BIOTECHNOLOGY

OF

ENDOPHYTIC FUNGI

OF GRASSES

Edited by

Charles W. Bacon

and

James F. White, Jr.

CRC Press

Boca Raton Ann Arbor London Tokyo

Library of Congress Cataloging-in-Publication Data

Biotechnology of endophytic fungi of grasses / edited by Charles W. Bacon,
    James F. White, Jr.
        p.   cm.
    Includes bibliographical references and index.
    ISBN 0-8493-6276-8
    1. Endophytic fungi. 2. Endophytic fungi—Biotechnology. 3. Grasses—
Ecophysiology. I. Bacon, Charles W. II. White, James F. (James Francis).
QK604.2.E53B56 1994
589.2′04524—dc20   93-47983
DNLM/DLC   CIP
for Library of Congress

No claim to original U.S. Government works
International Standard Book Number 0-8493-6276-8
Library of Congress Card Number 93-47983
Printed in the United States of America 1 2 3 4 5 6 7 8 9 0
Printed on acid-free paper

EDITORS

Charles W. Bacon is the research microbiologist, Toxicology and Mycotoxin Research Unit, U.S. Department of Agriculture, Agricultural Research Service, Russell Research Center, Athens, GA, and adjunct professor of plant pathology in the Department of Plant Pathology, University of Georgia, Athens.

Dr. Bacon graduated in 1965 from Clark College, Atlanta with a B.S. degree in biology and a minor in chemistry and obtained his Ph.D. degree in 1972 in botany and fungal physiology, from the University of Michigan. He was a research associate in biochemistry from 1972 to 1973 at the Department of Biological Chemistry, University of Michigan. It was in 1973 that he assumed his present positions.

Dr. Bacon is a member of the Mycological Society of America, the American Phytopathological Society, the American Society for Microbiology, the Agronomy Society of America, and the honorary society Sigma Xi. He has been the recipient of research grants from the Competitive Grants Program National Research Initiative. He received, among other awards, the U.S. Department of Agriculture Superior Service Award, for research on grass endophytes and for incorporating this finding into forage grass breeding research.

Dr. Bacon's research paper in 1977 first drew attention to endophytes of grasses as causes of pasture toxicity problems in livestock, within the 45-year-old history of forage grass toxicity. From this initial discovery, he has been at the forefront of several major important revelations of endophytic fungi and endophyte–grass relationships. Dr. Bacon has presented over 75 invited lectures at international and national meetings, including international meetings on grass endophytes. He has been a guest lecturer at more than 50 universities and at industry and veterinary service laboratories and has participated in numerous workshops and symposia dealing with mycotoxins and fungi. He has published more than 200 research papers, book chapters, and abstracts. His current research interests include the regulation and biosynthesis of mycotoxins, fungal endophyte–grass relationships, and the coevolution of fungal secondary products, primarily mycotoxins, with plants, as an adaptive strategy for mutualistic associations.

James F. White, Jr. is associate professor, Department of Biology, Auburn University at Montgomery, Montgomery, AL. He graduated in 1981 from Auburn University, Auburn, AL, with a B.S. degree in botany (with honors). In 1983 he obtained an M.S. degree in plant pathology from the same institution. In 1987 he received a Ph.D. degree in botany from the University of Texas, Austin. Postdoctoral research was conducted in the Department of Botany at the Ohio State University, Columbus, OH. In 1988 he assumed his present position.

Dr. White has received several research grants from the National Science Foundation and the Center for Interdisciplinary Studies in Turfgrass Science at Rutgers University. He has received the Junior Faculty Fellowship and Faculty Research Award at Auburn University at Montgomery. He is a member of the Honor Society of Phi Kappa Phi, the American Phytopathological Society, the Botanical Society of America, and the Mycological Society of America. He was an invited speaker at the first two international symposia on *Acremonium*–grass interactions and several other endophyte symposia.

Dr. White is the author of more than 60 research articles and book chapters. His research efforts helped to establish the widespread distribution of fungal endophytes in grasses throughout the world and emphasized the systematic diversity among endophytic taxa. His current research interests relate to systematics and ecology of the endophytic and epibiotic members of the Clavicipitaceae.

PREFACE

In only a few years, grass endophytes have emerged from relative obscurity to become a group of fungi of widespread interest and importance. Livestock toxicities from grasses provided the initial impetus for studies on the relationship of endophytic fungi and their grass hosts. Already two international symposia on *Acremonium* and other grass endophytes have been held to discuss developments in the area. Several major review articles in different disciplines of science have been written. Researchers have developed an endophyte information base that indicates that endophytes have significant impacts on grasses, plant ecology, and animal herbivory. Endophytes are now used in turf and forage grass breeding programs, to produce cultivars with enhanced stress tolerance.

It is the conviction of the editors, however, that the greatest discoveries on endophytes, and the most valuable applications, are yet to come. The area of endophyte physiology has received very little attention, and we know very little about the cellular relationships and compatibilities of endophytes and hosts. Perhaps the greatest applications lie in the potential for biotechnological manipulations of the endophytes. In this area genetic transformation of endophytes may provide the means to develop cost-efficient ways to produce drugs, nutrients, enzymes, and other products that currently must be extracted from rare organisms and organs or synthesized at high cost. For example, the endophytes may ultimately provide the vehicle by which grass crops may be modified into factories of important compounds, where the host provides all energy and raw material needed for the endophyte-directed process. This use of endophytes, along with many other aspects of endophytes, remains to be explored.

This book considers the biological, ecological, toxicological, and chemical aspects of research topics as they relate to endophytes of grasses. Several chapters reflect the very pragmatic applications of endophytes and endophyte-infected grasses. Other chapters offer future applications for endophytes and are therefore discussed from theoretical viewpoints. This book contains the collective writings of an international group of experts on fungal endophytes of grasses, all of whom are directed toward understanding, creating, and exploiting the positive aspects of endophytes. With this book we are attempting to stimulate and facilitate future explorations of the grass endophytes.

Charles W. Bacon
James F. White, Jr.

ACKNOWLEDGMENTS

This book is dedicated to all the early pioneers of endophyte biology and toxicology. Particularly, we acknowledge the dedicated industry of G. B. Garner and C. R. Funk, Jr., whose contributions to this exciting field extend beyond their personal research goals.

CONTRIBUTORS

Charles W. Bacon, Ph.D.
Research Microbiologist
USDA/ARS
Toxicology and Mycotoxin Research Unit
Richard B. Russell Research Center
Athens, Georgia

Faith C. Belanger, Ph.D.
Assistant Professor
Department of Plant Science
Cook College
Rutgers University
New Brunswick, New Jersey

Lowell P. Bush, Ph.D.
Professor
Department of Agronomy
University of Kentucky
Lexington, Kentucky

Keith Clay, Ph.D.
Associate Professor
Department of Biology
Indiana University
Bloomington, Indiana

Stephen L. Clement, Ph.D.
Research Entomologist
USDA/ARS
Regional Plant Introduction Station
Washington State University
Pullman, Washington

Herb Eichenseer, Ph.D.
Postdoctoral Research Assistant
Department of Entomology
Pennsylvania State University
University Park, Pennsylvania

C. Reed Funk, Ph.D.
Professor
Department of Plant Science
Cook College
Rutgers University
New Brunswick, New Jersey

George B. Garner, Ph.D.
Professor Emeritus
Animal Sciences Department
University of Missouri – Columbia
Columbia, Missouri

Nicholas S. Hill, Ph.D.
Associate Professor
Department of Crop and Soil Sciences
University of Georgia
Athens, Georgia

Walter J. Kaiser, Ph.D.
Research Plant Pathologist
USDA/ARS
Regional Plant Introduction Station
Washington State University
Pullman, Washington

Garrick C. M. Latch, Ph.D.
Doctor
New Zealand Pastoral Agriculture Research
     Institute Ltd.
Palmerston North, New Zealand

Adrian Leuchtmann, Ph.D.
Lecturer/Senior Research Associate
Geobotanisches Institut
Zurich, Switzerland

James A. Murphy, Ph.D.
Extension Specialist
Department of Plant Science
Cook College
Rutgers University
New Brunswick, New Jersey

Wayne A. Parrott, Ph.D.
Associate Professor
Department of Crop and Soil Sciences
University of Georgia
Athens, Georgia

James K. Porter, Ph.D.
Research Chemist
USDA/ARS
Toxicology and Mycotoxin Research Unit
Richard B. Russell Agricultural Research
     Center
Athens, Georgia

Daryl D. Rowan, Ph.D.
Doctor
The Horticulture and Food Research Institute of
     New Zealand Ltd.
Palmerston North, New Zealand

Christopher L. Schardl, Ph.D.
Associate Professor
Department of Plant Pathology
University of Kentucky
Lexington, Kentucky

Malcolm R. Siegel, Ph.D.
Department of Plant Pathology
University of Kentucky
Lexington, Kentucky

Frederick N. Thompson, Ph.D.
Professor
Department of Physiology and Pharmacology
University of Georgia
Athens, Georgia

Charles P. West, Ph.D.
Associate Professor
Department of Agronomy
University of Arkansas
Fayetteville, Arkansas

James F. White, Jr., Ph.D.
Associate Professor
Department of Biology
Auburn University at Montgomery
Montgomery, Alabama

TABLE OF CONTENTS

SECTION VI: UTILIZATION OF ENDOPHYTE-INFECTED GRASSES BASED ON AGRONOMIC CHARACTERISTICS

Section I

Principles of Classification and
Taxonomic Groups

Chapter 1

Taxonomic Relationships Among the Members of the Balansieae (Clavicipitales)

James F. White, Jr.

CONTENTS

I. INTRODUCTION

Numerous grasses have been discovered to be infected by endophytic or epibiotic Balansioid fungi.[5,8,20,39,41,46] These fungi are members of the Balansieae, one of three tribes classified in the ascomycete subfamily Clavicipitoideae (Clavicipitaceae, Clavicipitales). The other tribes are Clavicipiteae and Ustilaginoideae.[9] The systematic monograph *Balansia and the Balansieae in America*, written by W. W. Diehl in the 1950s,[9] is the main source by which potential biotechnologists and other biologists gain information about endophytes and their relatives. However, there is growing awareness that Diehl's monograph is incomplete. This chapter is written in an attempt to provide a general perspective on biological and systematic relationships in the Balansieae. It is also intended to be a preliminary step toward making systematic revisions.

Infection by many of the balansioid fungi affects host reproductive capacity, gives increased deterrence of herbivory by insects and, in some cases, by mammalian herbivores, and gives increased drought tolerance.[7,35,38] However, some species render colonized grass individuals sterile, or partially sterile, due to the development of external reproductive stages on host culms during flowering.[6,17] Other of the Balansieae have less destructive effects on host fitness, by producing reproductive structures on leaves or by relinquishing external reproduction and relying on seed transmission, a clonal means of propagation.[40,48] The latter are essentially benign, symptomless endophytes, although their presence in grasses may have a price in terms of energy drain on host food reserves.[12]

Some species, such as *Atkinsonella hypoxylon* (Peck) Diehl and *Epichloë typhina* (Pers.:Fr.) Tul., produce stromata on grass culms. The stroma is composed of both fungal mycelium and host tissues that are permeated with mycelium.[48,49] Other species, such as *Balansia obtecta* Diehl, produce a stroma that is modified to form a sclerified resting structure termed pseudothecium.[9] The developmental trigger for stroma formation is believed to be the presence of sugars or other energy compounds released from the inflorescence primordium at the onset of expansion.[48] On the surface of the stromata, spermatia (conidia) are produced. If these are transferred to a stroma of the opposite mating type, perithecia may develop there.[44]

The discovery of numerous asymptomatic asexual endophytes in grasses, all thought to have phylogenetic affinity to the Balansieae, have made it important that a more complete knowledge of systematics of Balansieae be developed. The realization that many of these asymptomatic endophytes establish mutualistic symbioses with their hosts has stimulated research on evolution of plant–fungus mutualism.[33] In recent years the status of such genera as *Atkinsonella* Diehl and *Balansiopsis* Höhnel has been questioned as to whether they should be retained as distinct from the genus *Balansia* Speg.[32] Within the genus *Epichloë* (Fr.) Tul., species concepts are not well defined and are in need of reevaluation.[33,42] It is clear that the Balansieae will require systematic attention to resolve these and other taxonomic problems. The eventual goal for systematic studies of Balansieae is to distinguish natural generic groupings and establish biological species concepts that are consistent with the tenants of population and evolutionary biology.

II. OVERVIEW: HISTORICAL DEVELOPMENTS IN THE SYSTEMATICS OF BALANSIEAE

The genus *Balansia* was established by Spegazzini, who recognized its affinity to *Claviceps* Tul.[36] Lindau classified *Balansia* in the Hypocreaceae, in the subfamily Clavicipiteae.[22] Höhnel recognized the Clavicipiteae as part of the Hypocreaceae and included his new genus, *Balansiopsis* Höhn., in it.[15] Gäumann also accepted this classification, but recognized three groups within the subfamily: *Oomyces-Ascopolyporus*, *Epichloë-Claviceps*, and *Cordyceps*.[11] Nannfeldt, however, removed the subfamily Clavicipiteae from the Hypocreales, elevating it to separate familial and ordinal levels as Clavicipitaceae and Clavicipitales.[29] Diehl[9] accepted the groups recognized by Gäumann and formally designated them as subfamilies within the Clavicipitaceae: Oomycetoideae, Clavicipitoideae, and Cordycipitoideae. He divided the subfamily Clavicipitoideae into the three tribes: Clavicipiteae, Balansieae, and Ustilaginoideae. The Clavicipiteae included only *Claviceps*, whereas four teleomorphic genera, *Atkinsonella*, *Balansia*, *Balansiopsis*, and *Epichloë*, were classified in the Balansieae. The Ustilaginoideae included two anamorphic genera: *Munkia* Speg. and *Ustilaginoidea* Bref. Diehl identified *Balansia chusqueicola* P. Henn. as a teleomorphic member of this tribe, but little more is known of the group. Luttrell and Bacon added *Myriogenospora* Atk. to the Balansieae.[24]

Diehl distinguished the tribes in the Clavicipitoideae on the basis of morphology of the conidial states.[9] In the Clavicipiteae, glutinous masses of microconidia are produced from usually short, compact conidiogenous cells arranged in a dense, convoluted palisade layer. In contrast, conidial states of Balansieae comprise two main morphology types: microconidia (Figure 1) produced from elongate attenuate *Acremonium*-type conidiogenous cells bearing small heads of conidia apically, and macroconidia (Figure 2) that are narrowly cylindrical to linear and are produced in moist masses on a hypothallus. The generic name *Ephelis* Fr. is based on the macroconidial state. The Ustilaginoideae is characterized by possession of smut-like conidia borne on closely packed parallel hyphae produced within pycnidium-like depressions of sclerified stromata. In addition to these differences, as recognized by Rykard et al.,[32] the Balansieae also differ from the Clavicipiteae in the extent of mycelial invasion of host tissues.

Species of *Claviceps*, including *C. paspali* Stevens & Hall on dallisgrass (*Paspalum dilatatum* Poir.) and *C. purpurea* (Fr.) Tul., the cause of ergot of numerous grasses, show characteristic, localized infections of grass ovaries only.[23] *Claviceps* species infect grasses by penetration of germ tubes between cells of stigmas or styles of their hosts. In contrast, members of the Balansieae generally cause systemic infection of grasses or sedges involving several different host tissues, although there is considerable variation in the precise organs involved. In *Balansia aristidae* (Atk.) Diehl, *B. epichloë* (Weese) Diehl, and *E. typhina* (Pers.: Fr.) Tul., systemic, intercellular mycelia occur endophytically in culm or leaf sheath tissues, whereas in *Atkinsonella hypoxylon* and *Balansia cyperi*, the mycelia are epibiotic and localized around meristems, young leaves, and inflorescences.[9,20,32,49] Fungal–host relationships are still poorly understood for most Balansieae.

III. GENERA OF BALANSIEAE

The genera of Balansieae are defined predominantly on the basis of the presence or absence of the microconidial and macroconidial states.[9] *Epichloë* has only microconidia, whereas *Balansia* and

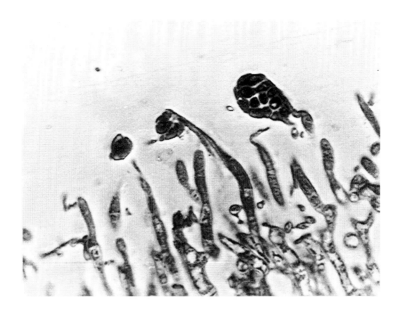

Figure 1 Heads of microconidia on *Acremonium*-type conidiogenous cells on stromata of an *Epichloë* sp. infecting *Elymus canadensis* L. (× 2600).

Figure 2 *Ephelis*-type macroconidia of *Atkinsonella hypoxylon* from *Danthonia spicata* L. (× 1000).

Myriogenospora have only macroconidia. *Atkinsonella* possesses both micro- and macroconidia, and *Balansiopsis* has neither. Recently, Rykard et al.[32] discovered a macroconidial state in specimens of *Balansiopsis pilulaeformis* (Berk. & Curt.) Diehl. The genus *Balansiopsis*, as defined by Diehl, requires careful scrutiny to assess its validity as a taxonomic unit.[9]

Cultures of many Balansieae exhibit production of the macroconidial (*Ephelis*) state. The form genus *Ephelis* has as its type species *E. mexicana* Fr., but it is uncertain whether this binomial is based upon the macroconidia of *Balansia claviceps* Speg. or *B. obtecta* Diehl. Binomials in *Ephelis* have been established for several other species, including *Balansia cyperi* and *Atkinsonella texensis* (Diehl)

Leuchtmann & Clay (*E. texensis* Ell. & Everh. and *E. borealis* Ell. & Everh., respectively), but not for others such as *B. aristidae* and *B. epichloë*. The macroconidia are produced holoblastically from a narrow, sometimes basally swollen, conidiogenous cell in which the tip buds conidia sympodially.[37] In most Balansieae, macroconidia are formed predominantly on the surface of a hypothallus.[32] They are produced on a flattened superficial layer in linear masses in *B. epichloë*, in crevices in the hypothallus in *A. texensis*, and in cup-shaped sporodochia in *A. hypoxylon* and *B. obtecta* (Figure 3).[21,27,32,49] Diehl used characteristics of the macroconidial conidiomata to distinguish two subgenera of *Balansia*: *Eubalansia*, with cupulate conidiomata, and *Dothichloe*, with conidiomata having any other form.[9]

The microconidial states have been referred to as sphacelial, typhodial, and *Cephalosporium*- or *Acremonium*-like by various authors,[2,9,26] however, these terms and corresponding Latin names have limited taxonomic value when referring to the microconidia on stromata, since these are always produced on stromata that are already classified in either *Epichloë* or *Atkinsonella*.

A convenient system of classifying asexual strains of some endophytes that do not form stromata has been developed using the cultural expressions of the isolated endophytes.[26,47] In this system endophytes are isolated and grown on standard laboratory media such as potato dextros agar (PDA). In culture, microconidiogenous cells are formed solitarily in aerial mycelia and resemble those of the form genus *Acremonium* Link. Because *Epichloë* is not closely related to the other ascomycete genera producing *Acremonium* states in culture (e.g., *Nectria* [Fr.] Fr., *Thielavia* Zopf, etc.), a separate section in the genus, sect. *Albo-lanosa* Morgan-Jones & W. Gams,[26] was proposed to accommodate the endophytes.

The system of classifying endophytes on the basis of cultural state is unavoidable because stromata are never formed on many hosts, and there is thus no basis for identification other than cultural expression. Proposing names for cultural expressions is only necessary when a given clone of an endophyte is present in a grass of economic importance or, alternatively, when large natural populations of endophyte-infected plants exhibit a distinctive endophyte.

IV. MAJOR SECTIONS OF BALANSIEAE BASED ON STRUCTURE AND HABIT

Presently, most data must be obtained from dried herbarium specimens, although studies have been conducted on living material of several species.[12,27,45] Examination of tissues of hosts subtending stromata in herbaria can be done to determine the presence or absence of endophytic colonization of host tissues. Also, conidial states, ascomata, ascus, and ascospore structure are ascertainable from dried specimens.

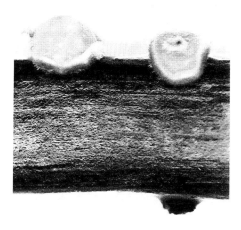

Figure 3 Cup-shaped sporodochia bearing masses of macroconidia on a pseudothecium of *Balansia obtecta*.

However, developmental patterns are difficult to determine, and cultural studies to assess physiological capacities and other *in vitro* characteristics cannot be conducted. Since herbarium work provides incomplete information on fungi, further research will be necessary to document groups. It thus seems prudent to organize Balansieae into structural groups whose taxonomic significance will await future confirmation using living material and appropriate molecular technology. The classification of fungi into homogenous groups based on possession of common features is a necessary precursor to more permanent taxonomic alterations.

Specimens of Balansieae were obtained and examined from the mycological herbarium in the Department of Plant Pathology at Cornell University (CUP), the New York Botanical Garden (NY), and mycological collections of the U.S. National Fungus Collections (BPI). The proposed groups outlined below include Balansieae collected in Eurasia and Africa that were excluded from Diehl's treatment of the tribe.[9] This grouping scheme relies on features of anamorphs to distinguish groups and, in this respect, builds on Diehl's systematic treatment. However, additional features of the fungi (e.g., endophytic vs. epibiotic) are employed to increase the level of precision and form more homogenous groupings.

It is notable that within the tribe, many of the species, as presently recognized, may represent aggregates of distinct biological species. Thus, for example, *B. claviceps* may actually be a plurality of similar-appearing species that may eventually be distinguished from one another after cultural, developmental, molecular, and reproductive studies have been conducted. Such a situation is evident in *Epichloë*, where fungi once believed to be assignable to the species *E. typhina* have been discovered to represent several distinct biological species that are reproductively incompatible with one another.[42] Because of the likelihood that other species of Balansieae will similarly prove to be aggregates, even individual species will require careful systematic attention to distinguish biological units. The proposed organization into groups should be regarded as an initial step in the development of a more holistic system of classifying Balansieae.

A. SECTION 1: *ECHINODOTHIS TUBERIFORMIS* GROUP

- Key features: epibiotic; forming stromata on inflorescences or culms; producing only microconidia; forming brown flat ascomata.
- Representative specimens examined: *Epichloë cinerea* on *Eragrostis tenuifolia* Hochst., Hebbal, Banglore, India, September, 1967, K. A. Lucy Channamma (BPI); *Epichloë warburgiana* on *Donax camaeforme*, Los Banos, Laguna Province, Philippines, December, 1915, C. F. Baker (NY); *E. tuberiformis* on *Arundinaria tecta*, Montgomery, AL, December, 1992, J. F. White, Jr. (BPI).

Fungi of this group include *Epichloë cinerea* Berk. & Br., *E. warburgiana* Magn., and *Echinodothis tuberiformis* (B. & Rav.) Atk. (Figure 4), which occur on grasses of both Asia and the Americas.[25,34] Hosts of the *E. cinerea* include panicoid genera *Eragrostis* Beauv., *Pennisetum* L. Rich., and *Sporobolus* R. Br. *Epichloë warburgiana* is known only on the bambusoid genus *Arundo* L. *Echinodothis tuberiformis* infects species of *Arundinaria*. Stromata of the first two species are formed on grass inflorescences, entirely imbedding them in a stroma, without inclusion of leaves. In *E. tuberiformis,* stromata develop on culms, often subtending leaves, and do not appear to include any meristematic tissues of the host.[43]

Stromata of all species included in this group are initially covered by an even layer of *Acremonium*-type conidiogenous cells that bear heads of microconidia apically. Following this stage, a brown ascoma forms evenly over the surface of the stroma. Perithecia are produced in flat ascomata with emergent perithecial necks, or in the case of *E. tuberiformis,* the entire perithecium is elevated to give a tuberculate ascoma.[43]

Asci of *E. warburgiana* are thick walled, with an apical tip that shows a pore around which there is no additional apical thickening. Multiseptate filamentous ascospores disarticulate to form numerous short hyaline one-celled cylindrical part-spores (Figure 5). *Epichloë cinerea* has asci that show thin lateral walls with a flattened truncate refractive thickening at the tip. In young asci the apical tip is thicker, but during maturation, tips appear to expand and flatten. Part-spores of this species are cylindrical and somewhat longer than those of *E. warburgiana*. Studies of the mating system of *E. tuberiformis* show that microconidia function as spermatia in initiating the development of the ascomata.[43] Asci in this species have a pronounced refractive tip. Filamentous ascospores were seen to become prominently multiseptate, but did not form part-spores. Living material of the other two species included in this group has not been examined.

8

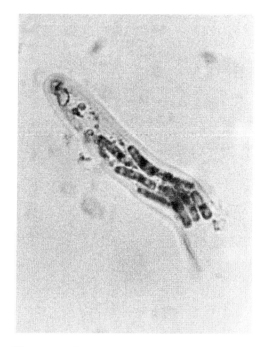

Figure 4 Stroma of *Echinodothis tuberiformis* on culm of *Arundinaria tecta*.

Figure 5 Part-spores in fragment of ascus of *Epichloë warburgiana* (× 1800).

B. SECTION 2: *ATKINSONELLA HYPOXYLON* GROUP

- Key features: epibiotic; forming stromata on inflorescences; producing both micro- and macroconidia; forming sessile black cushion-shaped ascomata.
- Representative specimens examined: *A. hypoxylon* on *Danthonia* sp., Highlands Biological Stations, Highlands, NC, July, 1961, R. H. Petersen and C. T. Rogerson (NY); *A. texensis* on *Stipa* sp., Greenville, TX, June, 1913, D. A. Saunders (NY); *Balansia andropogonis* on *Andropogon acicularis*, Luzon, Philippines, August, 1923, M. S. Clemens (NY); *Balansia gigas* on *Centotheca latifolia*, Sabaki, Sarawak, Borneo, 1929, J. and M. S. Clemens #6709 (NY).

Among the epibiotic Balansieae is a group characterized by possession of aseptate or 1-septate curved part-spores, microconidial and macroconidial synanamorphs, and black pulvinate sessile ascomata. Ascus walls are relatively thin, except at the tip, where a pronounced refractive thickening is evident. This group includes *Atkinsonella hypoxylon* (Figure 6), *A. texensis*, *Balansia gigas* Racib. (Figure 7), and probably *B. andropogonis* Lyd. & Butl.[13,21,27,28,37] The first two species produce stromata on inflorescences of species of cool-season grass, genera *Danthonia* L., *Stipa* L., and others. The latter two species form stromata on inflorescences of warm-season grasses, e.g., *Cyrtococcum, Centotheca, Andropogon* L., and others. During elongation of culms, an epibiotic mycelium proliferates on inflorescence primordia, forming the hypothallus. In *A. hypoxylon*, three leaves of the host are incorporated into stromata, but in the others only inflorescences are included in the stromata. On the surface of the hypothallus, a gray felty layer of tapering *Acremonium*-type conidiogenous cells with apical microconidia is produced. As the hypothallus matures, macroconidia begin to form at one, or several, sites on the stroma. In *A. texensis* and *B. andropogonis,* macroconidia exude in masses from what appears to be crevices in the hypothallus, while in *A. hypoxylon* and *B. gigas,* they are formed in cup-shaped sporodochia. In all species of this group, black cushion-shaped ascomata develop at macroconidial production sites, apparently forming only after heterothallic crossing. Developmental studies have been conducted on *A. hypoxylon*,[21] *A. texensis*,[27] and *B. gigas*,[37] but *B. andropogonis* has not been closely examined.

C. SECTION 3: *BALANSIA SCLEROTICA* GROUP

- Key features: epibiotic; forming black sclerified pseudomorphs on dwarfed tillers.
- Representative specimens examined: *Epichloë bambusae* on *Gigantochloae* apus, Tjikeumenh Buitenzorg,

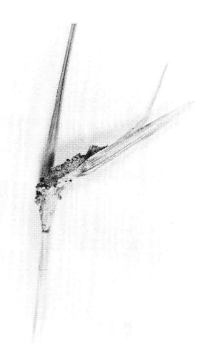

Figure 6 Stroma of *Atkinsonella hypoxylon* on culm of *Danthonia spicata*.

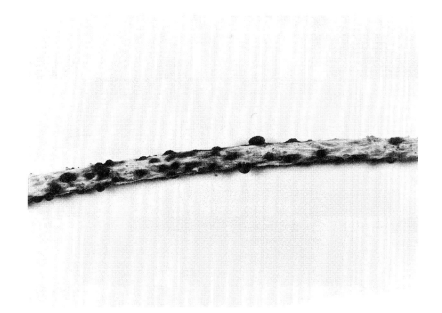

Figure 7 Stroma of *Balansia gigas* on culm of *Centotheca latifolia*.

Java, April, 1922, E. Gäumann (NY); *Epichloë cynodontis* on *Cynodon dactylon* (L.) Pers., Buffelspoort, district Rustenburg, Transvaal Province, South Africa, March, 1938, E. M. Dodge (BPI); *Epichloë omplismani* on *Omplismanis* sp., Kamerun, Victoria, Africa, 1900, G. F. Atkinson #2644 (CUP); *Epichloë sasae* on *Sasa paniculata* Mankino & Sieb, Mt. Nagi, Inaba, September, 1926, S. Tanaka #69 (BPI); *Epichloë schumanniana* on an unidentified grass, Masalu River, Nyassa, Africa, July, 1912, C. E. F. Allen (CUP); *Balansia sclerotica* on *Andropogon contortus*, India, August, 1904, E. J. Butler (BPI); *Epichloë volkensii* on *Hyparrhenia hirta* Stapf., Jimma, Ethiopia, June, 1955, R. B. Stewart #D-50 (NY).

Many African and Eurasian Balansieae produce exposed sclerotial-like pseudomorphs.[11,16] These Balansieae include, for example, *Epichloë bambusae* Pat., *E. cynodontis* Syd., *E. oplismani* P. Henn., *E. sasae* Hara, *E. schumanniana* P. Henn., *Balansia sclerotica* (Pat.) Höhn. (Figure 8), and *E. volkensii* P. Henn. These fungi are present on warm-season grasses, and although many are classified in *Epichloë*, they bear little structural similarity to fungi of the *Epichloë typhina* group widespread on cool-season grasses. They also appear distinct from other endophytic Balansieae (groups XII–IX) of the Americas. The often curved fusiform shape of their pseudomorphs is reminiscent of the similarly curved stromata of *Atkinsonella* spp. of North America, and both groups are epibiotic. However, the fungi are quite distinct otherwise. *Atkinsonella hypoxylon* (Pk.) Diehl produces soft stromata containing several leaves that are only slightly reduced in size, and has been demonstrated to be heterothallic.

In contrast, these sclerotial tiller epibionts completely embed a tiller bearing several dwarfed leaves of the host in the pseudomorph. These fungi produce a black rind, suggesting an ergot sclerotial structure similar to that formed by *Claviceps purpurea* (Fr.) Tul. In *Claviceps* the sclerotium is, however, further limited in the amount of host tissue involved, including only a single floret of the host. Whether these fungi are heterothallic or, like *C. purpurea,* homothallic is of additional interest.[23] *Claviceps* spp. produce a microconidial state; however, information on conidial states is lacking for the sclerotial tiller epibionts.[16] It may be that these fungi show an intermediate condition between *Balansia* and *Claviceps*. If this is proven by further study of living representatives of this group, the lines that presently distinguish tribes Clavicipiteae and Balansieae will become more diffuse. The eventual need is to evaluate the possible segregation of these sclerotial tiller epibionts from *Balansia* and *Epichloë* into a distinct genus.

D. SECTION 4: *MYRIOGENOSPORA ATRAMENTOSA* GROUP

- Key features: epibiotic; stroma develops in rolled or folded leaves; only macroconidia formed; ascomata black and linear; part-spores are fusoid.
- Representative specimens examined: *M. linearis* on *Pariana* sp., Aramanahy, Rio Tapajos, Para, Brazil, January, 1934, J. R. Swallen #3229 (BPI); *Myriogenospora atramentosa* on *Paspalum* sp., Montgomery, AL, August, 1992, J. F. White, Jr. (BPI).

Another group of epibiotic Balansieae includes *M. atramentosa* (Berk. & Curt.) Diehl and *M. linearis* (Rehm.) White & Glenn (Figure 9).[9,14,24,30,31] These fungi are characterized by possession of fusiform part-spores (Figure 10) that are produced by the disarticulation of linear ascospores, to form small ellipsoidal or rectangular segments that reinitiate determinate bipolar growth within the ascus. *Myriogenospora*

Figure 8 Pseudothecia of *Epichloë volkensii* on tillers of *Hyparrhenia hirta*.

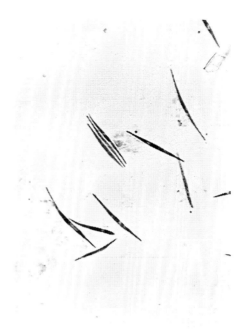

Figure 9 Linear stromata of *Balansia linearis* on rolled leaves of *Pariana* sp.

Figure 10 Fusiform part-spores of *Myriogenospora atramentosa* (× 1000).

atramentosa occurs on species of warm-season hosts, such as *Andropogon* L., *Paspalum* L., and others, while *M. linearis* infects only bambusoid grasses. Both species are limited to the Americas.

In these species macroconidia are formed prior to development of the ascoma, and microconidia are lacking. Asci have typical thin lateral walls, as in most Balansieae. In *M. linearis,* asci have typical thickened and refractive tips, although they truncate terminally. In *M. atramentosa,* ascus tips are modified, forming an apical dome in the top portion of the expanded ascus. Part-spores are ejected en masse from these asci by a circumscissile rupture of the ascus at the base of the apical dome. The presence of typical ascus tips in *M. linearis* may indicate that part-spores are ejected sequentially, as in other members of the Balansieae. Ascomata of these species are linear and black and occur predominantly on rolled or folded leaves of hosts, although inflorescences may sometimes be involved. In the ascoma of *M. atramentosa,* a single row of globose perithecia is embedded in ascomata emerging from crevices of folded leaves. In *M. linearis*, two rows of perithecia are embedded in linear ascomata that erupt through an overlying layer of the rolled leaf. Perithecia of this group are entirely embedded within the stroma, without emergent perithecial necks, a feature that is common in other groups of Balansieae.

In both *M. linearis* and *M. atramentosa,* host epidermal cells that interface with the mycelium of the hypothallus are modified, showing hypertrophy or hypotrophy, depending on the host, and reduced thickness of the cell wall layers over part of the epidermis.[30] This apparently enhances the flow of nutrients into the epibiotic mycelium. Information is unavailable on the mating systems employed by these species; however, the rapid production of perithecia by *M. atramentosa* may be an indication that it is homothallic.

E. SECTION 5: *BALANSIA CYPERACEARUM* GROUP

- Key features: epibiotic; forming stromata on leaf surfaces; only macroconidia formed; ascomata black and flattened.
- Representative specimens examined: *B. cyperacearum* on *Cyperus diffusus*, British Guiana, F. L. Stevens #325 (NY); *Epichloë kyllingiae* on *Fimbristylis* sp., Santiago, Bulacan Province, Luzon, Philippines, September, 1913, M. Ramos (BPI).

Another group of epibiotic Balansieae includes *Balansia cyperacearum* (Berk. & Curt.) Diehl and *E. kyllingiae* Rac.[9] The former species is limited to the Americas and parasitizes sedges, while the latter has been reported from Asian and American tropical areas on the grass genera *Bouteloua* Lag. and *Fimbristylis* L. Asci have thin lateral walls and typical hemispherical refractive tips. Part-spores are cylindrical and possess small guttules concentrating at septa. Exclusively macroconidia are produced by these species. Ascomata of *B. cyperacearum* and *E. kyllingiae* are black and flattened on leaves. Studies on reproductive biology of these species have not been conducted.

F. SECTION 6: *BALANSIOPSIS ASCLEROTIACA* GROUP

- Key features: epibiotic; stromata are formed on inflorescences; ascomata are black and pulvinate to stipitate.
- Representative specimens examined: *Balansia ambiens* on *Leersia grandiflora*, Viscosa-Escola, Minas Gerais, Brazil, March, 1933, A. S. Müller #415 (BPI); *Balansia cyperi* (type) on *Cyperus virens*, St. Gabriel, LA, August, 1917, A. T. Bell and C. W. Edgerton (NY); *B. asclerotiaca* (type) on *Orthoclada rariflora*, Amazonas, Jurua, Brazil, 1901, E. Ule (NY); *Balansiopsis pilulaeformis* on *Paspalum pubescens*, Savannah, GA, July, 1933, G. Crisfield (BPI).

Another group of epibionts includes *Balansia ambiens* Möller, *B. cyperi* Edg., *B. asclerotiaca* (P. Henn.) Diehl (Figure 11), and *B. pilulaeformis* (Berk. & Curt.) Diehl.[5,9,20] These species infect warm-season hosts, producing on inflorescences a thin hypothallus that initially bears discontinuous accumulations of macroconidia free of any fructification. The hosts for this group include predominantly panicoid and bambusoid grasses, although *B. cyperi* is parasitic of sedges. Where grasses are hosts, a thin hypothallus on inflorescences is usually partially covered by the terminal culm leaf. This apparently provides protection and optimal humidity during early development of the stroma. In *B. cyperi*, no such ensheathing leaf covers the hypothallus, which develops, entirely exposed, on the inflorescence. The ascomata arise from masses of macroconidia that appear at sites where the ensheathing leaf splits or is absent. *Balansia ambiens*, *B. cyperi*, and *B. pilulaeformis* are limited to the Americas. Numerous specimens examined in herbarium collections are identified as *B. asclerotiaca,* but were collected in Asian and American tropical sites. These American and Asian fungi show consistent differences in lengths of stipes on ascomata, and it thus seems probable that at least two distinct species are represented. Asci are thin-walled and have typical hemispherical refractive tips (Figure 12). The genus *Balansiopsis* was proposed to include species lacking conidial states;[9] however, Rykard et al.[32] have documented production of macroconidia by *B. pilulaeformis*. The macroconidia are ephemeral and shielded from view by an ensheathing leaf. It seems likely that *B. asclerotiaca* also produces macroconidia.

Figure 11 Stroma of *Balansiopsis asclerotiaca* on culm of *Orthoclada rariflora*.

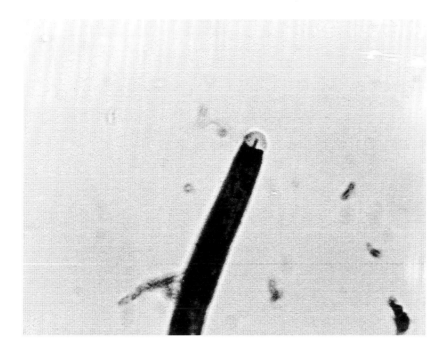

Figure 12 Ascus tip of *Balansiopsis asclerotiaca* (× 1800).

G. SECTION 7: *BALANSIA OBTECTA* GROUP

- Key features: endophytic; forming stromata on inflorescences; only macroconidia formed; ascomata are black and stipitate.
- Representative specimens examined: *Balansia claviceps* on *Sporobolus poiretii*, Guapiles, Costa Rica, September, 1952, J. B. Carpenter #573 (BPI); *B. obtecta* on *Cenchrus echinatus*, Gainesville, FL, November, 1932, G. F. Weber #9485 (BPI).

Several groups of Balansieae limited to the Americas are endophytic. Among these endophytic groups is one that may be characterized by formation of stromata on inflorescences of panicoid grasses where macroconidia and stipitate ascomata comparable to those of group VI are formed. *Balansia claviceps* (Figure 13) and *B. obtecta* Diehl are included in this group.[9] Asci of both species are thin walled, with typical refractive tips. Part-spores are cylindrical and show the tendency for small guttules to be concentrated at septa.

In *B. obtecta* the life cycle involves proliferation of the endophytic mycelium in meristematic tissues of inflorescence primordia, to produce a resistant mycelial structure referred to as the pseudothecium. This structure contains the inflorescence primordium, but no culm leaves. When immature, the pseudothecium is fully enclosed by a green, apparently healthy leaf blade that apparently serves as a moisture barrier to the environment and allows the rapidly growing mycelium to develop in a high-moisture microenvironment. When the pseudothecium is mature, it is released from the enveloping leaf. However, by this time it has developed a black impervious rind. Eventually, the pseudothecium detaches from the grass culm by a break that occurs on the dried grass culm, just below the pseudothecium.

Pseudothecia are like sclerotia in that they may lie dormant in the soil until temperature and moisture conditions are appropriate for germination. Prior to germination, moisture is absorbed through the culm remnant that appears to function as a moisture wick. Small cracks then appear at various sites on pseudothecia, from which macroconidia exude. After a few days, cup-shaped sporodochia bearing masses of macroconidia develop over the cracks on the pseudothecia. On some sporodochia, black stipitate ascomata develop, evidently only after heterothallic mating. Details of the mating process in nature require further evaluation.

The development of *B. claviceps* is similar, except that a resistant rind does not form on the stroma. The structural differences between these two species may affect their environmental tolerances and geographic distribution patterns. Moisture-conserving properties of *B. obtecta* appear to allow that

Figure 13 Stroma of *Balansia claviceps*.

Figure 14 Stroma of *Balansia strangulans* as node of culm of *Axonopus frucatus*.

species to colonize hosts, such as *Cenchrus* spp. and *Cynodon* spp., which are adapted to dry, temperate environments, without compromising the hosts ability to conserve moisture. *Balansia claviceps* is limited to the tropics, probably because of its inability to conserve moisture, which is abundantly available in the tropics.

H. SECTION 8: *BALANSIA STRANGULANS* GROUP

- Key features: endophytic; forming stromata at nodes; only macroconidia formed; ascomata black and flattened to short stipitate.
- Representative specimens examined: *Balansia aristidae* on *Aristida* sp., August, 1931, Auburn, AL, J. L. Seal (BPI); *Balansia discoidea* (as *Dothichloe discoidea* (P. Henn.) Diehl) on *Ichnanthus candidus* Nees, Vicosa, Escola, Brazil, April, 1933, A. S. Müller (CUP); *Balansia* sp. (as *B. linearis*) on *Chuquea* sp., Rio De Janeiro, Brazil, 1925, A. Chase #8496 (BPI); *b. strangulans* on *Panicum clandestinum* L., New York Botanic Garden, Bronx, NY, August, 1976, C. T. Rogerson #76-65 (NY); *Balansia subnodosa* on *Ichnanthus* sp., Trinidad, October, 1944, J. M. Todd #251 (BPI); *Balansiopsis gaduae* (as *Balansia regularis* A. Möll.) on *Gaduae* sp., Amazonas, Jurua, Brazil, 1901, E. Ule (NY).

Several endophytic Balansieae form stromata at the nodes of culms. Among these species are *Balansia aristidae*, *B. discoidea* P. Henn., *B. strangulans* (Mont.) Diehl (Figure 14), *B. subnodosa* Atk. ap. Chardon, and *Balansiopsis gaduae* (Rehm) Höhn. *Balansiopsis gaduae* infects bambusoid grasses, while the others colonize panicoid hosts. These species are distributed exclusively in the Americas. *Balansia discoidea* is similar to species of the *B. obtecta* group in that it forms macroconidia in cup-shaped sporodochia followed by stipitate ascomata. In contrast, macroconidia of *B. aristidae*, *B. strangulans*, and *B. subnodosa* are formed free of sporodochia, over the surface of the hypothallus, and the ascomata develop in a flat or raised layer. In *B. gaduae*, macroconidia have not been observed, and stromata appear at nodes opposite each leaf, rather than in a continuous circle around the culm, as seen in others. For these species, asci are thin-walled, and ascus tips are thickened and refractive, as typically observed. Part-spores are cylindrical and have numerous small guttules at septa (Figure 15). Ascospore development shows variability between species. In a process common to the endophytic *Balansia* species, filamentous ascospores disarticulate to form 1-septate cylindrical units prior to ejection from asci. Following ejection, the 1-septate units undergo a final disarticulation. The number of part-spores produced per filamentous ascospore, however, varies between species. In *B. strangulans* and several other species of endophytic Balansieae, including *B. claviceps*, *B. epichloë*, and *B. henningsiana* (Moell.) Diehl, filaments disarticulate to form only four 1-septate part-spores. However, in *B. aristidae,* eight 1-septate units form from a single filamentous ascospore. The way that part-spores infect host plants is not presently known.

I. SECTION 9: *BALANSIA HENNINGSIANA* GROUP

- Key features: endophytic; forming stromata on leaves without inclusion of inflorescences; only macroconidia produced; ascomata brown to black and flattened.
- Representative specimens examined: *B. epichloë* (as *Dothichloe limitata* Diehl) on *Eragrostis refracta*, Gainesville, FL, September, 1931, G. F. Weber (BPI); *B. henningsiana* (as *Dothichloe atramentosa* Berk. & Curt.) Atkinson, on *Andropogon* sp., unspecified collection location, 1900, G. F. Atkinson #1169 (CUP).

Balansia epichloë and *B. henningsiana* (Figure 16) produce stromata on leaves of several warm-season hosts.[9,46] Stromata of *B. epichloë* form on the upper surface of leaves of species in grass of genera *Chloris* Swartz, *Eragrostis* Beauv., *Panicum* L., *Sporobolus* R. Br., *Tridens* Roem. & Schult., and others, while stromata of *B. henningsiana* form on the lower surface of leaves of species of grass of genera *Andropogon* L., *Gymnopogon* Beauv., *Panicum*, and others. Stroma development in *B. epichloë* involves proliferation of endophytic mycelia in developing leaves, while hyphae are not present in inflorescence axes. In the mature stroma, an epibiotic hypothallus is connected, via hyphal bridges, to a network of the intercellular endophytic mycelium. It is through the hyphal bridges that water and nutrients are expected to pass to the stroma from within the leaf.

Endophytic mycelia of *B. henningsiana* are present in inflorescence axes, but apparently do not proliferate rapidly in those tissues. The endophytic mycelium develops rapidly in leaves and emerges to the surface, through breaks in the lower epidermis, to form the hypothallus. Sections through leaves of *Andropogon virginicus* L. bearing stromata of this species show abundant mycelia permeating vascular bundles, apparently to facilitate the transfer of food and moisture to the ectophytic stroma.

Seed transmission has not been demonstrated in these, or any other, species of endophytic *Balansia,* despite repeated attempts to document its occurrence. In *B. epichloë* the stroma is initially composed of

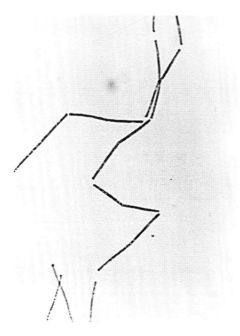

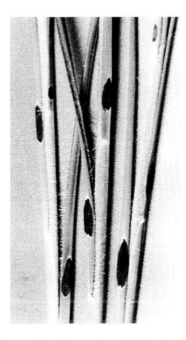

Figure 15 Part-spores of *Balansia aristidae* (× 1000).

Figure 16 Stromata of *Balansia henningsiana* on leaves of *Andropogon* sp.

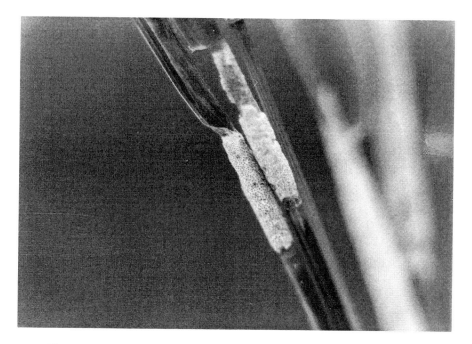

Figure 17 Stromata of *Epichloë baconii* on culms of *Agrostis capillaris.*

a white hypothallus on which a hymenial layer of cream-colored macroconidia is formed. A brown-to-black ascoma develops on the hypothallus after macroconidia of the opposite mating type are deposited on the stroma. Macroconidia are probably vectored by insects attracted to feed on the mycelia prominently displayed on leaf surfaces; however, more work is necessary to document specific groups of insect vectors involved. In *B. henningsiana* the hypothallus is initially off-white or pink and very rapidly becomes brown, as macroconidia form evenly over its surface. A slightly raised black ascoma forms on the hypothallus, apparently after heterothallic crossing. Asci have thin lateral walls and typical hemispherical refractive tips. Part-spores are cylindrical, showing an accumulation of small guttules at septa. They are ejected as 1-septate units. The process of infection is still unknown.

J. SECTION 10: *EPICHLOË TYPHINA* GROUP

- Key features: endophytic; forming stromata that include inflorescences and one or more culm leaves with a stromal leaf blade emergent from the apex of stroma; only microconidia formed; ascomata form in a flat layer and are yellow, orange, or brown in color; seed transmission frequently occurring.
- Representative specimens examined: *Epichloë baconii* (type) on *Agrostis* sp., Okehampton, Devon, England, July, 1992, J. F. White, Jr. (BPI); *Epichloë clarkii* (type) on *Holcus* sp., Okehampton, Devon, England, July, 1992, J. F. White, Jr. (BPI); *E. typhina* (type; as *Sphaeria typhina* Pers.) on *Dactylis glomerata* L., France, ca. 1790, C. H. Persoon (L 9312 Rijksherbarium, Leiden, The Netherlands).

This group includes *Epichloë baconii* White, *E. clarkii* White, and *E. typhina* (Figure 17).[42] These fungi form stromata on culms of cool-season grasses. Stromata bind inflorescence primordia and several stromal leaves, with a leaf blade emergent from the apex of the stroma. The emergent stromal leaf shows hypertrophy of several cell types, including mesophyll parenchyma and epidermis. Because of modifications to the emergent stromal leaf, evaporation of water from its surface is enhanced.[44] Increased moisture loss of the stromal stage of fungi of this group may explain why stromata are produced abundantly in cool, moist regions and are less common where moisture is lacking. The surface of the stromata is initially covered by a layer of microconidia produced in heads at the apices of *Acremonium*-type conidiogenous cells.

These *Epichloë* species are heterothallic, and ascomata do not form until conidia of one mating type are transferred to a stroma of the opposite mating type.[43] The role of the vector of microconidia is filled by Anthomyiid flies that consume conidia and mycelia of stromata.[4,18] Microconidia pass, undigested, through the alimentary tract of flies and are deposited on stromata in feces while flies are feeding and laying eggs.[3] From fertilization sites, ascomata begin to develop as a raised mycelium that becomes yellow to orange as perithecia differentiate. Asci have thin lateral walls and tips that show typical hemispherical refractive thickening. In this group, ascospore development differs between species.[42] For example, in *E. baconii* White, 1-septate cylindrical part-spores are formed from filamentous ascospores, while *E. clarkii* White forms 4- or 5-septate spear-shaped part-spores, and *E. typhina* ejects filamentous ascospores intact.

Ascospore development in *Epichloë* is variable, whereas in *Balansia* spp. ascospore development is a conserved feature. A possible explanation for this is that in *Epichloë*, ascospores germinate to form conidia in obligate iterative conidiation.[1] Conidia resulting from ascospores may play the dominant role in the infection of hosts, while in other Balansieae, part-spores may be directly involved in the infection process and thus are selectively stabilized. Further research is necessary to evaluate this possibility.

It is becoming clear that the task of defining species in *Epichloë* is in the beginning stages. For example, specimens of *Epichloë* also occur commonly on several other grass hosts, including *Agrostis hiemalis* (Walt.) B.S.P., *Calamagrostis* spp., *Sphenopholis* spp., and others. Preliminary examinations of ascomata in herbaria suggest that fungi on these hosts are distinct from those previously described.[42] For example, on *A. hiemalis*, *Calamagrostis* Adans., and *Sphenopholis* Scribn., perithecia are separated and superficial on the ascoma, while on some other hosts, the stromata have perithecia that are deeply immersed within the ascoma. Ascoma variations may be indications of species differences in *Epichloë*.

The nonstroma-forming *Acremonium* endophytes are known to be widespread in cool-season grasses.[47] These endophytes are believed to be stages of *Epichloë* that have lost the ability to form stromata on grasses. Several types of data support the affinity of *Epichloë* and the nonstroma-forming *Acremonium* endophytes. Cultures of *Epichloë* are usually identified as *Acremonium typhinum*; however, several grasses that are not known to bear stromata are found to be infected with endophytes that are identical to *A. typhinum*.[41] In addition, molecular studies of ribosomal DNA frequently indicate close affinity of the two groups and suggest that *Acremonium* endophytes may have been derived, on multiple occasions, from stroma-forming endophytes.[33]

To date, eight species of *Acremonium* endophytes have been proposed, including *A. chilense* Morgan-Jones, White, & Piontelli; *A. chisosum* White & Morgan-Jones; *A. coenophialum* Morgan-Jones & W. Gams; *A. huerfanum* White, Cole, & Morgan-Jones; *A. lolii* Latch, Christensen, & Samuels; *A. starrii* White & Morgan-Jones; *A. typhinum* Morgan-Jones & W. Gams; and *A. uncinatum* W. Gams, Petrini, & Schmidt.[10,19,26,47]

The species of endophytes have been proposed predominantly on the basis of the morphology of the conidia and conidiogenous cells. Additional cultural features have been used to propose varieties of stroma-forming endophytes. *A. typhinum* var. *bulliforme* White represents the cultural expression of *E. baconii* and is characterized by the production of hyphal swellings (bullulae), when cultured on glucose agar.[41] Bullulae are not typically produced by other endophytes. *A. typhinum* var. *ammophilae* White & Morgan-Jones accommodates the cultural expression of an endophyte in American beachgrass (*Ammophila breviligulata* Fernald.), producing tan colonies on potato dextrose agar and ephemeral nonfunctional stromata solely on inflorescences of the host.[46] This type of stroma is anomalus in that functional stromata of *E. typhina* include the inflorescence primordium, which fails to develop, leaves in varying stages of development, and the leaf sheath of a leaf that is emergent from the apex of the stromal mycelium. Perithecia do not develop on the anomalus stromata of *A. typhinum* var. *ammophilae*, since stromata desiccate soon after exposure from ensheathing leaves.

V. PROPOSED EVOLUTIONARY RELATIONSHIPS

Evolutionary relationships among groups of Balansieae are difficult to assess from this superficial evaluation of structural groups. However, if it is assumed that epibiotic groups are ancestral to endophytic groups and that morphological similarity reflects evolutionary relatedness, some generalizations are not unreasonable. The endophytic *Balansia obtecta*, *B. strangulans*, and *B. henningsiana* groups may have arisen from ancestral epibiotic forms related to either the *B. cyperacearum* or *B. asclerotiaca* groups. Members of these two groups produce only macroconidia and show similarities in teleomorph structure. In fact, the similar structures of stipitate ascomata in both epiphyte *B. asclerotiaca* and endophytes of the *B. obtecta* group, such as *B. claviceps*, are very suggestive of close evolutionary relatedness. *Balansiopsis asclerotiaca* may exhibit features similar to those of fungi that were ancestral to the endophytes. If this

Table 1 Isozyme diversity in grass endophytes of the genera *Atkinsonella*, *Balansia*, *Epichloë*, and *Acremonium*[a]

Genus Fungal Species or Host	*I* Between Species or Host Population	Percentage Polymorphism	Average Gene Diversity H_T or H_S	Sample Size	References
Atkinsonella	0.21–0.31	85.6	0.507	291	
A. hypoxylon		76.9	0.174	271	23
A. texensis		0.0	0.0	20	23
Balansia	0.00–0.24	100.0	0.672	21	
B. cyperi		0.0	0.0	7	unpub.[b]
B. epichloë		41.7	0.208	4	unpub.
B. henningsiana		8.3	0.042	2	unpub.
B. obtecta		8.3	0.042	8	unpub.
Epichloë (Type I)	0.02–0.76	100.0	0.645	60	
Agrostis stolonifera		33.3	0.153	14	unpub.
Anthoxanthum odoratum		16.7	0.093	3	unpub.
Bromus erectus		8.3	0.014	11	unpub.
Calamagrostis villosa		0.0	0.0	3	unpub.
Dactylis glomerata		50.0	0.229	18	unpub.
Glyceria striata		0.0	0.0	7	25
Holcus lanatus		0.0	0.0	1	25
Poa trivialis		25.0	0.102	3	unpub.
Epichloë (Type II)	0.21–0.89	100.0	0.544	224	
Agrostis hiemalis		54.5	0.257	13	25
A. perennans		54.5	0.182	6	25
Brachyelytrum erectum		27.3	0.053	18	25
Brachypodium sylvaticum		35.7	0.097	106	unpub.
Elymus villosus		0.0	0.0	1	25
E. virginicus		72.7	0.128	40	25
Hystrix patula		54.5	0.122	40	25
Acremonium (Type III)	0.05–0.99	100.0	0.617	392	
Agropyron caninum		14.3	0.071	5	unpub.
Bromus benekenii		35.7	0.102	22	unpub.
B. ramosus		7.1	0.017	21	unpub.
Elymus europaeus		57.1	0.290	41	unpub.
Festuca altissima		58.3	0.292	13	20
F. arundinacea		84.6	0.320	70	2, 25
F. gigantea		16.7	0.051	21	20
F. glauca		0.0	0.0	1	20
F. heterophylla		16.7	0.083	2	20
F. longifolia		16.7	0.070	8	20
F. obtusa		90.9	0.365	14	25
F. ovina		41.7	0.082	9	20
F. pratensis		30.8	0.137	17	2, 20
F. pulchella		8.3	0.035	20	20
F. rubra		41.7	0.087	29	20
F. valesiaca		0.0	0.0	1	20
Lolium perenne		69.2	0.199	29	2, 25
Melica ciliata		35.7	0.179	48	unpub.
Poa autumnalis		18.2	0.091	1	25

Table 1 (Cont.) **Isozyme diversity in grass endophytes of the genera** *Atkinsonella,* *Balansia, Epichloë,* **and** *Acremonium*[a]

Genus Fungal Species or Host	*I* Between Species or Host Population	Percentage Polymorphism	Average Gene Diversity H_T or H_S	Sample Size	References
P. sylvestris		72.7	0.243	13	25
P. wolfii		27.3	0.136	1	25
Sphenopholis nitida		63.6	0.330	2	25
S. pallens		0.0	0.0	3	25
Stipa robusta		9.1	0.045	1	25

[a] Data are compiled from several sources and reanalyzed, in part, based on allele frequencies of 11 to 14 putative loci. The range of Nei's genetic identity[31] *I* between endophyte species or host populations are given for genera, percentage of loci polymorphic, and Nei's average gene diversity[32] H_T or H_S, for genera and fungal species or host populations; [b] Leuchtmann, unpublished data.

The diversity of individual populations or species can be measured by the percentage of loci polymorphic, and Nei's average gene diversity, H.[32] Polymorphism was high within most species of *Epichloë* or *Acremonium* and reached 100% in endophyte genera or association types, with the exception of *Atkinsonella* (Table 1). However, not all species or endophyte host populations were variable. While some probably appeared to be monomorphic due to the small sample size, others showed no variation, despite multiple isolates from different areas or different hosts. For example, all isolates of *Balansia cyperi* Edg. from three different *Cyperus* hosts in the U.S., Peru, and Ecuador were identical. The absence of genetic variation in *B. cyperi* may indicate a recent genetic bottleneck, since other species in the same genus possess high levels of variation.

Isozyme data indicated further that there is no correlation between genetic diversity and sexual reproduction, in the endophytes studied here. In other fungi it has been noted that sexually reproducing populations tend to have higher H_S values than their asexual counterparts, because of continuing recombination of allozyme loci.[44] Several of the asexual *Acremonium* host populations were found to be highly variable in a small geographic area in woodland habitats.[25] This could indicate that endophytes from these hosts are not always strictly seedborne or that sexuality has ceased to occur only recently. The observation of a single plant of *Poa sylvestris* A. Gray bearing stromata, which otherwise always was asymptomatic, lends support to this conclusion.[5] Other endophyte host populations, such as *A. coenophialum* from *F. arundinacea,* or *A. lolii* from *L. perenne,* with a similar degree of diversity, were from larger geographic areas or were worldwide samples.[2]

The total average gene diversity (H_T) was almost equally high in all life-cycle types of *Epichloë* (Table 1). High differentiation among populations or species often indicates restricted gene flow due to reproductive isolation.[10] While it can be assumed that endophytes in type-III associations, where no reproductive structures are formed on the host plant, are reproductively isolated, this should not necessarily be the case in type-I and type-II associations. Nevertheless, genetic divergences were high among several stromata forming host populations, giving reason to believe that little or no gene flow occurs between them and that they might be adapted to specific hosts. This would support the concept of several biological species within *Epichloë,* which has been advocated recently by other researchers, based on molecular data and mating studies,[37,38] as well as on morphological characteristics.[49]

D. TAXONOMICAL APPLICATIONS

Traditionally, fungal taxonomy has been based primarily on morphological characters, which may be sufficient in most highly differentiated groups, including sexual species of *Balansia* or *Atkinsonella.* However, evolution at the molecular level is often uncoupled with morphological differentiation. This may particularly apply to fungal groups expressing simple morphological features, like seedborne *Acremonium* endophytes. Isozyme electrophoresis is a powerful tool to determine the genetic cohesiveness of morphologically defined species that are linked to a particular host. A recent study on the endophytes of *F. arundinacea, F. pratensis,* and *L. perenne,* including representative isolates from a large part of the hosts' geographical ranges, identified three distinct taxonomic groupings among isolates from

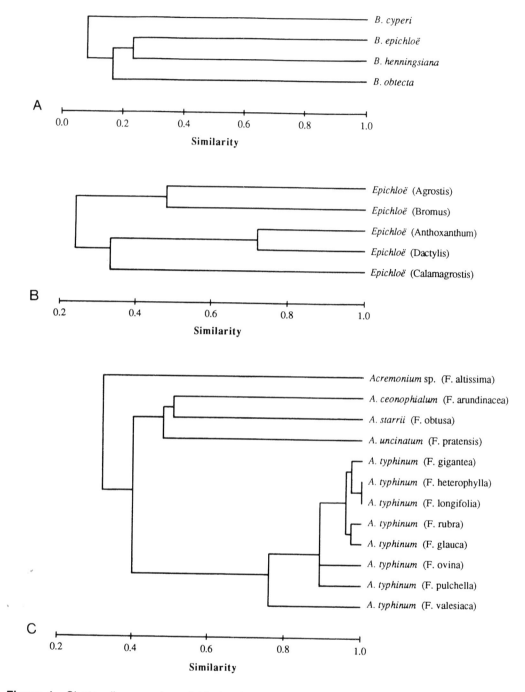

Figure 1 Cluster diagrams (unweighted pair-group method using arithmetic averages: UPGMA) showing the genetic similarities of Balansieae endophytes. The similarity measure is based on Nei's genetic identity, using 12 isozyme loci. (A, species of *Balansia*; B, *Epichloë typhina* from different grass hosts; C, symptomless *Acremonium* endophytes from 12 different *Festuca* host species.)

F. arundinacea, two among isolates from *L. perenne,* and one among isolates from *F. pratensis,* based on isozyme variation.[2] Furthermore, alkaloid profiles of four different alkaloid classes in the natural host grass/endophyte associations were consistent for all isolates within a single isozyme phenotype and for most isolates within a taxonomic grouping. Morphological characters, such as conidial length, showed considerable variation, with overlap among groupings, but were often linked to certain isozyme phenotypes or taxonomic groupings. Only one grouping among isolates from *F. arundinacea* that emerged from isozyme data conformed to the description of *A. coenophialum,* and only one from *L. perenne* conformed to that of *A. lolii.* It has been suggested that the additional groupings on these hosts, though morphologically less pronounced, may also deserve recognition at the rank of species.[2]

The genus *Atkinsonella* offers another example where isozyme analysis facilitated the recognition and delineation of a species. *A. texensis* was recently segregated from *A. hypoxylon* as a distinct species, based on isozyme divergence in addition to differences in stroma size and colony characteristics.[23,24] The new species possessed unique alleles at 9 out of 13 loci, which were not found in populations of *A. hypoxylon.*

E. PHYLOGENETIC RELATIONSHIPS AND COEVOLUTION

In many groups of organisms, isozyme data have proved helpful in inferring phylogenetic relationships, particularly at the species or below-species levels. However, in higher-ranking taxa, such as genera, this application has clear limitations. Isozyme divergence between distantly related taxa is often so excessive that too few, or no, alleles are shared, and those that are shared may be convergent.[30] A recent study looked at the *Acremonium* endophytes from 12 different *Festuca* species representing 5 subgenera.[20] Cluster analysis of isozyme data revealed at least five deeply rooted groupings among isozyme genotypes. The largest grouping comprised the genotypes from all six investigated host species of the subgenus *Festuca* (fine fescues) plus the endophytes of *F. gigantea* (L.) Vill. and *F. pulchella* Schrader, which both represented different subgenera. Distinct groupings were also formed by isolates from *F. arundinacea*, *F. altissima* All., *F. pratensis*, and *F. obtusa* Spreng. (Figure 1C). Assuming that subgeneric classification of *Festuca* species reflects phylogenetic divergence of the hosts, coherent endophyte clades associated with grass subgenera could indicate that they have coevolved with their host plants. Coherency was obvious among the fine fescue endophytes and in endophytes of two other subgenera. However, endophytes from grasses of the subgenus *Schedonorus,* including *F. arundinacea, F. gigantea,* and *F. pratensis,* did not appear to form a genetically coherent group, as several genotypes showed affinity with endophytes from other subgenera. Additional molecular phylogenetic tests are necessary on endophytes as well as on the *Festuca* hosts, to firmly establish evidence for coevolution.

F. ENDOPHYTE POPULATION STRUCTURE

Available data on the frequency and distribution of endophyte genotypes in sympatric populations of several infected hosts indicated that each host species usually harbors one or several genotypes strictly confined to one host. For example, in each of six sympatric woodland grasses in Switzerland, between 1 and 14 unique isozyme genotypes were found.[19] However, there are exceptions from this general observation, in few instances. The *Acremonium* endophytes from several fine fescue species and from *F. gigantea* were identical,[20] and the same genotypes of *E. typhina* were present in *Hystrix patula* Moench and in an *Elymus* species.[25] Identical genotypes of *A. hypoxylon* were also found on two different host species, although in unequal frequencies.[23] In all these cases occasional or regular sexual reproduction is known to occur, which may facilitate contagious spread of genotypes across host species.

As indicated earlier, isozyme variation was detected in most endophyte populations, when adequate isolate samples were analyzed. However, very often, alternate alleles occurred in low frequencies, and only one or two multilocus genotypes were prevalent on a host species. At least in part, this may be due to absent or infrequent sexual reproduction. Alternatively, different isozyme genotypes may not be selectively equivalent, contrary to what has often been assumed.[14] In *A. hypoxylon* from *D. spicata*, a rare genotype with unique alleles at six loci was found only once among 38 isolates of a single population.[23] This genotype was obviously reproductively isolated within an otherwise outcrossing population, since none of the alternate alleles occurred in the more common genotypes of the same, or other, population(s). It has been suggested that this genotype may be self-fertile, which could be the first step in a sympatric speciation process.

In another study isozyme genotypes and allelic frequencies in isolates of type-II *Epichloë* from stromata-forming and symptomless plants of *B. sylvaticum* were compared (Bucheli and Leuchtmann, unpublished data). Among the 129 isolates randomly sampled from a population with both types of plant, a highly significant correlation between isozyme genotypes and the original disease symptoms was observed. Moreover, several alleles of variable loci appeared to be confined to genotypes from stromata-bearing plants. These observations indicate that stromata formation on *B. sylvaticum*, at least in part, may be genetically controlled by the fungus and that allozymes could serve as markers.

III. COMPATIBILITY AND PERSISTENCE

A. EVIDENCE FOR HOST SPECIFICITY

Grass endophytes, like other plant parasitic fungi, are usually limited to one or several host species for which affinity is suggested. Indirect evidence for such a pattern in the balansioid fungi is provided by numerous reports on host ranges of both sexual species[6,8,16] and seedborne *Acremonium* endophytes.[19,47] In contrast, the spreading of the balansioid fungi to new or introduced hosts has been observed in *Myriogenospora atramentosa* (Berk. & Curt.) Diehl and *B. cyperi,* indicating a broad host range for these species.[3,36]

Another source of indirect evidence for host specificity of endophytes comes from the distribution of isozyme genotypes on different sympatric host species. Multiple isolates of *E. typhina* from two *Elymus* and one *Hystrix* species, which often occur intermixed in woodland habitats in southern Indiana, were found to be identical or very similar in their isozyme genotypes,[25] whereas *E. typhina* isolates from *Brachyelytrum erectum* (Schreb.) Beauv., also common in the same area, were genetically quite divergent. As indicated later by reciprocal inoculations, genetically similar strains were cross-compatible with all four hosts, while the divergent *B. erectum* strain was specific to the original host.[26] Host-specific genetic differentiation of endophytes has also been demonstrated in six sympatric woodland grass species in Switzerland, using isozyme analysis.[19] Genotypes of isolates from at least two host populations, each with up to 96 isolates per host, were always restricted to a single host. Endophyte isolates showed very little variation on the same host species, while genetic divergence among isolates from different host species was often considerable, with Nei's genetic identity ranging from 0.2 to 0.8.[19] Genotypes may represent host-specific physiological races that are reproductively isolated and adapted to a host species. However, there was only one host species, *B. sylvaticum,* on which stromata with conidia and ascospores were regularly produced. The other endophytes appeared to be strictly seedborne and no mechanism of contagious spread is known. Host specificity of seedborne endophytes, therefore, requires further clarification by inoculation experiments.

In a similar isozyme study, endophytes of 12 different *Festuca* species revealed a high degree of differentiation in some host species. However, in other hosts, including six species of fine fescue and two more distantly related species, very similar or identical genotypes were found.[20] This finding points to broadly compatible genotypes across host-species lines in fine fescues.

A direct approach to elucidate compatibility patterns in grass endophytes involves experimental inoculations. However, a reliable method to infect mature plants through stigmas or wounds is not available, and successful attempts in the past remained anecdotal.[6,36,46] More recently, techniques have been established to infect grasses with endophytes using young seedlings,[12,17,21] callus cultures,[12] somatic embryos,[13] or plantlets derived from meristem cultures.[33] All of these techniques rely on undifferentiated or young plant tissues to establish infection, suggesting that plants are predisposed to infection only during the initial stages of development.

The most widely applied method of artificial inoculation involves inserting mycelia from pure fungal cultures into the meristematic region, through a small incision at the junction of the mesocotyl and coleoptile of young seedlings (see Table 2). The seedlings should usually be between 5 and 10 days old to achieve successful infections in a compatible endophyte–host combination, whereby the infection rate can vary between 5 and 100%. A major disadvantage of the seedling inoculation technique is that it does not allow the introduction of multiple endophyte genotypes into a single plant genotype, which may be essential for research on plant–endophyte interactions. A most promising alternative is now offered by the recently described technique infecting plantlets derived from meristems of adult plants.[33] This technique does not require callus development and thus overcomes the possibility of somaclonal variation inherent in other tissue-culture techniques.[7] Furthermore, the effectiveness of the meristem method, yielding an infection rate of up to 56% in some endophyte–host combinations, appears to be comparable to, or even higher than, that of the seedling technique.

The seedling inoculation technique has been used to examine compatibility and host range in sexual species of the genera *Atkinsonella, Balansia,* and *Epichloë* as well as in a number of seedborne *Acremonium* endophytes, primarily from *Festuca* and *Lolium* grass hosts (Table 2). Several patterns emerged from these studies, which may have relevance for other groups not yet examined. The two known species of *Atkinsonella* naturally infecting different host genera (*Danthonia* or *Stipa*) in North America were not compatible with seedlings of the alternative host genus.[22] Moreover, *A. hypoxylon* was able to infect four host species of *Danthonia,* which grow in slightly different microhabitats, but in the same geographic area.[4] Experimental inoculations showed that isolates of *D. spicata* and *D. compressa* can

Table 2 **Compatibility of fungal endophytes with nonhost grass species assessed by seedling inoculations**

Fungus	Original Host	Inoculated Host	Compatibility	References
Atkinsonella hypoxylon	*Danthonia spicata*	*Danthonia compressa*	Long term, reciprocal	22
	Danthonia sericea	*D. spicata*	Long term, nonreciprocal	4
	D. sericea	*D. compressa*	Long term, nonreciprocal	4
	D. sericea	*D. epilis*	Long term, reciprocal	4
	D. epilis	*D. spicata*	Long term, nonreciprocal	4
	D. epilis	*D. compressa*	Long term, nonreciprocal	4
Balansia cyperi	*Cyperus virens*	*Cyperus rotundus*	Long term	21
Epichloë typhina	*Agrostis hiemalis*	*Festuca arundinacea*	Short term	19
	Elymus villosus	*Elymus virginicus*	Long term, reciprocal	26
	E. villosus	*Hystrix patula*	Long term, reciprocal	26
	E. villosus	*Brachyelytrum erectum*	Long term, nonreciprocal	26
	E. virginicus	*H. patula*	Long term, reciprocal	26
	E. virginicus	*B. erectum*	Long term, nonreciprocal	26
	H. patula	*F. arundinacea*	Short term	19
	Festuca longifolia	*F. arundinacea*	Persistence not known	39, 40
	F. longifolia	*Lolium perenne*	Persistence not known	39, 40
	Festuca rubra s. l.	*F. longifolia*	Persistence not known	40
	F. rubra s. l.	*F. arundinacea*	Long term, seed transmitted	17
	F. rubra s. l.	*L. perenne*	Long term, seed transmitted	17, 40
	Holcus lanatus	*F. arundinacea*	Short term	19
	L. perenne	*F. arundinacea*	Persistence not known	39
Acremonium sp.	*Festuca pulchella*	*F. arundinacea*	Persistence not known	19
	F. pulchella	*Festuca pratensis*	Persistence not known	19
A. coenophialum	*F. arundinacea*	*L. perenne*	Persistence limited	15, 39
A. lolii	*L. perenne*	*F. arundinacea*	Persistence not known	17, 39, 40
	L. perenne	*Lolium multiflorum*	Persistence not known	18
	L. perenne	*L. temulentum*	Persistence not known	18
A. starrii	*Festuca arizonica*	*F. arundinacea*	Long term	38, 40
	Poa sylvestris	*F. arundinacea*	Short term	19
	Sphenopholis nitida	*F. arundinacea*	Short term	19
A. uncinatum	*F. pratensis*	*F. arundinacea*	Persistence not known	G. C. M. Latch, personal communication

Methods for Cultivating and Detecting Grass/Endophyte Interactions

Chapter 3

In Vitro Approaches for the Study of *Acremonium-Festuca* Biology

Wayne A. Parrott

CONTENTS

I. INTRODUCTION

The economically important symbioses that exist among crop species, such as the *Rhizobium*–legume and the plant–mycorrhizal relationships, have been long recognized and studied. In contrast, the symbiotic relationship between some grasses and their endophytes has only recently been studied. Such an association is exemplified by tall fescue, *Festuca arundinacea* Schreb., and its endophyte, *Acremonium coenophialum* Morgan-Jones and W. Gams. The lack of external evidence for the existence of this fungal endophyte was a limiting factor to its discovery and study; the subtle nature of the physiological changes induced by the endophyte has made it difficult to discern the contributions made by the endophyte towards the fitness of its host. These studies have been further complicated by the reproductive biology of the fescues, related ryegrasses, and their endophytes, which has made it difficult to generate genetically defined materials for the critical studies necessary to define the interactions that occur between host and endophyte.

Firstly, only one partner in this relationship has the ability for sexual reproduction. Tall fescue and its near relatives are cross-pollinated species. The implication is that every single seed-derived plant is a genetically distinct individual. Secondly, the endophyte reproduces clonally, with no opportunity for meiosis. In addition, there is no known mechanism in nature for a tall fescue plant to be infected with a fungal endophyte of another strain; the possibility of karyogamy between strains is also precluded. Despite these reproductive features, the judicious application of *in vitro* techniques can help generate the genetically defined materials necessary for critical studies of host–endophyte symbiosis.

The life cycle of *A. coenophialum* is well defined.[3] Individual hyphae survive in meristematic and leaf sheath areas of the plant. When the flowering stem begins to elongate, the mycelium grows within the infloresesce meristem and eventually infects the maternal tissues of the ovule, where it remains until the seed is mature and the seed begins to germinate. The mycelium in the seed infects the embryo immediately, first infecting the scutellum and then the plumule. However, there is a 3- to 4-week window during seedling germination, when uninfected seedlings can be artificially infected with an endophytic mycelium. While young seedlings can be artificially infected with the endophyte, it is not possible to infect older or mature plants.[30] The inability to infect individual ramets from one tall fescue clone with different endophytes has been a major limitation to the study of endophyte–host interactions. The ability to culture the endophyte without its host, however, permits the dissection of some of these parameters and permits the generation of novel endophyte traits or endophyte–host combinations that may be of agricultural use.

II. CULTURE OF *ACREMONIUM*

Although *A. coenophialum* is an obligate symbiont, it may be cultured *in vitro* on a variety of media, where it grows at an extremely slow rate.[3] The ability of *Acremonium* to conidiate in culture is essential for its taxonomic classification.[31] Aside from maintenance of laboratory strains, the endophyte, once in culture, may be monitored for production of secondary metabolites, or the media may be manipulated to discern the nutritional requirements for either growth or production of secondary metabolites. Identification of the latter will be essential to help devise strategies to prevent accumulation of toxic alkaloids in pastures.

A. MEDIA

The components for endophyte media are summarized in Table 1. The endophyte may be isolated from tall fescue and cultured on potato dextrose agar (PDA) supplemented with 50 mg/l each of streptomycin and chloramphenicol,[2] or on cornmeal-malt extract agar (CMM). Subsequent growth has been compared in three liquid media formulations, two of which were undefined (M43 and M96) and the other was semidefined (M102). The best growth rate, as determined by mycelial dry weight, was obtained on the M43 medium.[2] *Acremonium* has subsequently been reported to grow on cornmeal dextrose soytone agar (CDSA). Growth on "medium No. 9" was reported to be superior to that on CDSA,[6] while growth on double- and triple-strength PDA was superior to that on standard PDA or M102.[36] The highest growth rate was achieved simply by doubling the amount of dextrose to 40 g/l in potato broth.[36] Medium No. 9 was later modified to GY agar,[29] which has improved buffering capacity. The endophyte may be cultured with a dialysis membrane between the endophyte and the medium. This prevents the endophyte from growing into the medium, facilitating the harvest of the mycelium freed of medium, for further study or chemical analysis.[36]

The endophyte has been reported to grow on a medium containing only glucose and yeast extract, with the optimal concentrations being 6 and 0.35%, respectively.[15] Higher concentrations were inhibitory.

Table 1 **Composition of semidefined media used for the culture of *Acremonium coenophialum***

	Media[a]					
Ingredient	**M43[b]**	**M96[b]**	**M102[b]**	**M104T[c]** (g/l)	**No. 9[e]**	**GY[d]**
Mannitol		10				
Sorbitol				100		
Sucrose	30	30	30			
Starch		15				
Dextrose					5	
Glucose				40		5
Malt extract	20	10	20			
Yeast extract	2.5	2	1	3	2	2
Peptone	1		2			
NH_4 succinate		0.68				
Glutamic acid				10		
Tryptophan				0.8		
$MgSO_4 \cdot 7H_2O$			0.5	0.3	0.5	0.5
KCl			0.5			
KH_2PO_4			1	0.5	5	3
$K_2HPO_4 \cdot 3H_2O$						2
pH	6.0	6.0	6.0	5.6	6.5	6.5

[a] Agar, if desired, is at 20 g/l; [b,c,d,e] References 2, 1, 6, and 29, respectively.

Such concentrations are low compared to those of other fungi; for example, *Aspergillus*.[15] This nutrient-poor solution may simulate the nutrient status found in the apoplasm of tall fescue. Tall fescue callus alone will support endophyte growth,[9] and the latter approach may have the potential to provide for greater long-term stability of cultured endophyte strains. Growth on callus would most likely approximate the physiological and nutritional conditions the endophyte encounters in its host. Furthermore, the medium for callus growth[33] also supports growth of the endophyte.[28,55]

B. NUTRITIONAL REQUIREMENTS

The nutritional requirements for one isolate of *A. coenophialum* were extensively studied by Kulkarni and Nielson,[29] using a series of chemically defined media. The endophyte was largely unable to utilize pentose sugars as a carbon source. Among the hexoses, it was able to use fructose, glucose, and mannose equally well, but unable to use galactose, sorbose, or rhamnose. Among disaccharides, the endophyte used sucrose and trehalose, but not maltose or lactose. It was able to use the oligosaccharide raffinose, but not the polysaccharides pectin and cellulose. It was marginally efficient at utilizing soluble starch as a carbon source. Carboxylic and uronic acids were generally not effective as carbon sources, while polyols were variable, with mannitol and sorbitol being the most effective. Among nitrogen sources, inorganic ammonium was as effective as arginine, asparagine, cysteine, glutamine, proline, and serine. The remaining amino acids were not very effective in supporting growth. Among the vitamins, an essential requirement was identified for thiamine. Interestingly enough, an optimized, defined medium was no better than standard undefined media, in supporting growth.

In a similar study evaluating the requirements of a wide range of *Acremonium* species, White et al.[55] confirmed the ability of *A. chisosum, A. coenophialum,* and *A. uncinatum* to use fructose, glucose, and sucrose. Arabinose was also an effective carbon source, while xylose supported little growth, and potassium acetate supported none at all. For all carbohydrates tested at two concentrations, 1% was as effective as 3%. In addition, *A. chisosum, A. starri,* and *A. typhinum* (= *Epichloë typhina* (Fries) Tul.) were tested for the ability to hydrolyze protein and utilize glycerol, soybean oil, and paraffin as carbon sources. The soybean oil supported growth of all the *Acremonium* species, while only *A. chisosum* and *A. starri* were able to hydrolyze protein.

Isolates of the stroma-forming *A. typhinum* show similar carbohydrate utilization patterns to those of the nonstroma-forming species, although they grew more rapidly and tended to grow better at the lower concentration of fructose than did the other *Acremonium*-type endophytes.[55] Even among isolates within the *A. typhinum* taxon, those able to form stroma grow faster in culture than isolates of the symptomless type, i.e., *A. coenophialum*.[57]

Finally, there is evidence that the remaining factors in the growth medium can affect the response to individual nutrients. In a defined medium consisting of mineral salts, the preference for carbon source was mannitol > fructose > mannose = sucrose > dextrose, for biomass production,[29] while in a potato broth background, the preference was mannitol > dextrose > sucrose > mannose > fructose.[36] As has been observed before,[3] such a wide diversity of strains and protocols have been used that it is impossible to draw definite conclusions as to the precise nutritional requirements of *Acremonium*-type endophytes. In addition, it is also likely that differences in substrate utilization exist among isolates of the same species.

C. METABOLITE PRODUCTION

The initial culture medium can have lasting effects on endophyte metabolism. Bacon,[1] working with 83 endophyte isolates growing on M104T medium, found that mean total ergot alkaloid production was always higher when the original isolation was on M102 medium, as compared to CMM. The mean ergot alkaloid concentration for the isolates cultured on CMM was 215 mg/l, while that for the isolates on M102 was 476 mg/l. In addition, the ability of cultured endophytes to produce ergot alkaloids can decrease over time. Alkaloid production during *in vitro* culture may not be a reliable indicator of an endophytic strain's ability to produce alkaloids *in planta*; thus, caution is warranted in strategies that rely on *in vitro* screening for alkaloid production, as a means to identify endophyte isolates with low capacity to produce ergot alkaloids. The ultimate verification of an isolate's ability for ergot alkaloid production should take place in several different host genotypes.

Metabolites whose synthesis was specifically attributed to *Acremonium* through *in vitro* culture include chanoclavine I,[37] 6,7-secoagroclavine, agroclavine, elymoclavine, penniclavine, festuclavine, ergovaline, ergovalinine,[38] ergosterol, ergosterol peroxide, and ergosta-4,6,8-(14),22-tetraen-3-one.[16] *In vitro* assays have also been used to determine that *A. coenophialum* has the ability to inhibit the growth of several soil saprophytic and pathogenic fungi[56] and has the ability to synthesize the auxin 3-indoleacetic acid, a plant growth regulator with the ability to affect the growth of the host plant.[17]

It is clear from results obtained thus far that *in vitro* culture of *Acremonium* may be used to distinguish endophyte traits from host traits, while other traits have been attributed to an interaction of the endophyte with its grass host (such as production of the lolines). The possibility cannot be excluded that these compounds are produced by the grass in response to the endophyte. With further refinement of media and culture protocols, it should become possible to identify the precursors that the fescue host must supply for the synthesis of secondary metabolites by the endophyte. Confounding the media-specific "preconditioning" effects discussed previously is the potential instability of the fungus in culture.[1] In addition, it may be possible that some endophytic traits require even more-defined culture conditions than those established to date or that a physical association between the endophyte and the fescue host is necessary for their expression and thus may never be expressed when the fungus is cultured alone. This may be the case for loline and perlolidine alkaloids, which are never found in noninfected plants. However, they are not found in cultured endophytes either.[3]

A. persicinum excretes β-glucans, an exopolysaccharide with several potential commercial applications, when cultured *in vitro* using 0.8 g/l total nitrogen in the form of KNO_3.[44] The level of exopolysaccharide can be increased by the addition of vegetable oils,[45] while it is decreased by the use of polypropylene glycol as an antifoaming compound.[47] While the yields of β-glucan are high in laboratory shake cultures, reaching levels as high as 15 g/l, scale-up attempts using bioreactors have had mixed success. Yields are decreased in stirred-tank designs unless the stirring speeds are a low 100 rpm. However, yields in air-lift reactors are comparable to those achieved in shake cultures. Part of these effects may be due to the lower sheer rates found in the air-lift design. In addition, the β-glucan accumulation began after a 10% drop in the original level of dissolved oxygen (0.25 l of air per liter of culture), and the nitrogen in the medium had been depleted.[46]

III. CULTURE OF THE GRASS HOST

A. SOMATIC EMBRYOGENESIS AND REGENERATION

Within the laboratory the endophyte is cultured as an intact organism, while the fescue host is cultured as isolated tissues or cells. Such *in vitro* cell cultures can be used to "rejuvenate" mature plants and allow them to recapitulate their embryogenic/seedling stages, which are susceptible to artificial infection by the endophyte. The ability to recover somatic embryos from a widespread variety of grass species is feasible, with the general pattern that mitotically active tissues (leaf basal tissue, intercalary meristems, embryos, or immature inflorescenses) are exposed to a synthetic auxin, usually 2,4-dichlorophenoxyacetic acid (2,4-D). Somatic embryos appear upon removal of the auxin in the medium or upon reduction of the auxin to a level that will not inhibit the development of somatic embryos.[53] The *in vitro* behavior of the fescues and ryegrasses is typical of this family.

Somatic embryos differ from zygotic embryos in two principle ways: first, somatic embryos have the same genotype as the plant that produced them; and second, a somatic embryo is essentially naked, lacking maternal tissues around it. However, zygotic and somatic embryos are functionally equivalent, and both may be infected during germination, with an *Acremonium* strain. Since virtually unlimited numbers of somatic embryos may be obtained from any one genotype of fescue, a mechanism is available whereby one genotype may be infected with a variety of *Acremonium* strains.

Murashige and Skoog basal medium[33] is the most common medium used for the culture of tall fescue tissues, although Schenk and Hildebrandt basal medium[41] has also been effective.[12,21] The use of N6 medium[14,27] has been discontinued. The auxin of choice to obtain somatic embryos is clearly 2,4-D, although both 2,4,5-trichlorophenoxyacetic acid (2,4,5-T),[21,28,51] 4-amino-3,5,6-trichloropicolinic acid (picloram),[28] and *p*-chlorophenoxyacetic acid (pCPA) can be used,[12,21] while the use of α-naphthaleneacetic acid is not at all effective.[51] If the production of callus, rather than embryos, is the goal, then the auxin of choice may be 2,4,5-T, which is reported to result in better callus growth than does the use of 2,4-D.[7,12] This could be important if callus itself is used as a medium for endophyte growth.

The concentrations of 2,4-D required for embryo induction are somewhat plastic, with results having been obtained within a range from 10 mg/l[14] down to 2 mg/l.[27] Both 2 mg/l[27] and 6.6 mg/l[28] have been reported as being optimal, a discrepancy that probably reflects the differences in plant genotypic responses to the inducing auxins. In fact, genotype effects have been described as being stronger than auxin effects, in tall fescue,[21] a phenomenon also true of annual and perennial ryegrasses.[10] A few reports have used low levels of the synthetic cytokinin 6-benzylaminopurine[13,14,56] during the regeneration process. However, given the success of regeneration protocols without exogenous cytokinins, it is unclear what role, if any, the cytokinin plays in the regeneration process.

As mentioned previously, induction of the original tissue into an embryogenic state is accomplished with exposure to an auxin, the concentration of which must be lowered to a permissive state if the induced tissue is to form somatic embryos. In the case of monocotyledonous species, the developing somatic embryos recapitulate the globular, scutellar, and coleoptilar stages of ontogeny.[42] Initially, it was thought that low levels of auxin were required after induction, to achieve continued cell division during embryogenesis in grasses.[52] Accordingly, protocols for tall fescue regeneration used 2,4-D levels ranging from 2 mg/l [18] to 0.5 mg/l [19,32] or less. It is now clear that no exogenous auxin is necessary for the recovery of somatic embryos and their subsequent germination into plants of red fescue[50,59,61] and tall fescue.[21,28] However, older callus may lose its embryogenic capacity; in which case both elevated sucrose levels[61] and activated charcoal[59] may restore regeneration capacity.

Thus far the most widely used explant for fescue tissue culture is the zygotic embryo. However, this is not a viable explant if the objective is to work on plants of a defined genotype; in which case, immature inflorescences,[13,18,21] elongating peduncles or internodes,[27] and elongating basal leaf sheaths[28] can be effectively used as explants. Once established, fescue cultures may be grown as suspensions.[14,39,60] Not only do suspension cultures greatly increase the scale-up culture potential, they have also served as a source of protoplasts capable of regeneration.[58]

B. ARTIFICIAL INOCULATIONS

As mentioned previously, somatic embryogenesis provides a method of inserting various endophyte isolates into different ramets of a given tall fescue clone. Such a technique was pioneered by Johnson et al.,[26] who took advantage of the endophyte's ability to grow on callus. Callus cultures were inoculated with endophyte mycelia and allowed to grow for 10 weeks, after which the 2,4-D concentration was lowered in the medium to permit the formation of somatic embryos. Approximately 17% of the resulting plants were infected. Kearney et al.[28] later identified the time of 2,4-D removal from the medium, when somatic embryos begin to elongate, as the best time for inoculation with endophyte mycelia (Figure 1).

Plants recovered from inoculations of somatic embryos have revealed that the level of ergopeptide alkaloid production is mostly controlled by the tall fescue host, even though the alkaloids themselves are produced by the endophyte.[25] In retrospect, given the ability of culture media to modulate alkaloid production, as described previously, it is not surprising that the host genotype has a similar effect. The implication is that research directed towards the amelioration of fescue toxicosis cannot ignore the effects of the host genotype and concentrate on modifications of the endophyte as the sole strategy to reduce ergopeptide alkaloid production. The selection of low ergot-producing endophyte isolates should be coupled to the use of recurrent selection to reduce the host plants' ability to permit high levels of ergot alkaloid synthesis by the endophyte.

The possibility of deriving new host–endophyte combinations through the transfer of endophytes between species has been raised.[3] Limited progress towards this goal was made by Latch and Christianson,[30] who managed to transfer *A. lolii* from ryegrass into tall fescue, although the reciprocal transfer was not successful. This approach is unlikely to be successful to any large extent, as successful symbiosis requires a compatibility between the genome of an endophyte and that of each individual host. Whereas a given endophyte may be compatible with a range of host genotypes, it is improbable that a given endophyte will be compatible with every host genotype of a given species, or that every endophyte/host combination will be mutually beneficial. In our own work,[28] it was impossible to recover certain fescue genotype–endophyte combinations, again suggesting that even within one species, not all genotypes are compatible with certain strains.

C. GENETIC TRANSFORMATION
1. Host Transformation

The advent of genetic transformation technology offers a more precise method to modify the host–endophyte combination, with the possibility of either or both partners may be subject to transformation. Tall fescue has been transformed through the electroporation of protoplasts.[24] However, this technique for grass transformation will very likely be replaced by easier, less mutagenic techniques such as those used for maize,[11,22,23] rice,[4,5] or wheat.[54] One example of what may be accomplished is the transformation

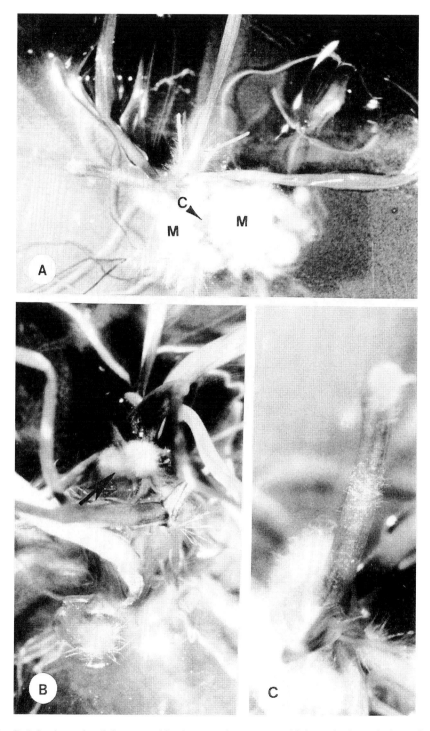

Figure 1 Reinfection of tall fescue with *Acremonium coenophialum* via inoculation of somatic embryos with mycelium. (A) A somatic embryo that has germinated into a seedling with multiple leaves emerging from a callus, C, that has almost been completely overgrown with mycelium, M, 61 days postinoculation. (B) A germinated somatic embryo ringed with mycelium (arrow) around its crown 84 days postinoculation. (C) *A. coenophialum* mycelium growing out of the leaves of a germinated somatic embryo inside a culture vessel 61 days after inoculation.

of fescue to produce "plantibodies", i.e., antibodies produced in a plant, against the ergot alkaloids responsible for fescue toxicosis. The production of functional antibodies has been demonstrated in plants,[48] and monoclonal antibodies active against various ergot alkaloids are already available,[43] including one against the lysergic acid ring common to the ergopeptine and clavine alkaloids.[49]

2. Endophyte Transformation

Alternatively, the endophyte itself may be transformed, then the transformed endophyte reinserted into its host plant, a procedure pioneered with *Acremonium lolii*, which was transformed and reinserted into its ryegrass host[34] (see also Chapter 11). Endophyte, rather than host, transformation can have several advantages. First, fungal transformation remains far simpler than grass transformation, making it easier to deploy genes more rapidly via fungal transformation. Second, elite genotypes (e.g., the parental clones for a synthetic cultivar) containing an engineered endophyte may be rid of their endophyte and replaced with another engineered strain, perhaps containing a more efficient version of a transgene. Finally, deployment of the transgene would be more easily managed, as spread would be limited to seed, and not pollen, dispersal.

D. SOMACLONAL VARIATION

A final consideration for the use of any *in vitro* technique is the occurrence of somaclonal variation, a term that refers to a wide variety of mutations that can occur in cultured tissues. If the goal is to assess host–endophyte interactions, it is difficult to determine if any variability is the result of somaclonal variation or of the endophyte on the host. There is no information available as to the extent of somaclonal variation that may occur in cultured *Acremonium* species. As mentioned previously, cultured endophytes can lose the ability to produce ergot alkaloids, although this may be epigenic in nature, rather than a result of mutations.

In contrast, the incidence of somaclonal variation in cultured fescue tissues has been extensively documented.[18,19,40] The DNA content in fescue callus has been found to increase with time,[8] and aneuploidy may be present in regenerated plants.[12] The amount of somaclonal variation appears to be a function of time in culture, with increasing somaclonal variation with increasing time in culture.[20,27] Accordingly, protocols may be modified to minimize somaclonal variation by decreasing the time in culture to a minimum and minimizing callus production prior to somatic embryo formation.[28] The use of alternative regeneration schemes, such as micropropagation from meristems, does not prevent somaclonal variation, and if anything, regeneration via somatic embryogenesis appears to be less subject to somaclonal variation[35] than is regeneration via other pathways.

IV. CONCLUSIONS

In vitro technology for the study of fungal endophytes and their grass hosts has advanced rapidly over the past few years. These can be used to further characterize the endophyte, define the endophyte–host relationship, and generate novel materials. Nevertheless, there is still a substantial need for additional research effort. With the continued refinement of these techniques, the range of issues that can be addressed and the types of materials that can be generated can only continue to increase.

REFERENCES

1. **Bacon, C. W.,** Procedure for isolating the endophyte from tall fescue and screening isolates for ergot alkaloids, *Appl. Environ. Microbiol.,* 54:2615–2618, 1988.
2. **Bacon, C. W., J. K. Porter, J. D. Robbins, and E. S. Luttrell,** *Epichloë typhina* from toxic tall fescue grasses, *Appl. Environ. Microbiol.,* 34:576–581, 1977.
3. **Bacon, C. W. and M. R. Siegel,** Endophyte parasitism of tall fescue, *J. Prod. Agric.,* 1:45–55, 1988.
4. **Cao, J., X. Duan, D. McElroy, and R. Wu,** Regeneration of herbicide resistant transgenic rice plants following microprojectile-mediated transformation of suspension culture cells, *Plant Cell Rep.,* 11:586–591, 1992.
5. **Christou, P., T. L. Ford, and M. Kofron,** Production of transgenic rice (*Oryza sativa* L.) plants from agronomically important indica and japonica varieties via electric discharge particle acceleration of exogenous DNA into immature zygotic embryos, *Biotechnology,* 9:957–962, 1991.

6. **Clark, E. M., J. F. White, and R. M. Patterson,** Improved histochemical techniques for the detection of *Acremonium coenophialum* in tall fescue and methods of *in vitro* culture of the fungus, *J. Microbiol. Methods,* 1:149–155, 1983.

7. **Conger, B. V., J. V. Carabia, and K. W. Lowe,** Comparison of 2,4-D and 2,4,5-T on callus induction and growth in three *Gramineae* species, *Environ. Exp. Bot.,* 18:163–168, 1978.

8. **Conger, B. V., K. W. Lowe, and J. V. Carabia,** Relative DNA content of cells in *Festuca arundinacea* and *Dactylis glomerata* calli of different ages, *Environ. Exp. Bot.,* 20:401–408, 1980.

9. **Conger, B. V. and J. K. McDaniel,** Use of callus cultures to screen tall fescue seed samples for *Acremonium coenophialum, Crop Sci.,* 23:172–174, 1983.

10. **Creemers-Molenaar, J., J. P. M. Loeffen, and P. van der Valk,** The effect of 2,4-dichlorophenoxyacetic acid and donor plant environment on plant regeneration from immature inflorescence-derived callus of *Lolium perenne* L. and *Lolium multiflorum* L., *Plant Sci.,* 57:165–172, 1988.

11. **D'Halluin, K. D., E. Bonne, M. Bossut, M. De Beuckeleer, and J. Leemans,** Transgenic maize plants by tissue electroporation, *Plant Cell,* 4:1495–1505, 1992.

12. **Dahleen, L. S. and G. C. Eizenga,** Meiotic and isozymic characterization of plants regenerated from euploid and selfed monosomic tall fescue embryos, *Theor. Appl. Genet.,* 79:39–44, 1990.

13. **Dale, P. J. and S. J. Dalton,** Immature inflorescence culture in *Lolium, Festuca, Phleum* and *Dactylis, Z. Pflanzenphysiol.,* 111:39–45, 1983.

14. **Dalton, S. J.,** Plant regeneration from cell suspension protoplasts of *Festuca arundinacea* Schreb. (tall fescue) and *Lolium perenne* L. (perennial ryegrass), *J. Plant Physiol.,* 132:170–175, 1988.

15. **Davis, N. D., E. M. Clark, K. A. Schrey, and U. L. Diener,** In vitro growth of *Acremonium coenophialum,* an endophyte of toxic tall fescue grass, *Appl. Environ. Microbiol.,* 52:888–891, 1986.

16. **Davis, N. D., R. J. Cole, J. W. Dorner, J. D. Weete, P. A. Backman, E. M. Clark, C. C. King, S. P. Schmidt, and U. L. Diener,** Steroid metabolites of *Acremonium coenophialum,* an endophyte of tall fescue, *J. Agric. Food Chem.,* 34:105–108, 1986.

17. **De Battista, J. P., C. W. Bacon, R. Severson, R. D. Plattner, and J. H. Bouton,** Indole acetic acid production by the fungal endophyte of tall fescue, *Agron. J.,* 82:878–880, 1990.

18. **Eizenga, G. C.,** Cytogenetic and isozymic characterization of anther-panicle culture derived tall fescue aneuploids, *Euphytica,* 36:175–179, 1987.

19. **Eizenga, G. C.,** Meiotic analysis of tall fescue somaclones, *Genome,* 32:373–379, 1989.

20. **Eizenga, G. C. and P. L. Cornelius,** Comparison of the isozyme variation in tall fescue parents and their somaclones, *Euphytica,* 51:249–256, 1991.

21. **Eizenga, G. C. and L. S. Dahleen,** Callus production, regeneration and evaluation of plants from cultured inflorescences of tall fescue (*Festuca arundinacea* Schreb.), *Plant Cell Tissue Org. Cult.,* 22:7–15, 1990.

22. **Fromm, M. E., F. Morrish, C. Armstrong, R. Williams, J. Thomas, and T. M. Klein,** Inheritance and expression of chimeric genes in the progeny of transgenic maize plants, *Biotechnology,* 8:833–839, 1990.

23. **Gordon-Kamm, W. J., T. M. Spencer, M. L. Mangano, T. R. Adams, R. J. Daines, W. G. Start, J. V. O'Brien, S. A. Chambers, W. R. Adams, Jr., N. G. Willetts, T. B. Rice, C. J. Mackey, R. W. Krueger, A. P. Kausch, and P. G. Lemaux,** Transformation of maize cells and regeneration of fertile transgenic plants, *Plant Cell,* 2:603–618, 1990.

24. **Ha, S.-B., F.-S. Wu, and T. K. Thorne,** Transgenic turf-type tall fescue (*Festuca arundinacea* Schreb.) plants regenerated from protoplasts, *Plant Cell Rep.,* 11:601–604, 1992.

25. **Hill, N. S., W. A. Parrott, and D. D. Pope,** Ergopeptine alkaloid production by endophytes in a common tall fescue genotype, *Crop Sci.,* 31:1545–1547, 1991.

26. **Johnson, M. C., L. P. Bush, and M. R. Siegel,** Infection of tall fescue with *Acremonium coenophialum* by means of callus culture, *Plant Dis.,* 70:380–382, 1986.

27. **Kasperbauer, M. J., R. C. Buckner, and L. P. Bush,** Tissue culture of annual ryegrass x tall fescue F1 hybrids: callus establishment and plant regeneration, *Crop Sci.,* 19:457–460, 1979.

28. **Kearney, J. F., W. A. Parrott, and N. S. Hill,** Infection of somatic embryos of tall fescue with *Acremonium coenophialum, Crop Sci.,* 31:979–984, 1991.

29. **Kulkarni, R. K. and B. D. Nielsen,** Nutritional requirements for the growth of a fungus endophyte of tall fescue grass, *Mycologia,* 78:781–786, 1986.

30. **Latch, G. C. M. and M. J. Christensen,** Artificial infection of grasses with endophytes, *Ann. Appl. Biol.,* 107:17–24, 1985.

31. **Latch, G. C. M., M. J. Christensen, and G. J. Samuels,** Five endophytes of *Lolium* and *Festuca* in New Zealand, *Mycotaxon,* 20:535–550, 1984.

32. **Lowe, K. W. and B. V. Conger,** Root and shoot formation from callus cultures of tall fescue, *Crop Sci.,* 19:397–400, 1979.

33. **Murashige, T. and F. Skoog,** A revised medium for rapid growth and bioassays with tobacco tissue cultures, *Physiol. Plant.,* 15:473–497, 1962.

34. **Murray, F. R., G. C. M. Latch, and D. B. Scott,** Surrogate transformation of a perennial ryegrass, *Lolium perenne* using genetically modified *Acremonium* endophyte, *Mol. Gen. Genet.,* 233:1–9, 1992.

35. **Ozias-Akins, P. and I. K. Vasil,** In vitro regeneration and genetic manipulation of grasses, *Physiol. Plant.,* 73:565–569, 1988.

36. **Pope, D. D. and N. S. Hill,** Effects of various culture media, antibiotics, and carbon sources on growth parameters of *Acremonium coenophialum*, the fungal endophyte of tall fescue, *Mycologia,* 83:110–115, 1991.

37. **Porter, J. K., C. W. Bacon, and J. D. Robbins,** Ergosine, ergosinine, and chanoclavine I from *Epichloë typhina*, *J. Agric. Food Chem.,* 27:595–598, 1979.

38. **Porter, J. K., C. W. Bacon, J. D. Robbins, and D. Betowski,** Ergot alkaloid identification in Clavicipitaceae systemic fungi of pasture grasses, *J. Agric. Food Chem.,* 29:653–657, 1981.

39. **Rajoelina, S. R., G. Alibert, and C. Planchon,** Continuous plant regeneration from established embryogenic cell suspension cultures of Italian ryegrass and tall fescue, *Plant Breed.,* 104:265–271, 1990.

40. **Reed, J. N. and B. V. Conger,** Meiotic analyses of tall fescue *Festuca arundinacea* plants regenerated from callus cultures, *Environ. Exp. Bot.,* 25:277–284, 1985.

41. **Schenk, R. U. and A. C. Hildebrandt,** Medium and techniques for induction and plant growth of monocotyledonous and dicotyledonous plant cell cultures, *Can. J. Bot.,* 50:199–204, 1972.

42. **Senaratna, T.,** Artificial seeds, *Biotech. Adv.,* 10:379–392, 1992.

43. **Shelby, R. A. and V. C. Kelley,** Detection of ergot alkaloids in tall fescue by competitive immunoassay with a monoclonal antibody, *Food Agric. Immunol.,* 3:169–177, 1991.

44. **Stasinopoulos, S. J. and R. J. Seviour,** Exopolysaccharide formation by isolates of *Cephalosporium* and *Acremonium*, *Mycol. Res.,* 92:55–60, 1989.

45. **Stasinopoulos, S. J. and R. J. Seviour,** Stimulation of exopolysaccharide production in the fungus *Acremonium persinicum* with fatty acids, *Biotech. Bioeng.,* 36:778–782, 1990.

46. **Stasinopoulos, S. J. and R. J. Seviour,** Exopolysaccharide production by *Acremonium persinum* in stirred-tank and air-lift fermentors, *Appl. Microbiol. Biotechnol.,* 36:465–468, 1992.

47. **Stasinopoulos, S. J., R. J. Seviour, and D. F. Auer,** Inhibition of fungal exopolysaccharide production by chemical antifoams, *Lett. Appl. Microbiol.,* 8:91–93, 1989.

48. **Swain, W. F.,** Antibodies in plants, *Tibtech,* 9:107–109, 1991.

49. **Thompson, F. N., N. S. Hill, D. L. Dawe, and J. A. Stuedemann,** The effects of passive immunization against lysergic acid derivatives on serum prolactin in steers grazing endophyte-infected tall fescue, In: D. E. Hume, G. C. M. Latch, and H. S. Easton, Eds., *Proc. 2nd Int. Symp. on* Acremonium/*Grass Interactions*, AgResearch, Grasslands Research Centre, Palmerston North, New Zealand, 1993, 135.

50. **Torello, W. A., R. Rufner, and A. G. Symington,** The ontogeny of somatic embryos from long-term callus cultures of red fescue, *Hortscience,* 20:938–942, 1985.

51. **Torello, W. A., A. G. Symington, and R. Rufner,** Callus initiation, plant regeneration, and evidence of somatic embryogenesis in red fescue, *Crop Sci.,* 24:1037–1040, 1984.

52. **Vasil, I. K.,** Somatic embryogenesis and its consequences in the gramineae, In: R. R. Henke, K. W. Hughes, M. J. Constantin, and A. Hollaender, Eds., *Tissue Culture in Agriculture and Forestry*, Vol. 31, Plenum Press, New York, 1985.

53. **Vasil, I. K.,** Developing cell and tissue culture systems for the improvement of cereal and grass crops, *J. Plant Physiol.,* 128:193–218, 1987.

54. **Vasil, V., A. M. Castillo, M. E. Fromm, and I. K. Vasil,** Herbicide resistant fertile trasgenic wheat plants obtained by microprojectile bombardment of regenerable embryogenic callus, *Biotechnology,* 10:667–674, 1992.

55. **White, J. F., Jr., J. P. Breen, and G. Morgan-Jones,** Substrate utilization in selected *Acremonium*, *Atkinsonella* and *Balansia* species, *Mycologia,* 83:601–610, 1991.

56. **White, J. F., Jr. and G. T. Cole,** Endophyte-host associations in forage grasses. III. *In vitro* inhibition of fungi by *Acremonium coenophialum*, *Mycologia,* 77:487–489, 1985.

57. **White, J. F., Jr., A. C. Morrow, G. Morgan-Jones, and D. A. Chambless,** Endophyte-host associations in forage grasses. XIV. Primary stromata formation and seed transmission in *Epichloe typhina*: developmental and regulatory aspects, *Mycologia,* 83:72–81, 1991.
58. **Zaghmout, O. M. F. and W. A. Torello,** Isolation and culture of protoplasts from embryogenic suspension cultures of red fescue (*Festuca rubra* L.), *Plant Cell Rep.,* 9:340–343, 1990.
59. **Zaghmout, O. M. F. and W. A. Torello,** Enhanced regeneration from long-term callus cultures of red fescue by pretreatment with activated charcoal, *Hortscience,* 23:615–616, 1988.
60. **Zaghmout, O. M. F. and W. A. Torello,** Somatic embryogenesis and plant regeneration from suspension cultures of red fescue, *Crop Sci.,* 29:815–817, 1989.
61. **Zaghmout, O. M. F. and W. A. Torello,** Restoration of regeneration potential of long-term cultures of red fescue (*Festuca rubra* L.) by elevated sucrose levels, *Plant Cell Rep.,* 11:142–145, 1992.

Chapter 4

Stains, Media, and Procedures for Analyzing Endophytes

Charles W. Bacon and James F. White, Jr.

CONTENTS

I. INTRODUCTION

The techniques for examining, culturing, and using endophytes have been slowly developing over the few years that we have been actively studying them. These techniques vary from simple to complex and have been shown to be fundamental for *in vitro* and *in planta* studies of grass endophytes. In some instances several techniques or media formulations are unavailable in published form. Furthermore, to our knowledge there is no publication where all the common procedures are assembled for easy reference.

   This chapter is intended to provide researchers with the basic information and techniques for examining, culturing, and manipulating grass endophytes. This information is intended as a supplement to specific protocols presented in several chapters of this book. Additional general references dealing with techniques include those on ELISA detection of endophytes,[10,20] artificial infection of grasses with endophytes,[16] tissue culture of grasses,[14] methods for analysis of ergot alkaloids[21,25] and peramine analysis.[9,12,22]

II. STAINS

A. ANILINE BLUE-LACTIC ACID STAIN

Lactic acid (85%)	50.0 ml
Aniline blue stain	0.1
Water	100.0 ml

To prepare this stain, mix aniline blue powder with water, and once the stain is dissolved, add lactic acid. This stain is frequently used to examine scrapings of plant tissues from fresh and dried culms and leaf sheaths, for the presence of endophytic mycelia.[6,7,18] Tissues from sheaths or culms are removed, macerated if necessary, placed on a slide, covered with stain (three or more drops), and gently heated for a few seconds to aid penetration of the stain. Excess stain is then removed, water is added, and the tissues are examined with a light microscope at ×400 or more magnification, for presence of nonbranching, intercellular, blue-stained hyphae. If dried culms are examined for endophytic mycelia, a stain mixture of 0.1% aqueous aniline blue without lactic acid may be used, and heating of the slides is omitted. Slides may be stored for 2 to 4 weeks by sealing the coverslip with clear fingernail polish. Grass tissue containing some species of endophytes, e.g., *Balansia* Speg., should be cleared in order to examine endophytic hyphae (see Section C).

B. ROSE BENGAL STAIN

Ethanol (absolute)	5.0 ml
Rose bengal stain (Fisher certified)	0.5 g
Distilled water	95.0 ml

To prepare this stain, mix rose bengal powder with distilled water; then add alcohol. This stain is used to examine endophytic mycelia in grasses.[23] Tissue is scraped from culms or leaf sheaths onto a slide. Several drops of rose bengal stain are applied to tissue that is then macerated. After 2 or 3 min at room temperature, a coverslip is applied. The tissue is examined using a light microscope fitted with a green interference filter. Endophytic mycelia appear red, while plant tissues of culms appear green.

C. CLEARING OF GRASS TISSUE FOR DETECTING FUNGAL ENDOPHYTES

This procedure was developed by Hignight et al.[11] and is used to clear grass tissues. In this procedure methyl salicylate is employed as a clearing agent, while the endophyte is stained with aniline blue. This procedure has been used to visualize endophytes in tall, hard, and red fescue species, as well as in perennial ryegrass; it should also work with the *Balansia*-infected grasses. The use of this technique does not require the peeling of epidermal tissue. Grass samples, 2 to 3 cm long, are processed according to the following protocol.

Killing and Fixing Procedure:

1. All samples are placed in Carnoy's solution (6:3:1 ethyl alcohol:chloroform:85% glacial acetic acid) for 24 h (samples can remain in this solution for at least 12 weeks).
2. The samples are transferred to 70% aqueous ethyl alcohol for 24 h to remove chlorophyll.
3. The tissues are cut to 0.5- to 1.0-cm lengths and stained with aniline blue. The recommended stain is a modification of the aniline blue solution above and consists of 2:1 aniline blue in 70% aqueous ethyl alcohol:85% lactic acid. The minimum recommended staining times are as follows: tall fescue, 12 min; read and hard fescue, 5 min; and perennial ryegrass, 8 min.

Clearing Procedure:

1. The stained tissue is cleared during the sequential transferring of tissue in the following solutions:
   a. 100% ethyl alcohol for 60 min
   b. 100% ethyl alcohol for 60 min
   c. 1:3 methyl salicylate:ethyl alcohol for 60 min
   d. 1:1 methyl salicylate:ethyl alcohol for 60 min
   e. 3:1 methyl salicylate:ethyl alcohol for 60 min
   f. 100% methyl salicylate for 60 min

2. Finally, all specimens are placed ventral side up on a microscope slide, mounted in methyl salicylate, and examined at ×200, using bright-field microscopy.

III. MEDIA

A. GLUCOSE-CITRATE AGAR

Murashige and Skoog Basal Salt Mixture	4.3 g
(Sigma Chemical Company, St. Louis, MO; product number M5524)	
Potassium citrate	5.0 g
Glucose	10.0 g
Agar	10.0 g
Distilled water	1000.0 ml

To prepare this medium, all components are mixed in 1 l of distilled water and autoclaved. This defined medium has been found to be useful for conducting developmental studies of *Atkinsonella hypoxylon* (Peck) Diehl (White, unpublished). When isolates of this species are cultured on the above medium in which potassium citrate is omitted, a smooth, white, undifferentiated mycelium develops that produces no conidia. However, as the citrate concentration in the medium is increased, colonies differentiate, showing forms that approximate the gray color and fealty texture of stromata on the host grass *Danthonia spicata* (L.) Beauv. Sporodochia producing masses of pink macroconidia may develop on these differentiated colonies. Acetate has also been found to induce a partial differentiation of colonies of *A. hypoxylon*.

B. PARAFFIN AGAR

Murashige and Skoog Basal Salt Mixture	1.0 g
(Sigma Chemical Company, St. Louis, MO; product number M5524)	
Paraffin embedding compound	0.5 g
(melting pt. 60–62°C; Will Scientific, Inc., Rochester, NY)	
Agar	3.0 g
Distilled water	200.0 ml

To prepare this medium, all components are mixed and autoclaved. Following autoclaving, the media is cooled, vigorously agitated, and poured into plates. As the agar solidifies, paraffin forms small beads on the surface of the agar.

This medium is used to evaluate the potential of epibiotic Balansieae to hydrolyze wax.[27] Culturing fungi on paraffin agar for several weeks, followed by microscopic examination of wax beads, is necessary to assess bead colonization. Formation of thick mantles of mycelia around wax beads is taken as evidence of wax degradation.

C. STARCH-MILK AGAR

Murashige and Skoog Basal Salt Mixture	4.3 g
(Sigma Chemical Company, St. Louis, MO; product number M5524)	
Nonfat powdered milk	5.0 g
Fine corn starch	10.0 g
Agar	10.0 g
Distilled water	1000.0 ml

To prepare this medium, basal salts are first dissolved in distilled water, other components are then added, and the mixture is autoclaved. The medium must be agitated before pouring into plates, to prevent settling of starch.

This semidefined medium employs inorganic nitrogen sources in the salts mixture, along with undefined sources in powdered milk. Starch is used as the primary carbon source, although some carbon is supplied in the milk. Starch-milk agar has been used to distinguish different groups of endophytes and evaluate the starch degradation capacity among Balansieae.[28] A small percentage of the slow-growing endophytes isolated from *Festuca* L. and *Lolium* L. species are unrelated to the balansoid endophytes. The *p*-endophytes[1] and *Phialophora*- and *Gliocladium*-like endophytes[17] fall into this category. On potato

To soften seeds for microscopic examination to determine the presence of endophytes, pour a 10-ml volume of seeds into 100 ml of the sodium hydroxide solution in a beaker, stir, and let stand at room temperature for approximately 8 h. Pour off the sodium hydroxide solution, and cover the beaker with cheese cloth or fine screen mesh. Rinse the seeds for 20 min by use of a constant flow of tap water into the beaker. To assess the seeds for endophytes, deglume a seed, using forceps. Place the seed on a slide, in a drop of aniline blue-lactic acid stain or rose bengal stain. Squash the seed beneath a coverglass, and examine the periphery of the squash preparation for evidence of a mycelium associated with the densely staining aleurone layer of the seed.[7]

B. ENDOPHYTE ISOLATION PROCEDURE

Clorox® (6.25% NaOCl)	50 ml
Tap water	50 ml

1. Seed Isolation

Mix Clorox® and tap water to make a 1.25% solution of NaOCl. To isolate endophytes from seeds, deglume the seeds by rubbing them vigorously between the hands for several minutes. Periodically collect the seeds that are freed of any adhering glumes. After 100 or more seeds have been collected, place them in a 250-ml beaker, and pour a 50% Clorox® solution into the beaker. Cover the beaker with parafilm, and agitate continuously for 15 to 20 min. Pour off the Clorox® solution, and replace it with 100 ml of sterile distilled water. Agitate it for 5 min, then pour off the water, and replace it with another 100 ml of sterile distilled water. After agitating the seeds for 3 to 4 min more, pour off most of the water. Using sterile forceps, remove a seed, and press it into potato dextrose agar, yeast extract-glucose agar, cornmeal malt extract medium, or another suitable medium. A minimum of 20 plates with three seeds per plate is recommended. Seal the plates using strips of parafilm, masking tape, or other material, and incubate them at room temperature. Colonies emerging from the seeds in the first week are likely to belong to surface contaminants of the seeds. These colonies may be removed from the plates as they develop. The characteristic white colonies of *Acremonium* endophytes may become visible after 2 to 4 weeks.

2. Leaf or Stem Isolation

To isolate endophytes from leaf sheaths or culms, the procedure is similar. Young tissue should be obtained for isolation, as older tissues often contain many additional fungi that make isolation of slow-growing *Acremonium* endophytes difficult. Pieces of tissue approximately 5 mm or less in size should be obtained and placed in the 50% Clorox® solution, as described above. Tissue should be agitated continuously for 15 min. However, after the first 5 min, two or three pieces of tissue may be removed from the Clorox® solution every couple of minutes and rinsed vigorously in sterile distilled water. These pieces are then pressed into agar media, as above. Four or five different surface disinfection times should be represented among the resulting plates. The plates are sealed and incubated, as above.

3. Ascospore Isolation

Several of the Balansiae, (i.e., *Balansia epichloë* Weese) Diehl and *Epichloë typhina* (Fr.) Tul., produce fertile stromata that can be used as sources of isolates. The stromata must be mature and capable of discharging ascospores. Leaves or stems bearing mature stromata of the desired species are collected and used fresh; dried material usually will give poor results. The stromata are washed in sterile water, excess water is removed, and the stromata are taped to the lid of petri plates containing the desired medium. A satisfactory medium for this purpose is cornmeal malt extract. An antibiotic should be added to this medium. If the stroma is dry, a small amount of water is placed on its surface, and the lid is replaced on the dish. Ascospores discharge should proceed within a few minutes. The area under the leaf-bearing stroma may be marked on the bottom of the dish, to identify the area to be viewed. Discharged ascospores may be observed under ×200 of a light microscope. When enough ascospores are discharged, a new lid without stromata is placed on the plate, and the petri plate is sealed with parafilm. Plates are incubated at 24 to 26°C for a 4- to 6-week period. Colonies of endophytes are identified (usually this is based on their conidia and conidiophore morphology[16]); this should be evident within at least 5 weeks after discharge. Colonies of *Myriogensopora atramentosa* (Berk. & Curt.) Diehl will not be visible until 6 to 7 weeks have passed, but this is host specific.

C. *EPICHLOË* CROSSING PROCEDURES

It has been shown that *Epichloë* species are heterothallic.[26] This finding has led to the development of methods for evaluating reproductive compatibility between individuals of *Epichloë*.[28] The simplest method to make a cross is to rub a stroma on which spermatia are located, against another stroma also bearing spermatia. If stromata are of opposite mating types and the two individuals of *Epichloë* are genetically close enough to be reproductively compatible, fertile perithecia will mature on both stromata 30 to 60 days following the cross. Since it is necessary to make crosses on intact plants, often under field conditions, stromata must be protected from insects that feed on stromata. This is accomplished by inserting the freshly inoculated stroma into a plastic drinking straw that is narrow enough to exclude most potential insect visitors.

Unfortunately, attempts to make crosses on agar cultures have not been successful; thus, perithecia have never been obtained in culture. However, several means have been found to cross cultures to stromata on plants. These methods involve either use of conidia of monokaryotic cultures or use of mycelia of previously dikaryotized or mixed cultures. Schardl and Tsai[24] used enzymes to form proto-plasts that were then placed on water agar and induced to produce conidia. These conidia were then used to inoculate stromata. Conidia may also be obtained directly from colonies growing on agar media.[28] Here the entire colony is removed from a plate and rubbed on a stroma.

Another procedure used to make crosses employs cultures that were obtained from stromata on which perithecia were developing. Presumably these cultures are already dikaryotic. This method may facilitate the determination of reproductive compatibility between populations of *Epichloë*, because compatibility may be assessed regardless of the mating type of the recipient stroma. Such cultures produce perithecia on both mating types of stromata if fungi are similar enough to be reproductively compatible (White, unpublished). To conduct crosses using this method, colonies growing on potato dextrose agar are used. Holes are burned into plates at extreme margins. The stroma is then inserted through the holes so that the mycelia of the stroma and colony are in contact. The plate may be immobilized by fixing it with tape to a rod pressed into soil. Stromata should remain in contact with colonies for 2 or 3 days. After the inoculation period, the plates should be removed, and the stromata should be covered using straws. Nothing is known about the relative contributions to the mature perithecia of the three sets of nuclei involved in these crosses.

D. ANTISERUM PRODUCTION

Antisera from endophytes may be used to study the taxonomic relatedness among endophytes, and may be used in one of several procedures for detecting endophytes in grass tissue, i.e., the tissue print-immunoblot.[10] The following procedure is that of An et al.[1] and, according to M. R. Siegel, is a modification of the combination of the procedures of Johnson et al.[13] and Reddick and Collins.[20] This procedure can be scaled up to produce gram quantities of hyphae, which suggests that it may be modified for the large-scale isolation of specific proteins. The fungus is cultured for 7 to 14 days on cellophane disks on top of potato dextrose agar, and an inoculum is prepared from these by grinding the tissue in an Omni Mixer Homogenizer in 10 to 15 ml of sterile water. The inoculum, 1 to 2 ml, are placed in 300-ml flasks containing 50 ml of potato dextrose broth. Six flasks are required for each rabbit to be inoculated.

The broth cultures were grown at 21°C on an orbital shaker for 2 to 3 weeks. The mycelia were collected from the required number of flasks by vacuum filtration and washed with 2 l of phosphate-buffered saline minus potassium (PBNa) (20 mM sodium phosphate buffer, pH 7.3, and 150 mM NaCl). The washed mycelia were resuspended in 75 ml of PBNa and homogenized in a Brinkmann Polytron PT20 for 45 sec at setting 7. The homogenate was centrifuged for 10 min at $8000 \times g$, and the supernatant was decanted into a 250-ml beaker and cooled to 4°C. Solid polyethylene glycol (8000 molecular weight) was slowly added, with stirring, until a final concentration of 10% (wt/vol) was reached. The protein was precipitated from this solution overnight at 4°C, collected by a 20-min centrifugation at $12,000 \times g$, resuspended in 5.0 ml of PBNa containing 0.1% sodium dodecyl sulfate, and then warmed to 60°C and incubated for 5 min. Insoluble material was repelleted and discarded. The supernatant was cooled on ice, and 4 volumes of cold (−20°C) acetone were added. Protein was precipitated overnight at −20°C, repelleted, resuspended in 5.0 ml of PBNa, and divided into 1-ml portions. The concentration of protein was estimated by A_{280}, and the portions were freeze-dried for storage. For rabbit inoculations, protein extracts were dissolved in 0.5 ml of water, thoroughly emulsified with 0.5 ml of Freund's incomplete adjuvant, and injected subcutaneously or intramuscularly. The rabbits are injected three times at 2-week intervals and then bled 2 weeks following the last injection.

E. GENERATION OF ENDOPHYTE-FREE CLONES

1. Propiconazole Method

Endophyte-free and endophyte-infected ramets are used to study physiological and morphological responses of plants, in order to determine if there is an endophyte effect on these responses. The following procedure uses the fungicide propiconazole to remove the fungus. The method was developed for tall fescue by M. R. Siegel,[8] but should work for other grasses as well.

Three to five young tillers are removed from a ramet of the desired genotype, and each is planted in 300-ml styrofoam cups without draining holes, which contain 300 g of washed sand to which was added enough of the propiconazole (1-[{2-(2,4-dichlorophenyl)-4-propyl-1,3-doxolan-2-yl}methyl]-1H-1,2,4-triazole) solution (6 µg/ml) to completely inundate the seedling. These cups were weighed, and every 2 days, water or nutrient solution was added to equal the original weight, in order to maintain the initial concentration of propiconazole. These plants were grown in a greenhouse for 5 weeks.

After this period each ramet was removed, washed free of sand-propiconazole solution, and replanted into pots containing greenhouse soil mixture [50% soil and 50% peak-vermiculite, (v/v)]. After 3 months of growth in the greenhouse, three tillers per plant were checked for endophyte infection.

2. Folicure Method

G. M. C. Latch (unpublished) finds it much easier to kill endophytes in ryegrass than in tall fescue. Benomyl[16] or propiconazole usually work with perennial ryegrass, but with tall fescue, there are mixed results following the use of propiconazole, and benomyl will not work at all with this grass. Folicur (terbuconazole) appears to be better than benomyl or propiconazole, for killing endophytes, especially in tall fescue (Latch personal communication), and is recommended for any grass. The method described below uses three to five tillers per genotype, which usually is not necessary, since success is usually achieved with one tiller. With this procedure endophyte-free grasses are produced in less than half the time required with the propiconazole method, and the procedure is simpler.

1. Split the plants into single tillers, wash off the soil, and trim the leaves. Select three to five tillers for each plant genotype, to be sure to obtain at least one freed of the endophyte.
2. Place the tillers in a beaker containing a solution of 0.1 ml/l of Folicur (250 g ai/l), and let them soak for 6 h.
3. Plant the treated tillers singly in soil in small pots with bottom drainage.
4. Grow the plants for 3 to 4 weeks or until several new tillers have been produced.
5. Examine the leaf sheath of the new tillers, and if there is no endophytic mycelium present, remove the newly produced endophyte-free tillers and plant them in fresh pots of soil. It is important to remove the originally treated tiller and not use it for any studies.
6. Over the next few months, continue to check the new tillers to make sure that they are still free of endophyte mycelia.

V. SUMMARY

This annotated list of procedures includes those that the authors have used or that the authors and contributors feel should be incorporated into other complex protocols, which if properly designed, should help to define other biotechnological uses and biological relevance of endophytes and endophyte-infected grasses. Lacking among these procedures are physiological and biochemical techniques designed to study the independent and interactive nature of the symbioses and their components. Thus, this chapter should bring to the attention of the experienced scientist and novice of endophyte biology the diversity of research methods needed to define the diverse aspects of endophytes. We hope that this chapter will serve as a guide and starting point for the development of new and refined methods and techniques for endophyte research.

ACKNOWLEDGMENTS

The authors are grateful to the following people for providing information about stains, media, or procedures: F. Belanger and P. Halisky, Rutgers University, New Brunswick, NJ; R. Shelby, Auburn University, Auburn, AL; M. R. Siegel, University of Kentucky, Lexington; and G. C. M. Latch, AgResearch, Grasslands Research Center, Palmerston North, New Zealand.

REFERENCES

1. **An, Z.-Q. M. R. Siegel, W. Hollin, H.-F. Tsai, D. Schmidt, and C. L. Schardl,** Relationship among non-*Acremonium* sp. fungal endophytes in five grass species, *Appl. Environ. Microbiol.,* 59:1540–1548, 1993.

2. **Bacon, C. W.,** Laboratory production of ergot alkaloids by species of *Balansia, J. Gen. Microbiol.,* 113:119–126, 1979.

3. **Bacon, C. W.,** A chemically defined medium for the growth and synthesis of ergot alkaloids by the species of *Balansia, Mycologia,* 77:418–423, 1985.

4. **Bacon, C. W.,** Procedure for isolating the endophyte from tall fescue and screening of it for ergot alkaloids, *Appl. Environ. Microbiol.,* 54:2615–2618, 1988.

5. **Bacon, C. W.,** Isolation, culture, and maintenance of endophytic fungi of grasses, In: D. P. Labeda, Ed., *Isolation of Biotechnological Organisms from Nature,* McGraw-Hill, New York, 1990, 259.

6. **Bacon, C. W., J. K. Porter, J. D. Robbins, and E. S. Luttrell,** *Epichloe typhina* from toxic tall fescue grasses, *Appl. Environ. Microbiol.,* 34:576–581, 1977.

7. **Clark, E. M., J. F. White, Jr., and R. M. Patterson,** Improved histochemical techniques for the detection of *Acremonium coenophialum* in tall fescue and methods of in vitro culture of the fungus, *J. Microbiol. Methods,* 1:149–155, 1983.

8. **De Battista, J. P., J. H. Bouton, C. W. Bacon, and M. R. Siegel,** Rhizome and herbage production of endophyte-removed tall fescue clones and populations, *Agron. J.,* 82:651–654, 1990.

9. **Fannin, F. F., L. P. Bush, and M. R. Siegel,** Analysis of peramine in fungal endophyte-infected grasses by reverse phase think-layer chromatography, *J. Chromatogr.,* 503:288–292, 1990.

10. **Gwinn, K. D., M. H. Shepard-Collins, and B. B. Reddick,** Tissue print-immunoblot: an accurate method for the detection of *Acremonium coenophialum* in tall fescue, *Phytopathology,* 81:747–748, 1991.

11. **Hignight, K. W., G. A. Muilenburg, and A. J. P. van Wijk,** A clearing technique for detecting the fungal endophyte *Acremonium* sp. in grasses, *Biotech. Histochem.,* 68:87–90, 1993.

12. **Hill, N. S., G. E. Rottinghaus, C. S. Agee, and L. M. Schultz,** Simplified sample preparation for HPLC analysis of ergovaline in tall fescue, *Crop Sci.,* 33:331–333, 1993.

13. **Johnson, M. C., T. P. Pirone, M. R. Siegel, and D. R. Varney,** Detection of *Epichloe typhina* in tall fescue by means of enzyme-linked immunosorbent assay, *Phytopathology,* 72:647–650, 1982.

14. **Kasperbauer, M. J., R. C. Buckner, and L. P. Bush,** Tissue culture of annual ryegrass x tall fescue F_1 hybrids: callus establishment and plant regeneration, *Crop Sci.,* 19:457–461, 1979.

15. **Lam, C., F. Belanger, J. F. White, Jr., and J. Daie,** Sugar uptake by *Acremonium typhinum,* an endophytic fungus infecting *Festuca rubra, Mycologia,* in press.

16. **Latch, G. C. M. and M. J. Christensen,** Artificial infection of grasses with endophytes, *Ann. Appl. Biol.,* 107:17–24, 1985.

17. **Latch, G. C. M., M. J. Christensen, and G. J. Samuels,** Five endophytes of *Lolium* and *Festuca* in New Zealand, *Mycotaxon,* 20:535–550, 1984.

18. **Niell, J. C.,** The endophyte of ryegrass, *Lolium perenne, N.Z. J. Sci. Technol.,* 22:280–290, 1940.

19. **Porter, J. K.,** Ergosine, ergosinine, and chanoclavine I from *Epichloe typhina, J. Agric. Food Chem.,* 27:595–598, 1979.

20. **Reddick, B. B. and M. H. Collins,** An improved method for detection of *Acremonium coenophialum* in tall fescue plants, *Phytopathology,* 78:418–420, 1988.

21. **Rottinghaus, J. D., G. B. Garner, C. N. Cornell, and J. L. Ellis,** HPLC method for quantitating ergovaline in endophyte-infected tall fescue: seasonal variation of ergovaline levels in stems with leaf sheaths, leaf blades, and seed heads, *J. Agric. Food Chem.,* 39:112–115, 1991.

22. **Rowan, D. D. and D. L. Gaynor,** Isolation of feeding deterrents against argentine stem weevil from ryegrass infected with the endophyte, *J. Chem. Ecol.,* 12:647–658, 1986.

23. **Saha, D. C., M. A. Jackson, and R. L. Tate,** A rapid staining method for detection of endophytic fungus in turfgrasses, *Phytopathology,* 74:812, 1984.

24. **Schardl, C. L. and H. F. Tsai,** Molecular biology and evolution of the grass endophyte, *Nat. Toxins,* 1:171–184, 1992.

25. **Shelby, R. A. and V. C. Kelley,** Detection of ergot alkaloids in tall fescue by competitive immunoassay with a monoclonal antibody, *Food Agric. Immunol.,* 3:169–177, 1991.

26. **White, J. F., Jr. and T. L. Bultman,** Endophyte-host associations in forage grasses. VIII. Heterothallisim in *Epichloe tyhinum, Am. J. Bot.,* 74:1716–1721, 1987.

27. **White, J. F., Jr., J. P. Breen, and G. Morgan-Jones,** Substrate utilization in *Acremonium, Atkinsonella,* and *Balansia* species, *Mycologia,* 83:601–610, 1991.

28. **White, J. F., Jr.,** Endophyte-host associations in grasses. XIX. A systematic study of some sympatric species of *Epichloe* in England, *Mycologia,* 85:444–455, 1993.

Section III

Ecology of Endophyte-Infected Grasses

Chapter 5

Ecological Relationships of Balansiae-Infected Graminoids

Nicholas S. Hill

CONTENTS

I. INTRODUCTION

Mutualistic relationships often are between partners from different phyla within, or between, kingdoms. Insects, for example, are important for the dispersal of pollen or seeds and are therefore considered mutualists in plant communities. The necessity for such interactions is so essential for some plant species that the concept of mutualism appears to be overtly obvious. More subtle ecological relationships occur between graminoids and clavicipitaceous fungi, which are members of the tribe Balansiae.[22] Balansiae fungi vary in their mechanisms of reproduction and dissemination, and their different life cycles may or may not favor conditions for reinfection and sustenance of the association. Within the Balansiae tribe, fungi have different strategies of cohabitation with the host, which range from the parasitic *Myriogenospora atramentosa* (Berk. and Curt.) Diehl[44] to the obligately host-dependent and nonsexual species of *Acremonium*.[47]

The maintenance cost of the association between the two organisms is different for each endophyte–plant association and requires an in-depth examination of how they interact. Some endophytes affect the reproductive biology of their host plant such that seedling plants are asexual, larger, and more vigorous than those without endophytes,[11] while others produce plants that are smaller and less vigorous.[44] In other associations it is the endophyte that is asexual,[4] but it increases the fecundity and vegetative proliferation of the host.[28] The sexual nature of the fungal organisms within the Balansiae tribe, and their effects on the reproductive biology of their hosts, suggests that those that are not currently mutualists may be in transition from parasite to mutualist.

This chapter examines the sexual behavior of plants and their endophytes to assess which have evolved towards mutualism, discuss the apparent strategies that the cosymbionts are using to evolve towards mutualism, and compare the benefits of the association of the endophyte with the cost to their host plants.

II. CONCEPTS OF SYMBIOSIS AND MUTUALISM

Terminology used to describe mutually beneficial associations between organisms has different meanings to different scientific groups. Therefore, it is important to define the terminology that is used in this chapter. Often the term "symbiosis" is used as a contrast to parasitism.[42] While correct, using symbiosis to describe all nonparasitic associations is simplistic because it does not detail specific interactions of the where w is the fitness of the population containing mutualistic fungi (mf), g_o is the growth rate of the endophyte-free form of the plant, p is the probability a plant will encounter the endophyte, b_m is the benefit of fungal products (i.e., alkaloids, osmoticum, etc.) received by the plant from the interaction with the endophyte, c_m is the cost of the endophyte's association to the plant, s_m are structural changes that increase the efficiency of the plant–endophyte association, and f_m is a feedback loop that represents the increased growth of the plant due to the association. Therefore, $b_m + s_m + f_m \geq c_m$, or otherwise, the frequency of the association depends on virulence of the organism towards its host, and the fungal organism would be parasitic rather than mutualistic. By reviewing the life cycles of the Balansiae and how each affect their host, a determination may be made as to which are parasitic and which are mutualistic.

V. FUNGAL–HOST ASSOCIATIONS

A. *MYRIOGENOSPORA ATRAMENTOSA*

Of the species within the Balansiae tribe, the least understood is *Myriogenospora atramentosa* (Berk. & Curt.) Diehl. This is a fungus that infects *Paspalum notatum* Flugge and is associated with livestock toxicity.[44] It also infects four other C_4 grass species.[59,65] The fungus grows superficially, but affects the host by inhibiting expansion of immature leaves inside the whorl of the plant. The fungus is epibiotic and colonizes the leaf surface. As growth continues, the fungus develops elongated stroma on the leaf surface, and the unexpanded leaf entombs the stroma, thereby developing a refuge for the fungus. As the leaf continues to develop, the fungus behaves like a cementing agent and maintains the surrounding protective structure by preventing the leaf lamina from expanding. Tips from younger leaves may adhere to the stroma on another leaf blade, causing a looped-leaf appearance. Although the leaf blades are never infected with the fungus, the plant epidermis cells that are subcuticular to the fungus become enlarged and remain alive with intact chloroplasts. The fungus engulfs the inflorescence of affected tillers and prevents them from emerging from the flag leaf. Although the infected tillers are essentially asexual, healthy tillers are often present within individuals that are infected by *M. atramentosa,* and can be cultured in their noninfected form. Plants without the fungus are more vigorous than, and can out compete, their affected clones. *Myriogenospora atramentosa* is therefore noninvasive, but reduces vegetative and reproductive effort within the plants, by causing stunting and preventing inflorescence development.

With *M. atramentosa* it is rather obvious that this is a pathogenic organism. Other than possibly providing benefits to the plant by reducing livestock and insect grazing,[17] there appears to be no other benefit (b_m) to the plant from its association with the fungus. The association between the two organisms appears to be opportunistic and is therefore dependent upon the trilateral relationship between virulence of the fungus, susceptibility of the host, and appropriate environmental conditions for infection to occur. Therefore, the probability of association between the two (p) is likely to vary among years, locations, plant populations, and fungal strains. The significance of the enlarged epidermis cells in response to the fungus has not been determined, but could be an initial step towards the mutualistic association between the plant and fungus. In tetraploid plants, enlarged mesophyll cells are more photosynthetically efficient than in smaller cells of diploid plants.[72,73] It seems logical that increased photosynthetic efficiency and/ or capacity of plant cells adjacent to the fungus would be necessary to provide nutrition to a noninvasive pathogen, and epidermis cell enlargement is the mechanism by which that demand is met (s_m). But the cost (c_m) of maintaining the association with the fungus is prohibitive. Carbon is exported from the plant (likely from enlarged epidermis cells) and utilized by the fungus.[65] Some of the carbon is returned to the plant as low molecular weight compounds and is incorporated back into the sugar fraction of the plant and exported from one leaf to another, for metabolism. Formation of stroma on the leaf surface and around the developing seedhead impedes expansion of leaf area, emergence of new leaves from the whorl of the plant, and seedhead production.[44] Consequently, the infected plants are less vigorous and abundant, than noninfected plants, within a population. Because only noninfected tillers can produce seedheads, it appears that if a mutualistic association were ever to develop between these two organisms, it will likely come from infection through the leaf tissue and not through the seed. However, it is interesting that not all tillers of the plant are infected by the fungus and that the plant is still capable of sexual reproduction. As noted above, genetic recombination is a necessary component of the host life cycle, for coevolution of the species to continue.

B. *BALANSIA* SPECIES

Species of *Balansia* Speg. appear to have the greatest diversity of hosts and mechanisms of host inhabitation of the Balansiae. As a group this genus of fungi was originally thought to be entirely endophytic,[22] but recent investigations have demonstrated that some are entirely epibiotic,[38] and others have combined endophytic/epibiotic inhabitation of the plant.[14,58,78] Detailed microscopic examination of fruiting structures and infection sequences have not been conducted on all plant species infected with the genus, but it is unlikely that any are totally endophytic in their transmission from one plant generation to the next.

Generally, the life cycles of *Balansia* species are similar. They develop stroma that require transfer of conidia from one stroma to another, as a prerequisite for production of ascospores.[58,78] Ascospores or conidia associated with leaf blade, sheath, stem, or inflorescence, germinate and survive either epibiotically or endophytically with the plant. If conidia or ascospores of the epibiotic fungi fall on a leaf surface of a seedling plant, they germinate, and the developing mycelium must actively or passively migrate to meristematic regions for active growth.[39] When the leaves expand at rapid rates, the epiphyte becomes disjointed and survives as "patches" of latent or inactive mycelium[38] that initiate asexual reproductive structures on the stem or the inflorescence.[12,22,58] The epibiotic fungi totally or partially inhibit inflorescence development or engulf the developing inflorescence after stem elongation. With endophyte/epibiotic fungal organisms, the reproductive structures are on the leaves.[78] While the exact mechanisms of invasion of endophyte/epibiotic *Balansia* species are unknown, it has been demonstrated that they do not invade the developing inflorescences and are inhibited by sugars present in the developing seedhead, especially fructose. Therefore, it is assumed that their mechanism of infection of the host is through seedling plants.[78]

Over 100 C_4 grass and sedge species have been identified that are infected with species of *Balansia*,[14] and no species has been associated with C_3 grasses. In-depth experimentation has not been conducted on the majority of these plant–endophyte associations, and therefore, the exact cost:benefit analysis cannot be performed for each at this time. From those that have sufficient data to begin a cost:benefit analysis, it is evident that varying sexual and asexual reproductive strategies of the hosts confound a general cost:benefit model, and they must be considered individually.

1. Epibiotic *Balansia* Species

Balansia cyperi Edg. infects nine species of sedges,[22] including *Cyperus virens* Michx. and *C. rotundus* L.[11,12,17,67] *C. virens* is adapted to wetland conditions in the marshlands of Louisiana. It tillers from corky rootstock and produces inflorescences in early spring. Although only 50% of *C. virens* is infected with *B. cyperi*, plants that are infected are more likely to have infected plants as neighbors than noninfected plants.[12] The fungus lives epiphytically on the plant.

B. cyperi changes the reproductive capacity of specific host plants. In *C. virens* the epibiotic fungus causes the developing seedhead to become viviparous and develop epibiotic-infected plantlets from vegetative propagation. Not all infected plants are viviparous. Nonviviparous plants can pass the epibiotic fungus onto the seed, but most will abort their inflorescences. Epibiotic-infected plants, viviparous and nonviviparous, have greater numbers of tillers than do noninfected plants, but those that are viviparous are more vigorous than the infected nonviviparous plants.[11] The viviparous plantlets root quickly and easily become established in wet soil. Because they are larger seedling plants, viviparous seedlings have a competitive advantage over noninfected or nonviviparous infected seedlings, during the establishment period.

Clay[11] found variability in the number of viviparous, nonviviparous, and noninfected individuals among unique populations of *C. virens*, and variability in the number of viviparous offspring within and among the populations. This suggests that specific interactions occur between the host and endophyte, which may complement one another for vegetative viviparous reproduction. Auxin production varies among members of the Balansiae tribe when they are cultured *in vitro*,[21,51] and is necessary to initiate undifferentiated cellular growth and somatic embryogenesis during events leading to vivipary.[70,71] During somatic embryogenesis, exposure to auxin may induce somaclonal variation and provide a source of genetic variability in the host. Therefore, the genetic variability of the host is maintained in viviparous plants and is necessary for the continuation of evolutionary events leading to a mutualistic association.

Another sedge, *C. rotundus,* has nearly 100% abortion of developing inflorescences when infected by *B. cyperi*. As in *C. virens*, the epibiotic fungus engulfs the developing seedhead, but does not induce vivipary in its host.[12] Not all inflorescence development is inhibited by the epibiotic fungus. Some inflorescences are partially engulfed by the fungus and produce normal flowers. The fungus can be transmitted from one plant to another by growing on rhizomes attached to underground tubers.[67] Tubers of infected plants tend to be smaller than those of noninfected plants, and the percent that sprout and develop new plants may be less for infected plants than for noninfected plants, depending upon environmental conditions. Although the tubers are smaller, the number of tillers produced by infected plants is 80% greater than noninfected plants, and the total weight of storage organs is greater. While tiller initiation and development is delayed among epibiotic-infected plants, the number and weight of tillers is greater.

The life cycle of the epibiotic *B. cyperi* is similar to that of epiphytic *M. atramentosa*, but it is obvious that *B. cyperi* has developed into a mutualistic organism. It would appear that the carbon costs to the plant for maintenance of the epibiotic fungi are similar. The difference between the cost of the two organisms is likely to be the efficiency with which they fix the carbon through photosynthesis. Leaf rolling caused by the mycelium cementation of plant epidermal cells by *M. atramentosa* and inhibition of leaf expansion, by the stroma, are likely to result in reduced leaf lamina exposed to the sun combined with reduced CO_2 exchange through stomata from the rolled leaves. The cost (c_m) of supporting the *B. cyperi* is probably negligible and similar to that of other Balansiae organisms that are mutualistic. The structural changes that increase the efficiency of the association (s_m) are the increased numbers of tubers and rhizomes in *C. rotundus* and viviparous seedling plants of *C. virens*. These affect the competitiveness of the two plant species differently. In *C. rotundus* a strategy of investment into vegetative reproductive effort through rhizomes and bulbs appears to provide the infected plant with a competitive advantage for local demography. Production of plant hormones (b_m) is likely to be the explanation for differences in the numbers and sizes of tubers and the numbers and mass of tillers (f_m) from those tubers. In *C. virens* the increased robustness of the infected plants appears to be a consequence of auxin (b_m)-induced viviparous seedlings (s_m) that are larger than for noninfected plants. Not only are they larger, but they often have a ready-made root system prior to falling from the stem to the soil. The viviparous plantlets become physically removed from the parent plant and provide an opportunity for transport, by water, insects, or birds, from one location to another. Having a transport mechanism may enable them to expand their occupation of a community, rather than concentrating on local demographics, as in *C. rotundus*. Being larger and having a root system enables *B. cyperi*-infected *C. virens* to establish themselves and expand their leaf area at a faster rate than the noninfected plants, increasing the feedback loop for association with host plants in future generations (f_m).

While species of *Cyperus* differ in their responses to *B. cyperi*, both retain the ability to sexually reproduce, a precondition for coevolution according to Law.[36] However, they are not obligate mutualists, as they require infection of the seedling plants either through vivipary or association of the mycelium with the meristem of seedling plants. Although their association is not obligate, the frequency with which they infect their hosts indicates they have evolved to be more host compatible than *M. atramentosa*. If *B. cyperi* were a pathogen, the sexual nature of the host would gradually eliminate plant susceptibility, and frequency of infection within natural populations would be minimal. The regularity of infection suggests that the plant makes little or no attempt to defend itself from the fungus. While no acquired function of the epibiotic fungus has been documented, it is likely it exhibits similar characteristics that provide benefits to other host organisms from their association with Balansiae fungi. However, the epiphytic *B. cyperi* is likely to be comensalistic in its association with its hosts, if not mutualistic.

2. Endophytic/Epibiotic Fungus *Balansia* Species

Two species, *B. epichloë* (Weese) Diehl and *B. henningsiana* (Moell.) Diehl, have similar life cycles, although they inhabit different hosts. *Balansia epichloë* infects species of *Ctenium* Panzer, *Sporobolus* R. Br., *Eragrostis* Beauv., *Aristida* L., *Panicum* L., and *Triodia* R. Br. (*Tridens* Roem & Shult.). These fungi live endophytically within the plant during cool climatic conditions, but produce stromata on the upper and lower leaf surfaces, respectively, of their hosts when climatic conditions are warm.[22] The stroma are limited to the leaves and do not develop on seedheads if they are present on the plant.[22,59,78] The stromata or endophytic mycelia have no apparent detrimental effect on the vegetative tissues of the host plant, with the exception of halos surrounding the stroma.[22,59] The stroma produce both conidia and ascospores. A unique feature of *B. epichloë* is, with the exception of infection of *Eragrostis*, it permits its host plant to flower only on rare occasions. Reduced flowering is presumed to be a consequence of hormonal production by the endophyte,[51] which alters plant meristematic growth and differentiation. Therefore, there is no mechanism for obligate association of an endophyte with its host, from one plant generation to the next. It is apparent that infection of the subsequent plant generation is through infection by deposition of mycelia on florets during seed maturation or by invasion of seedling plants.

B. henningsianna has the same life cycle as *B. epichloë*, but its effect on the host plant differs dramatically. When infecting *Panicum agrostoides*, inflorescence development was not inhibited, but rather, vegetative tiller production was stimulated.[14] Inflorescences develop from tillers of plants, but show no evidence of infection by the endophyte, suggesting that not all tillers are infected within a plant. Microscopic examination of seeds from the inflorescences revealed no associated mycelium or viable spores, suggesting that infection of subsequent generations occurs from infection of seedling plants. As with the epiphytic *Balansia* species, the frequency of infection of *B. henningsianna* averaged 50% of naturally occurring plant populations and was as high as 62%. Therefore, the association between the host and endophyte is not by coincidence, but must be an event directed by the benefit to either the host or endophyte. Plant morphological changes and acquired defenses to pests suggest that the endophyte is a mutualist. Clay[14] found that infected plants produced 50% more tillers, and 25% more biomass, than noninfected plants and had fewer lesions from the fungal pathogen *Alternaria triticina*. Evaluations for insect herbivory found that there was no benefit to having the endophyte in the plant, but equally important, they noted that the noninfected host plant was not a good substrate for those insects studied.

C. *ATKINSONELLA* SPECIES

Atkinsonella hypoxylon (Peck) Diehl is known to infect three species: *Danthonia compressa* Austin, *D. spicata* (L.) Beauv., and *D. sericea* Nutt.[22] A second species, *A. texensis* (Diehl) Leuchtmann & Clay, infects *Stipa leucotricha* (Trin. & Rupr.).[22] These grass genera produce two types of flowers: chasmogamous cross-pollinated flowers on the terminal spikelets of the panicles, and cleistogamous self-pollinated flowers in spikelets located in the axils of lower leaf sheaths.[9,26] The fungi were originally believed to be endophytic,[22] but recently they have been documented to grow entirely epiphytically on the host.[38] They have a life cycle essentially identical to the epiphytic species of *Balansia,* in that they grow at the meristem of the plant. When intercalary growth of the leaf blade and stem occurs, the fungus becomes disjointed into clusters of mycelia resting on the cuticle of the host plant. When seedheads develop, reproductive fructifications develop around the flowering culm, initially producing conidia and, later, ascospores.[18] Seedheads of infected plants change their reproductive mechanism when infected with *A. hypoxylon* and only produce cleistogamous flowers.[10] Therefore, seeds from infected plants are self-pollinated only. Cleistogamous seeds have *A. hypoxylon* mycelia in association and produce infected plants.[18] Seeds derived from chasmogamous flowers become infected with *A. hypoxylon* when spores land on the leaves of seedlings and are funneled towards the meristematic regions by wind, water, or insects.[38]

Plants infected by *A. hypoxylon* appear dwarfed, because of changes in the growth characteristics of the culm,[22] but vegetative reproduction is greater in infected plants than in noninfected plants,[10,22] even though infected tillers developed into sterile inflorescences.[22] The uninfected plants produce approximately four times the number of seeds than the infected plants produce, but infected plants produce over twice as many tillers as noninfected plants. Although not clearly defined, infected plants have adaptive features that make them more competitive in mixed-plant communities than are noninfected plants. Much of the improved competitiveness can be attributed to phytohormone production or to insect deterrence. The rate of infection of *A. hypoxylon* within a plant species is dependent upon the age of the population.[22] Infection rates of young stands are approximately 15%, but the rate of infection of older stands is 50% or greater. This is evidence of the superior fitness of infected plants over their noninfected siblings.

Infection of cleistogamous seeds appears to be an adaptive advantage to the plant–fungus association, because cleistogamous seeds are produced at less photosynthetic cost than are chasmogamous seed.[8] Cleistogamous flowers are favored over chasmogamous flowers, during periods of stress, when chasmogamous flowers may completely abort. Generally, cleistogamous seeds mature earlier than do the chasmogamous seeds and, in *D. spicata,* have heavier caryopses, greater germination rates, and higher survival rates.

The probability (p) of a plant being infected with an epibiotic fungus is age dependent within a population, but within established populations, it will be approximately 0.5, and approximately 0.15 in newly established populations. As with the *Balansia* species, the cost (c_m) of maintenance of the fungal association is likely to be negligible to the plant, and the benefit from fungal products (b_m) are hormonal production and regulation of plant meristems and/or alkaloids associated with insect deterrence.

Several structural changes (s_m) occur that appear to be beneficial to the plant. First, the infected plant produces more leaf area and more vegetative tillers than does the noninfected plant. This provides the infected plants with more photosynthate for allocation to sustenance of the plant, making it more competitive with its noninfected neighbors. Second, the preference of infected plants to produce cleistogamous seeds guarantees association of the fungus with the next generation of the plant (p) and with seeds that, because of their larger size (f_m), have a better chance of survival than if that association were with chasmogamous seeds. If, for some reason, passage of the epibiotic fungus from one generation to the next is unsuccessful, the mechanism still exists for sporal infection of seedling plants (p).

At first glance one might assume that preference for cleistogamous reproduction by the host may have adverse genetic consequences. Inbreeding depression often occurs in plants that are cross-pollinators. However, cleistogamous species do not express inbreeding depression.[8] In addition, cleistogamous seeds are genetically unique individuals, as they are progeny from often cross-pollinated species.[9] Therefore, cleistogamous seeds could place added selection pressure on both the endophyte and the host, for evolution towards a perpetual and obligatory mutualistic association. Selfing of a host permits recombination of nuclear genes responsible for making the parent host suitable. Recombination provides opportunities for new mixes of nuclear genes, with probabilities that at least some of the offspring will be better hosts than was the parent.

Plants vary in the ability to be suitable hosts for specific *A. hypoxylon* strains, and some fungal strains appear to be more virulent to their potential hosts than do others.[40] Sexual reproduction within the endophyte results in a genetic diversity among *A. hypoxylon*.[41] If Law's theory[36] of reproductive behavior of mutualists is accurate, then *A. hypoxylon* would not have a sexual stage, and it would have no effect on its host's ability to sexually reproduce. Yet the ability of the fungus to impart benefits to the plant suggests the fungus and its hosts are, in all respects, mutualists. Therefore, evolution of the two species to reduce sexual reproduction of the fungus, yet maintain sexual reproduction of the host, would serve only to increase the probability of association of the fungus with a suitable host (p).

D. *EPICHLOË TYPHINA* AND *ACREMONIUM* SPECIES

Epichloë typhina (Fr.) Tul. infects approximately 100 C_3 grass species.[14] The fungus is similar to other stroma developing Balansiae fungi; however, some species of plants infected with this fungus are asymptomatic or have latent production of stromata.[4,61] (See Chapter 1 for discussion of two other species that have been recently delineated.) Asymptomatic forms of *E. typhina* have been taxonomically classified as species of *Acremonium* (Morgan-Jones and Gams), but molecular sequence data suggest that they may be hybrid anamorphs of *E. typhina*[47] (see Chapter 11); they are referred here and elsewhere in this volume as the *Epichloë–Acremonium* complex.

1. *Epichloë Typhina*

The life cycle of *E. typhina* differs from that of the other Balansiae fungi in that open wounds are necessary for its transmission to and infection of noninfected plants.[75] Conidia or ascospores that fall on cut stems germinate and travel down the pith of the stem to infect the meristematic region of the plant. The fungus survives as an endophyte within the primordia of the infant leaves and is restricted to the sheaths of older leaves.[61,75,77] In some host species the endophyte is passed from generation to generation through the seeds.[33,61] In other hosts the meristem of the tiller housing the endophyte initiates floral development; the fungus invades the flower primordium and "chokes" floral production in one of three ways: (1) the panicles develop normally, but do not produce viable seeds; (2) the panicle develops, but the fungus binds it into a stroma as it emerges from the flag leaf, and further growth is prevented; or (3) floral primordia are invaded prior to complete differentiation, preventing normal development and elongation of the reproductive tiller.[33]

In Europe *E. typhina* is found on several cool-season (C_3) pasture grasses.[7] The frequency of occurrence is greater in older pastures than in younger pastures, suggesting that grazing by livestock is one mechanism of transmittance of the fungus.[7,61] Populations of *E. typhina*-infected grasses that are prevented from flowering do not cease tillering and root production, during and following floral initiation, as do noninfected plants.[7] By not investing energy into seed production, the growth of vegetative components of the grass are at a premium, making the grass aggressive in the acquisition and occupance of areas within the community void of plant life. Chinch bug resistance has been documented in three grass species,[60] and phenolic fungitoxins are found in the stromata of plants infected with *E. typhina*.[35]

While additional biological comparisons between *E. typhina*-infected and noninfected populations of plants are needed, it is obvious that there are benefits imparted to the plant by the fungus. As with the other Balansiae fungi, it is presumed that the carbon cost (c_m), to the plant, for maintenance of the endophyte is minimal because of the apparent increase in the vigor of infected plants. There appear to be benefits of several fungal products (b_m), because of decreased insect grazing and antifungal products found in infected plants. Fungal metabolites may be present to provide disease or insect resistance in the

C_4 grass and sedge species infected with Balansiae fungi, but the hosts are not likely to be as palatable or susceptible to insects and pathogens as the cool-season species. Although reduced flowering is apparently disadvantageous to the plant, for genetic reasons, increased rooting and vegetative tillering (s_m) make the plant more competitive and aggressive for nutrients and water (f_m), especially under limiting conditions.

Complete sterilization within specific host plants is a demographic strategy, on the part of the mutualists, to perpetuate existing infected plants rather than produce competitors (seeds) for occupation of voids in a sward. Increased infection rates in older stands suggest the strategy is successful. Host sterilization is not complete within all grass populations, and some host plants maintain their ability to produce viable seeds from tillers that have not been infected with *E. typhina*. Therefore, the host species retains the capacity for genetic recombination, which is necessary for a mutualistic coevolutionary process to continue.

2. *Acremonium* Species

The economic significance of the biological association between *Acremonium* species and their hosts has resulted in intense investigation and a better understanding of how this genus interacts with its hosts. Two in particular, *A. coenophialum* (Morgan-Jones and Gams) and *A. lolii* (Latch, Christensen, and Samuels), are associated with livestock disorders that have initiated new awareness of the structure and functions of the pastoral agriculture ecosystem foodwebs.

The life cycle of *Acremonium* species are simple. They are intercellular endophytes inhabiting leaf sheaths and stems of plants, never invading the cells or becoming pathogenic. They produce no stromata and are passed from generation to generation within the host plants, through the seeds in a clonal form.[4,48] These fungi are in obligatory association with their hosts, as they do not naturally survive outside of their hosts, nor is there a known mechanism of transmittance from one plant to another in nature. Therefore, *Acremonium*-infected hosts beget infected seeds, and the life cycle perpetuates.

Defensive chemicals are produced by the endophyte, or the host plant in response to the endophyte, that are disruptive to numerous animal species that are capable of utilizing endophyte-free grasses for their substrate. Ergopeptine alkaloids and tremorgenic neurotoxins are responsible for causing the livestock conditions known as fescue toxicosis and ryegrass staggers, respectively. Sheep grazing *A. lolii*-infected perennial ryegrass (*Lolium perenne* L.) pastures suffer from the stiff-jointed staggering condition elicited by lolitrem A, lolitrem B, and paxilline alkaloids.[23,46] Animals grazing *A. coenophialum*-infected tall fescue ingest ergopeptine and clavine alkaloids, which induce various metabolic disorders that decrease reproduction and growth of the grazing animal.[45,68] The ergopeptine alkaloids reduce serum prolactin,[68] a growth hormone that regulates feed intake in mammalian herbivores.[49]

Ergopeptine, peramine, and pyrollizidine-based loline alkaloids are biologically active on numerous species of insects, including aphids,[30,53,64] sod webworms,[25] leaf hoppers,[34] chinch bugs,[60] fall army-worm,[17] and Argentine stem weevil.[52,53] The alkaloids not only serve as feeding deterrents, but also decrease the reproduction and growth of the insects as well. Nematode reproduction and growth are decreased when the *Acremonium* species is present in the plant.[32,74]

Compelling evidence that deterrence of primary consumers occurs when host species are infected by *Acremonium* species illustrates that mechanisms for solar energy capture and retention are a premium priority for the association. Reduced appetite by grazers (insects and livestock) means photosynthetically active leaf area is more abundant when the plant is in association with the fungus. The result is greater energy capture by the plant, with less demand for partitioning that energy between expenditure for new leaf-area development and storage as energy reserves.[27] Plants with high energy reserves have more vigor than do those whose energy reserves have been depleted by defoliation. When reserve energy is high, plants are capable of withstanding environmental stresses and defoliation. Reduced nematode grazing and greater root masses of plants infected with *Acremonium* species undoubtedly enable the host plant to scavenge more water and nutrients. Consequently, infected plants invariably outpersist their noninfected counterparts, in pastures, especially during drought.[55]

It has been demonstrated that *A. coenophialum* mediates an accumulation of osmotica in meristems and elongating leaves of tall fescue.[74] While osmotic adjustment is likely to be one mechanism of survival during drought, others have found no osmotic adjustment among limited comparisons between genetically identical *A. coenophialum*-infected and noninfected tall fescue plants.[79] This has led to speculation that drought-surviving capabilities of the host plant may be avoidance or escape mechanisms, rather than the endophyte serving as an active osmotic pump. Regardless of the mechanism, *Acremonium*-infected grasses have improved capabilities to survive drought.

Acremonium-infected plants are more competitive in populations mixed with noninfected plants. Even when infection rates of the seed are low (3%), the resulting population of plants increases to 40% after establishment as pastures, regardless of whether they are subjected to drought or pest stresses. Established plants have more tillers and biomass and produce more seeds than do genetically identical plants without endophytes,[27,28,56] and are more competitive when grown under field conditions.[28] However, as was noted with tolerance to abiotic stress, changes in endophyte-mediated plant morphology and competitiveness are not consistent from one plant–endophyte combination to the next, and a specific plant–endophyte combination may respond differently to changes in the environment.[27,28] Therefore, *Acremonium*-infected populations express greater phenotypic diversity than do their noninfected counterparts, without increased plant genotypic variation. Increased phenotypic variability makes plant populations more plastic.[62] Together with their resistances to biotic stress, *Acremonium*-infected grass populations have a wider region of adaptation than do noninfected populations.

Acremonium-infected populations have diverted from the pattern of host sterility common to the other Balansiae fungi and therefore represent a major change in the cost:benefit equation. Differences in the maintenance respiration of *Acremonium*-infected plants are barely detectable when compared with genetically identical noninfected clones.[6] With *Acremonium* species the probability of the fungus finding a suitable host is virtually guaranteed by the obligatory association between the two species ($p = 1.0$). Not only is the association guaranteed, but increases in the numbers of tillers and seeds from infected plants (s_m) increases the probability of successful recruitment of unoccupied space by the infected component of a population. Greater root mass increases the probability of persistence of the plants once they become established. Fungal products (b_m), which benefit the plant, are numerous and designed specifically to reduce the removal of leaf area by grazing mammals and insects, reduce root pruning or feeding by insects or pathogens, provide tolerance to adverse environmental conditions, and increase the growth rate of infected plants (f_m). When considering the benefit to the plant, the asexual life cycle of the endophyte, and the sexual life cycle of the host plant, one can only conclude that *Acremonium*–plant associations are model organisms for mutualistic associations.

VI. CONCLUSIONS

Balansiae fungi affect the competitive and reproductive capacity of their plant hosts differently and induce varying effects on local and community demographies. They range from the parasitic *M. atramentosa*, whose stroma engulf reproductive tillers and reduce the photosynthetic capacity of the plant, to the obligatory mutualistic associations of the *Acremonium* species, which provide stress tolerances and enhance the fecundity of their hosts. The other species of Balansiae, *Balansia, Atkinsonella,* and *Epichloë*, are less host specific and are more diverse in their effects on their hosts. While these species benefit their hosts in a manner whereby they fit the mathematical model for mutualism (Equation 1),[31] their sexual reproductive capabilities and inhibitory effect on plant sexual reproduction are a contrast to the mutualistic reproductive model.[36] To resolve the argument, the two models can be integrated into a larger model, rather than viewing each as independent models for mutualism. A combined model is useful in assessing the effects of the organisms on their hosts and in determining how far they have travelled on the evolutionary path to a final and stable mutualistic association. This can be justified by examining the frequency and distribution of the organisms within naturally occurring host populations. Those fungal organisms that appear more primitive have widespread associations among mutualists. While initial infection rates among plant populations may be low, the rate increases with the age of the population. For the frequency of infection to increase, each organism must benefit from the association; they are, therefore, mutualists. Fungi that are components of more advanced mutualistic associations are ones that are transmitted from generation to generation in, or on, the seeds, yet occasionally reproduce sexually. Their initial infection rates are higher among host populations than the more primitive associations and are maintained through seed transmission. While incidence of sexual reproduction by the fungus is low among these associations, it is evident that it is a predisposition before a final obligatory association can occur. Regardless of the degree of coevolutionary development, the Balansiae fungi provide multifaceted benefits to the plant, including insect and disease resistance, drought tolerance, and growth regulation.

REFERENCES

1. **Agee, C. S. and N. S. Hill,** Ergovaline variability in *Acremonium*-infected tall fescue due to environment and plant genotype, *Crop Sci.*, in press.
2. **Arachevaleta, M., C. W. Bacon, C. S. Hoveland, and D. E. Radcliffe,** Effect of the tall fescue endophyte on plant response environmental stress, *Agron. J.*, 81:83–90, 1989.
3. **Austin, M. P.,** Community theory and competition in vegetation, In: B. Grace and D. Tillman, Eds., *Perspectives on Plant Competition*, Academic Press, San Diego, 1990, 215.
4. **Bacon, C. W., J. K. Porter, J. Robbins, and E. S. Luttrell,** *Epichloe typhina* from oxic tall fescue grasses, *Appl. Environ. Microbiol.*, 34:576–581, 1977.
5. **Bayliss, M. W.,** Chromosomal variation in plant tissue culture, *Environ. Exp. Bot.*, 21:325–332, 1980.
6. **Belesky, D. P., O. J. Devine, J. E. Pallas, Jr., and W. C. Stringer,** Photosynthetic activity of tall fescue as influenced by a fungal endophyte, *Photosynthetica*, 21:82–87, 1987.
7. **Bradshaw, A. D.,** Population differentiaion in *Agrostis tenuis* Sibth. II. The incidence and significance of infection by *Epichloe typhina*, *New Phytol.*, 58:310–315, 1959.
8. **Campbell, C. S., J. A. Quinn, G. P. Cheplick, and T. J. Bell,** Cleistogamy in grasses, *Ann. Rev. Ecol. Syst.*, 14:411–441, 1983.
9. **Clay, K.,** Environmental and genetic determinants of cleistogamy in a natural population of the grass *Danthonia spicata*, *Evolution*, 36:734–741, 1982.
10. **Clay, K.,** The effect of the fungus *Atkinsonella hypoxylon* (Clavicipitaceae) on the reproductive system and demography of the grass *Danthonia spicata*, *New Phytol.*, 98:165–175, 1984.
11. **Clay, K.,** Induced vivipary in the sedge *Cyperus virens* and the transmission of the fungus *Balansia cyperi* (Clavicipitaceae), *Can. J. Bot.*, 64:2984–2988, 1986a.
12. **Clay, K.,** New disease (*Balansia cyperi*) of purple nutsedge (*Cyperus rotundus*), *Plant Dis.*, 70:597–599, 1986b.
13. **Clay, K.,** Clavicipitaceous fungal endophytes of grasses: coevolution and the change from parasitism to mutualism, In: K. A. Pirozinski and D. L. Hawksworth, Eds., *Coevolution of Fungi with Plants and Animals*, Academic Press, San Diego, 1988a, 79.
14. **Clay, K.,** Clavicipitaceous endophytes of grasses: their potential as biocontrol agents, *Mycol. Res.*, 92:1–12, 1989.
15. **Clay, K.,** Fungal endophytes of grasses, *Ann. Rev. Ecol. Syst.*, 21:275–297, 1990.
16. **Clay, K., G. P. Cheplick, and S. Marks,** Impact of the fungus *Balansia henningsiana* on *Panicum agrostoides*: frequency of infection, plant growth and reproduction, and resistance to pests, *Oecologia*, 80:374–380, 1989.
17. **Clay, K., T. N. Hardy, and A. M. Hammond, Jr.,** Fungal endophytes of grasses and their effects on an insect herbivore, *Oecologia*, 66:1–6, 1985.
18. **Clay, K. and J. P. Jones,** Transmission of *Atkinsonella hypoxylon* (Clavicipitaceae) by cleistogamous seed of *Danthonia spicata* (Gramineae), *Can. J. Bot.*, 62:2893–2895, 1984.
19. **Cooke, R.,** *The Biology of Symbiotic Fungi*, John Wiley & Sons, London, 1977.
20. **Day, P. R.,** *Genetics of Host-Parasite Interaction*, W. H. Freeman, San Francisco, 1974.
21. **De Battista, J. P., C. W. Bacon, R. Severson, R. D. Plattner, and J. H. Bouton,** Indole acetic acid production by the fungal endophyte of tall fescue, *Agron. J.*, 82:651–654, 1990.
22. **Diehl, W. W.,** *Balansia* and the Balansiae in America, U.S. Department of Agriculture Monogr. 4. U.S. Government Printing Office, Washington, D.C., 1950.
23. **Fletcher, L. R. and B. L. Sutherland,** Liveweight change in lambs grazing perennial ryegrasses with different endophytes, In: D. E. Hume, G. C. M. Latch, and H. S. Easton, Eds., *Proc. 2nd Int. Symp. Acremonium/Grass Interactions*, AgResearch, Grasslands Research Centre, Palmerston North, New Zealand, 1993, 125.
24. **Francis, S. M. and D. B. Baird,** Increase in the proportion of endophyte-infected perennial ryegrass plants in overdrilled pastures, *N.Z. J. Agric. Res.*, 32:437–440, 1989.
25. **Funk, C. R., P. M. Halisky, M. C. Johnson, M. R. Siegel, and A. V. Stewart,** An endophytic fungus and resistance to sod webworms, *Biotechnology*, 1:189–191, 1983.
26. **Gould, F. W. and R. B. Shaw,** *Grass Systematics*, 2nd ed., Texas A & M University Press, College Station, 1983.

27. **Hill, N. S., W. C. Stringer, G. E. Rottinghaus, D. P. Belesky, W. A. Parrott, and D. D. Pope,** Growth, morphological, and chemical component responses of tall fescue to *Acremonium coenophialum, Crop Sci.,* 30:156–161, 1990.

28. **Hill, N. S., D. P. Belesky, and W. C. Stringer,** Competitiveness of tall fescue as influenced by *Acremonium coenophialum, Crop Sci.,* 31:185–190, 1991.

29. **Hooker, A. L.,** Breeding to control pests, In: D. R. Wood, K. M. Rawal, and M. N. Wood, Eds., *Crop Breeding,* American Society Agronomy, Madison, WI, 1983, 199.

30. **Johnson, M. C., D. L. Dahlman, M. R. Siegel, L. P. Bush, and G. C. M. Latch,** Insect feeding deterrents in endophyte-infected tall fescue, *Appl. Environ. Microbiol.,* 49:568–571, 1985.

31. **Keeler, K. H.,** Cost:benefit analysis of mutualism, In: D. H. Boucher, Ed., *The Biology of Mutualism: Ecology and Evolution,* Croom Helm, London, 1985, 100.

32. **Kimmons, C. A., K. D. Gwinn, and E. C. Bernard,** Nematode reproduction on endophyte-infected and endophyte-free tall fescue, *Plant Dis.,* 74:757–761, 1990.

33. **Kirby, E. J. M.,** Host-parasite relations in the choke disease of grasses, *Trans. Br. Mycol. Soc.,* 44:493–503, 1961.

34. **Kirfman, G. W., R. L. Brandenburg, and G. B. Garner,** Relationship between insect abundance and endophyte infestation level in tall fescue in Missouri, *J. Kansas Entomol. Soc.,* 59:552–554, 1986.

35. **Koshino, H., S. Terada, T. Yoshihara, S. Sakamura, T. Shimanuki, T. Sato, and A. Akitoshi,** Three phenolic acid derivatives from stromata of *Epichloe typhina* on *Phleum pratense, Phytochemistry,* 27:1333–1338, 1988.

36. **Law, R.,** Evolution in a mutualistic environment, In: D. H. Boucher, Ed., *The Biology of Mutualism: Ecology and Evolution,* Croom Helm, London, 1985, 145.

37. **Law, R. and D. H. Lewis,** Biotic environments and the maintenance of sex. Some evidence from mutualistic symbioses, *Biol. J. Linnean Soc.,* 20:249–276, 1983.

38. **Leuchtmann, A. and K. Clay,** *Atkinsonella hypoxylon* and *Balansia cyperi,* epiphytic members of the Balansiae, *Mycologia,* 80:192–199, 1988a.

39. **Leuchtmann, A. and K. Clay,** Experimental infection of host grasses and sedges with *Atkinsonella hypoxylon* and *Balansia cyperi* (Balansiae, Clavicipitaceae), *Mycologia,* 80:291–297, 1988b.

40. **Leuchtmann, A. and K. Clay,** Experimental evidence for genetic variation in compatibility between the fungus *Atkinsonella hypoxylon* and its three host grasses, *Evolution,* 43:825–834, 1989a.

41. **Leuchtmann, A. and K. Clay,** Isozyme variation in the fungus *Atkinsonella hypoxilon* within and among populations of its host grasses, *Can. J. Bot.,* 67:2600–2607, 1989b.

42. **Lewin, R. A.,** Symbiosis and parasitism. Definitions and evaluations, *Biosci. Rep.,* 32:254–259, 1982.

43. **Lewis, D. H.,** Symbiosis and mutualism: crisp concepts and soggy semantics, In: D. H. Boucher, Ed., *The Biology of Mutualism: Ecology and Evolution,* Croom Helm, London, 1985, 9.

44. **Luttrell, E. S. and C. W. Bacon,** Classification of *Myriogenospora* in the Clavicipitaceae, *Can. J. Bot.,* 55:2090–2097, 1977.

45. **Lyons, P. C., R. D. Plattner, and C. W. Bacon,** Occurence of peptide and clavine ergot alkaloids in tall fescue, *Science,* 232:487–489, 1986.

46. **diMenna, M. E., P. H. Mortimer, R. A. Prestidge, A. D. Hawkes, and J. M. Sprosen,** Lolitrem B concentrations of *Acremonium lolii* hyphae, and the incidence of ryegrass taggers in lambs on plots of *A. lolii*-infected perennial ryegrass, *N.Z. J. Agric. Res.,* 5:211–217, 1992.

47. **Morgan-Jones, G. and W. Gams,** Notes on hyphomycetes. XLI. An endophyte of *Festuca arundinacea* and the anamorph of *Acremonium, Mycotaxon,* 15:311–318, 1982.

48. **Neill, J. C.,** The endophytes of *Lolium* and *Festuca, N.Z. J. Sci. Technol.,* 23A:185–195, 1941.

49. **Noel, M. B. and B. Woodside,** Effects of systemic and central prolactin injections on food intake, weight gain, and estrous cyclicity in female rats, *Physiol. Behav.,* 54:151–154, 1993.

50. **Odum, E. P.,** *Fundamentals of Ecology,* W.B. Saunders, Philadelphia, 1953.

51. **Porter, J. K., C. W. Bacon, H. G. Cutler, F. F. Arrendale, and J. D. Robbins,** In vitro auxin production by *Balansia epichloe, Phytochemistry,* 24:1429–1431, 1985.

52. **Pottinger, R. P., G. M. Barker, and R. A. Prestige,** A review of the relationships between endophytic fungi of grasses (*Acremonium* spp.) and Argentine stem weevil (*Listronotus bonarienses* Kuschel), In: *Proc. 4th Australasian Conf. on Grassland Invertebrate Ecology,* Lincoln College, Canterbury, New Zealand, 1985, 322.

53. **Prestige, R. A., R. P. Pottinger, and G. M. Barker,** An association of *Lolium* endophyte with ryegrass resistance to Argentine stem weevil, In: *Proc. N.Z. Weed Pest Control Conference,* 1982, 119.

54. **Read, C. P.,** *Parasitism and Symbiology,* Ronald Press, New York, 1970.

55. **Read, J. C. and B. J. Camp,** The effects of the fungal endophyte *Acremonium coenophialum* in tall fescue on animal performance, toxicity, and stand maintenance, *Agron. J.,* 78:848–850, 1986.

56. **Rice, J. S., B. W. Pinkerton, W. C. Stringer, and D. J. Undersander,** Seed production in tall fescue as affected by fungal endophyte, *Crop Sci.,* 30:1303–1305, 1990.

57. **Russel, G. E.,** *Plant Breeding for Pest and Disease Resistance,* Butterworths, London, 1978.

58. **Rykard, D. M., E. S. Luttrell, and C. W. Bacon,** Conidiogenesis and conidiomata in the clavicipitoideae, *Mycologia,* 76:1095–1103, 1984.

59. **Rykard, D. M., C. W. Bacon, and E. S. Luttrell,** Host relations of *Myriogenospora atramentosa* and *Balansia epichloe* (Clavicipitaceae), *Phytophathology,* 75:950–956, 1985.

60. **Saha, D. C., J. M. Johnson-Cicalese, P. M. Halisky, M. I. Van Heemstra, and C. R. Funk,** Occurance and significance of endophytic fungi in the fine fescues, *Plant Dis.,* 1:1021–1024, 1987.

61. **Sampson, K.,** The systematic infecton of grasses by *Epichloe typhina* (Pers.) Tul., *Trans. Br. Mycol. Soc.,* 18:30–47, 1933.

62. **Schmid, B.,** Clonal growth in grassland perennials. III. Genetic variation and plasticity between and within *Bellis perennis* and *Prunella vulgaris, J. Ecol.,* 73:819– 830, 1985.

63. **Shelby, R. A. and L. W. Dalrymple,** Incidence and distribution of tall fescue endophytes in the United States, *Plant Dis.,* 71:783–786, 1987.

64. **Siegel, M. R., G. C. M. Latch, L. P. Bush, N. F. Fannin, D. D. Rowen, B. A. Tapper, C. W. Bacon, and M. C. Johnson,** Alkaloids and insecticidal activity of grasses infected with different endophytes, *J. Chem. Ecol.,* 16:3301–3315, 1991.

65. **Smith, K. T., C. W. Bacon, and E. S. Luttrell,** Carbohydrate movement between host and fungus in bahiagrass infected with *Myriogenospora atramentosa, Phytopathology,* 75:407–411, 1985.

66. **Stanley, S. M.,** Clades versus clones in evolution: why we have sex, *Science,* 190:382–383, 1975.

67. **Stovall, M. E. and K. Clay,** The effect of the fungus, *Balansia cyperi* Edg., on growth and reproduction of purple nutsedge, *Cyperus rotundus* L., *New Phytol.,* 109:351–359, 1988.

68. **Stuedemann, J. A. and F. N. Thompson,** Management strategies and potential opportunities to reduce the effects of endophyte-infested tall fescue on animal performance, In: D. E. Hume, G. C. M. Latch, and H. S. Easton, Eds., *Proc. 2nd Int. Symp.* Acremonium/*Grass Interactions: Plenary Papers,* AgResearch, Grasslands Research Centre, Palmerston North, New Zealand, 1993, 103.

69. **Thompson, F. N., J. A. Stuedemann, J. L. Sartin, D. P. Belesky, and O. J. Devine,** Selected hormonal changes with summer fescue toxicosis, *J. Anim. Sci.,* 65:727–733, 1987.

70. **Vasil, I. K.,** Developing cell and tissue culture systems for the improvement of cereal and grass crops, *J. Plant Physiol.,* 128:193–218, 1987.

71. **Vasil, I. K.,** Progress in the regeneration and genetic manipulaiton of cereal crops, *Biotechnology,* 6:397–402, 1988.

72. **Warner, D. A. and G. E. Edwards,** Effects of polyploidy on photosynthetic rates, photosynthetic enzymes, contents of DNA, chlorophyll, and sizes and numbers of photosynthetic cells in the C_4 dicot *Atriplex confertifolia, Plant Physiol.,* 91:1143–1151, 1989.

73. **Warner, D. A., M. S. B. Ku, and G. E. Edwards,** Photosynthesis, leaf anatomy, and cellular constituents in the polyploid C_4 grass *Panicum virgatum, Plant Physiol.,* 4:461–466, 1987.

74. **West, C. P., E. Izekor, A. Elmi, R. T. Robbins, and K. E. Turner,** Endophyte effects on drought tolerance, nematode infestation and persistence of tall fescue, In: C. P. West, Ed., *Proc. Arkansas Fescue Toxicosis Conf.,* University of Arkansas Agriculture Experiment Station, Fayetteville, 1989, 23.

75. **Western, J. H. and J. J. Cavett,** The choke disease of cocksfoot (*Dactylis glomerata*) caused by *Epichloe typhina* (Fr.) Tul., *Trans. Br. Mycol. Soc.,* 42:298–307, 1959.

76. **White, J. F.,** Endophyte-host associations in forage grasses. XI. A proposal concerning origin and evolution, *Mycologia,* 80:442–446, 1988.

77. **White, J. F. and D. A. Chambless,** Endophyte-host associations in forage grasses. XV. Clustering of stromata-bearing individuals of *Agrostis hiemalis* infected by *Epichloe typhina, Am. J. Bot.,* 78:527–533, 1991.

78. **White, J. F. and J. R. Owens,** Stromal development and mating system of *Balansiae epichloe,* a leaf-colonizing endophyte of warm-season grasses, *Appl. Environ. Microbiol.,* 58:513–519, 1992.

79. **White, R. H., M. C. Engelke, S. J. Morton, J. M. Hohnson-Cicalese, and B. A. Ruemmele,** *Acremonium* endophyte effects on tall fescue drought tolerance, *Crop Sci.,* 32:1392–1396, 1992.

Chapter 6

The Potential Role of Endophytes in Ecosystems

Keith Clay

CONTENTS

I. INTRODUCTION

Grasses are our dominant plant family in terms of their abundance and dominance in both natural and agricultural ecosystems. They carpet the Earth and feed its people. The large volume of research on interactions between grasses and clavicipitaceous endophytes, reviewed in detail elsewhere in this volume, has provided basic information on the distribution of endophytes in different groups of grasses and different habitats, the effects of endophyte infection on individual host plants, and the consequences of these effects on community-level processes such as competition and herbivory. The objective of this paper is to consider the potential role of endophytes in a broad context of community- and ecosystem-level processes. Four general topics will be considered in particular: (1) the effect of endophytes on plant biodiversity, (2) the effect of endophytes on food chains, (3) the ecological consequences of the release of novel grass–endophyte associations, and (4) the response of endophyte-infected grasses and grasslands to increased atmospheric CO_2. Human impacts on all of these phenomena are considerable and increasing.

While there has been a great deal of research on grass–endophyte interactions, little of it has focused directly on the four topics above. Therefore, much evidence for the possible role of endophytes is indirect and extrapolary. It is hoped that this contribution will raise awareness and discussion of issues that have generally not been considered in endophyte research.

II. GRASSES AND GRASSLANDS

Nearly 10,000 species of grasses exist, making them one of the largest of all plant families.[80] Grasses occur in virtually all terrestrial habitats, from Arctic tundra to tropical rainforests and from alpine habitats

27. **Ellstrand, N. C. and C. A. Hoffman,** Hybridization as an avenue of escape for engineered genes, *Bioscience,* 40:438–442, 1990.

28. **Finn, G. and M. Brun,** Effects of atmospheric CO_2 enrichment on growth, nonstructural carbohydrate content and root nodule activity in soybean, *Plant Physiol.,* 69:327–331, 1982.

29. **Fomba, S. N.,** Rice disease situation in mangrove and associated swamps in Sierra Leone, *Trop. Pest Manage.,* 30:73–81, 1984.

30. **Francis, S. M. and D. B. Baird,** Increase in the proportion of endophyte-infected perennial ryegrass plants in over-drilled pastures, *N.Z. J. Agric. Res.,* 32:437–440, 1989.

31. **Funk, C. R., R. H. White, and J. B. Breen,** Importance of *Acremonium* endophytes in turf-grass breeding and management, *Agric. Ecosyst. Environ.,* 44:215–232, 1993.

32. **Funk, C. R., P. M. Halisky, M. C. Johnson, M. R. Siegel, A. V. Stewart, S. Ahmad, R. H. Hurley, and I. C. Harvey,** An endophytic fungus and resistance to sod webworms: association in *Lolium perenne, Biotechnology,* 1:189–191, 1983.

33. **Gould, F. W. and R. B. Shaw,** *Grass Systematics,* Texas A&M University Press, College Station, 1983.

34. **Govindu, H. C. and M. J. Thirumalachar,** *Ephelis* on *Sorghum halepense* in Mysore, *Science,* 119:288–289, 1954.

35. **Harberd, D. J.,** Note on choke disease of *Festuca rubra, Scottish Plant Breeding Stat. Rep.,* Pentlandfield, Midlothian, Scotland, 1961, 47.

36. **Hill, N. S., W. C. Stringer, G. E. Rottinghaus, D. P. Belesky, W. A. Parrot, and D. D. Pope,** Growth, morphological, and chemical component responses of tall fescue to *Acremonium coenophialum, Crop Sci.,* 30:156–161, 1990.

37. **Hoveland, C. S.,** Importance and economic significance of *Acremonium* endophytes to performance of animals and grass plant, *Agric. Ecosyst. Environ.,* 44:3–12, 1993.

38. **Johnson, M. C., L. P. Bush, and M. R. Siegel,** Infection of tall fescue with *Acremonium coenophialum* by means of callus culture, *Plant Dis.,* 70:380–382, 1986.

39. **Kirfman, G. W., R. L. Brandenburg, and G. B. Garner,** Relationship between insect abundance and endophyte infestation level in tall fescue in Missouri, *J. Kansas Entomol. Soc.,* 59:552–554, 1986.

40. **Knoch, T. R., S. H. Faeth, and D. S. Arnott,** Fungal endophytes: plant mutualists via seed predation and germination, *Bull. Ecol. Soc. Am.,* 74(Abstr.):313, 1993.

41. **Langevin, S., K. Clay, and J. B. Grace,** The incidence and effects of hybridization between cultivated rice and its related weed red rice (*Oryza sativa* L.), *Evolution,* 44:1000–1008, 1990.

42. **Large, E. C.,** Surveys for choke (*Epichloe typhina*) in cocksfoot seed crops, 1951–53, *Plant Pathol.,* 3:6–11, 1954.

43. **Latch, G. C. M.,** Physiological interactions of endophytic fungi and their hosts. Biotic stress tolerance imparted to grasses by endophytes, *Agric. Ecosyst. Environ.,* 44:143–156, 1993.

44. **Latch, G. C. M. and M. J. Christensen,** Artificial infections of grasses with endophytes, *Ann. Appl. Biol.,* 107:17–24, 1985.

45. **Latch, G. C. M., M. J. Christensen, and D. L. Gaynor,** Aphid detection of endophytic infection in tall fescue, *N.Z. J. Agric. Res.,* 28:129–132, 1985.

46. **Latch, G. C. M., L. R. Potter, and B. F. Tyler,** Incidence of endophytes in seeds from collections of *Lolium* and *Festuca* species, *Ann. Appl. Biol.,* 111:59–64, 1987.

47. **Leuchtmann, A.,** Taxonomy and host relations of fungal endophytes of grasses, *Nat. Toxins,* 1:150–162, 1992.

48. **Leuchtmann, A. and K. Clay,** Experimental evidence for genetic variation in compatibility between the fungus *Atkinsonella hypoxylon* and its host grasses, *Evolution,* 43:825–834, 1989.

49. **Leuchtmann, A. and K. Clay,** Nonreciprocal compatibility between the fungus *Epichloe typhina* and four host grasses, *Mycologia,* 85:157–163, 1993.

50. **Levin, D. A. and H. W. Kerster,** Gene flow in seed plants, *Evol. Biol.,* 7:139–220, 1974.

51. **Lewis, G. C. and R. O. Clements,** A survey of ryegrass endophyte (*Acremonium loliae*) in the U.K. and its apparent ineffectuality on a seedling pest, *J. Agric. Sci.,* 107:633–638, 1986.

52. **Mackintosh, C. G., M. B. Orr, R. T. Gallagher, and I. C. Harvey,** Ryegrass staggers in Canadian wapiti deer, *N.Z. Vet. J.,* 36:106–107, 1982.

53. **Madej, C. W. and K. Clay,** Avian seed preference and weight loss experiments: the role of fungal endophyte-infected tall fescue seeds, *Oecologia,* 88:296–302, 1991.

54. **Marks, S. and K. Clay,** Effects of CO_2 enrichment, nutrient addition, and fungal endophyte-infection on the growth of two grasses, *Oecologia*, 84:207–214, 1990.

55. **Marks, S., K. Clay, and G. P. Cheplick,** Effects of fungal endophytes on interspecific and intraspecific competition in the grasses *Festuca arundinacea* and *Lolium perenne, J. Appl. Ecol.*, 28:194–204, 1991.

56. **May, R. M.,** *Stability and Complexity in Model Ecosystems,* Princeton University Press, Princeton, NJ, 1973.

57. **Mooney, H. A. and J. A. Drake,** *Ecology of Biological Invasions of North America and Hawaii,* Springer-Verlag, New York, 1986.

58. **Moore, C. W. E.,** Distribution of grasslands, In: C. Barnard, Ed., *Grasses and Grasslands,* St. Martin's Press, New York, 1966, 182.

59. **Murray, F. R., G. C. M. Latch, and D. B. Scott,** Surrogate transformation of perennial ryegrass, *Lolium perenne,* using genetically modified *Acremonium* endophyte, *Mol. Gen. Genet.*, 23:1–9, 1992.

60. **Nobindro, U.,** Grass poisoning among cattle and goats in Assam, *Indian Vet. J.,* 10:235–236, 1934.

61. **Norby, R. J.,** Nodulation and nitrogenase activity in nitrogen-fixing woody plants stimulated by CO_2 enrichment of the atmosphere, *Physiol. Plant.*, 71:77–82, 1987.

62. **Norby, R. J., E. G. O'Neill, and R. J. Luxmoore,** Effects of atmospheric CO_2 enrichment on the growth and mineral nutrition of *Quercus alba* seedlings in nutrient poor soil, *Plant Physiol.*, 82:83–92, 1987.

63. **Odum, E. P.,** *Fundamentals of Ecology*, W.B. Saunders, Philadelphia, 1971.

64. **O'Neill, E. G., R. J. Luxmoore, and R. J. Norby,** Elevated atmospheric CO_2 effects on seedling growth, nutrient uptake, and rhizosphere bacterial populations in *Liriodendron tulipifera* L., *Plant Soil,* 104:3–11, 1987.

65. **O'Neill, E. G., R. J. Luxmoore, and R. J. Norby,** Increases in mycorrhizal colonization and seedling growth in *Pinus echinata* and *Quercus alba* in an enriched CO_2 atmosphere, *Can. J. For. Res.,* 17:878–883, 1987.

66. **Ou, S.,** *Rice Diseases*, Eastern Press, London, 1972.

67. **Prestidge, R. A., M. E. di Menna, and S. van der Zijpp,** An association of *Lolium* endophyte with ryegrass resistance to Argentine stem weevil, *Proc. N.Z. Weed Pest Control Conf.,* 35:119–122, 1985.

68. **Pshedetskaya, L. I.,** Distribution of the agent "choke" of cereals *Epichloe typhina* Tul. in the USSR, *Vestnik LGU,* 3:118–120, 1984 (in Russian).

69. **Read, J. C. and B. J. Camp,** The effect of fungal endophyte *Acremonium coenophialum* in tall fescue on animal performance, toxicity, and stand maintenance, *Agron. J.,* 78:848–850, 1986.

70. **Rice, J. S., B. W. Pinkerton, W. C. Stringer, and D. J. Undersander,** Seed production in tall fescue as affected by fungal endophyte, *Crop Sci.,* 30:1303–1305, 1990.

71. **Ricklefs, R. E.,** *The Economy of Nature,* Chiron Press, New York, 1983.

72. **Sadler, K.,** Of rabbits and habitat, a long term look, *Missouri Conservationist,* March:4–7, 1980.

73. **Schardl, C. L.,** Molecular biology and evolution of grass endophytes, *Nat. Toxins,* 1:171–184, 1992.

74. **Schmidt, S. P. and T. G. Osborn,** Effects of endophyte-infected tall fescue on animal performance, *Agric. Ecosyst. Environ.,* 44:233–262, 1993.

75. **Shaw, J.,** On the changes going on in the vegetation of South Africa, *Bot. J. Linnean Soc.,* 14:202–208, 1873.

76. **Shelby, R. A. and L. W. Dalrymple,** Incidence and distribution of the tall fescue endophyte in the United States, *Plant Dis.,* 71:783–785, 1987.

77. **Siegel, M. R., G. C. M. Latch, L. P. Bush, N. F. Fannin, D. D. Rowan, B. A. Tapper, C. W. Bacon, and M. C. Johnson,** Fungal endophyte-infected grasses: alkaloid accumulation and aphid response, *J. Chem. Ecol.,* 16:3301–3315, 1990.

78. **Simmonds, N. W.,** *Evolution of Crop Plants,* Longmans, London, 1976.

79. **Smith, J. E., A. W. Cole, and V. H. Watson,** Selective smut-grass control and forage quality response in bermudagrass-dallisgrass pastures, *Agron. J.,* 66:424–426, 1974.

80. **Soderstrom, T. R., K. W. Hilu, C. Campbell, and M. E. Barkworth,** *Grass Systematics and Evolution,* Smithsonian Institution Press, Washington, D.C., 1987.

81. **Stebbins, G. L.,** *Chromosomal Evolution in Higher Plants*, Edward Arnold, London, 1971.

82. **Stebbins, G. L.,** Coevolution of grasses and herbivores, *Ann. Missouri Bot. Gard.,* 68:75–86, 1981.

83. **Stewart, A. V.,** Perennial ryegrass seedling resistance to Argentine stem weevil, *N.Z. J. Agric. Res.,* 28:403–407, 1985.

84. **Steudemann, J. A. and C. S. Hoveland,** Fescue endophyte: history and impact on animal agriculture, *J. Prod. Agric.,* 1:39–44, 1988.

85. **Stovall, M. E. and K. Clay,** The effect of the fungus *Balansia cyperi* on the growth and reproduction of purple nutsedge, *Cyperus rotundus, New Phytol.,* 109:351–359, 1988.

86. **Strain, B. R. and J. D. Cure,** Direct effects of CO_2 enrichment on plants and ecosystems: a bibliography with abstracts. Oak Ridge National Laboratory Publication ORNL CDIC-13, Oak Ridge, TN, 1986.

87. **Sutherland, B. L. and J. H. Hogland,** Effect of ryegrass containing the endophyte *Acremonium lolii* on the performance of associated white clover and subsequent crops, *Proc. N.Z. Grasslands Assoc.,* 50:265–269, 1989.

88. **Telson, M. A., I. A. Leone, and F. B. Flower,** The role of an ectomycorrhizal fungus *Pisolthus tinctorius* in the survival and growth of scots pine subjected to landfill conditions, *Phytopathology,* 70:470, 1980.

89. **Trabalka, J. R., J. A. Edmonds, J. Reilly, R. H. Gardner, and L. D. Voorhees,** Human alterations of the global carbon cycle and the projected future, In: Trabalka, J. R. Ed., Atmospheric carbon dioxide and the global carbon cycle, DOE publication ER-0239, Washington, D.C., 1985.

90. **Walker, J.,** Systemic fungal parasite of *Phalaris tuberosa* in Australia, *Search,* 1:81–83, 1970.

91. **Welty, R. E., M. D. Azevedo, and T. M. Cooper,** Influence of moisture content, temperature, and length of storage on seed germination and survival of endophytic fungi in seeds of tall fescue and perennial ryegrass, *Phytopathology,* 77:893–900, 1987.

92. **West, C. P., E. Izekor, D. M. Oosterhuis, and R. T. Robbins,** The effect of *Acremonium coenophialum* on the growth and nematode infestation of tall fescue, *Plant Soil,* 112:3–6, 1988.

93. **White, J. F.,** Widespread distribution of endophytes in the Poaceae, *Plant Dis.,* 71:340–342, 1987.

94. **White, J. F., P. M. Halisky, S. Sun, G. Morgan-Jones, and C. R. Funk,** Endophyte-host associations in forage grasses. XVI. Patterns of endophyte distribution in species of the tribe Agrostideae, *Am. J. Bot.,* 79:472–477, 1992.

95. **White, J. F., A. C. Morrow, G. Morgan-Jones, and D. A. Chambless,** Endophyte-host associations in forage grasses. XIV. Primary stromata formation and seed transmission in *Epichloe typhina*: developmental and regulatory aspects, *Mycologia,* 83:72–81, 1991.

96. **White, J. F., G. Morgan-Jones, and A. C. Morrow,** Taxonomy, life cycle, reproduction and detection of *Acremonium* endophytes, *Agric. Ecosyst. Environ.,* 44:13–37, 1993.

97. **Wilson, A. D., S. L. Clement, W. J. Kaiser, and D. G. Lester,** First report of clavicipitaceous anamorphic endophytes in *Hordeum* species, *Plant Dis.,* 75:215, 1991.

98. **Wray, S. M. and B. R. Strain,** Response of two old field perennial to interactions of CO_2 enrichment and drought stress, *Am. J. Bot.,* 73:1486–1491, 1986.

99. **Zavos, P. M., B. Salim, J. A. Jackson, D. R. Varney, M. R. Siegel, and R. W. Hemken,** Effect of feeding tall fescue seed infected by endophytic fungus *Acremonium coenophialum* on reproductive performance in male rat, *Theriogenology,* 25:281–290, 1986.

Chapter 7

Physiology and Drought Tolerance of Endophyte-Infected Grasses

Charles P. West

CONTENTS

I. INTRODUCTION

Grasses have evolved various structural and physiological means of adapting to stress conditions, to ensure population stability in natural environments. Agricultural scientists have exploited such traits, in breeding for improved seed yield of annual cereal crops and for improved forage and turf characteristics of perennial grasses. It has been recently discovered that the endophytic fungus, *Acremonium coenophialum* Morgan-Jones & Gams, enhances the drought tolerance, competitiveness, and broad-range pest resistance of its host, tall fescue (*Festuca arundinacea* Schreb.). *Acremonium lolii* Latch, Christensen & Samuels also imparts insect resistance to perennial ryegrass (*Lolium perenne* L.). Such mutualistic symbioses may provide a means of supplementing conventional plant breeding, to further improve abiotic and biotic stress resistances in grasses. Even though endophytic fungi are widely distributed in the Poaceae,[61] short-term progress in exploiting endophytes is most likely with species of *Acremonium* Link ex Fr. in *Festuca* L. and *Lolium* L. grasses, inasmuch as the agronomic benefits of these mutualisms are well described.[4,27,32,56]

Endophyte–grass mutualisms are complex associations involving integrated physiological responses on the part of each organism in terms of growth, water relations, nutrient acquisition and use, and secondary metabolite synthesis. This chapter reviews current knowledge of the effects of the *Acremonium* endophytes on host physiology, relating to growth and drought stress tolerance, and discusses approaches for using such endophytes to improve drought tolerance. Use of the term "endophyte" will refer to *Acremonium* species, with emphasis on the *A. coenophialum*–tall fescue association. Furthermore, this association may be referred to as symbiotic or, in the case of uninfected plants, nonsymbiotic.

II. GROWTH

A. CONTROLLED ENVIRONMENTS

Although several studies have shown that endophyte presence tends to increase shoot mass,[1,17,38] tillering,[1,17,58] and seed production[14,45] in tall fescue, the effects are not consistent across host genotypes[6,45] and populations.[17] Belesky et al.[8] compared endophyte-infected and endophyte-free clones of four genotypes selected from a 15-year-old "Kentucky 31" population, for growth characteristics in a growth chamber study. Genotypes varied in their responses, in final leaf length, tiller number, pseudostem mass, and root mass, to endophyte infection status or water regime. Using the same genotypes in a greenhouse trial, however, Hill et al.[25] found that infected genotypes tended to produce larger, but fewer, tillers than noninfected isolines in one trial (Table 1). In a repeat of this trial, most infected genotypes produced greater shoot biomass than did their noninfected isolines, owing to greater mass per tiller instead of greater tiller number. Greater shoot growth of infected isolines was not explained by higher nonstructural carbohydrate concentrations, suggesting that the endophyte increases the efficiency of carbohydrate utilization for regrowth. Interestingly, infected plants had consistently lower crown depth below the soil surface than did noninfected isolines, a trait that the authors suggested may buffer the plant from temperature extremes.

Growth of other grasses is also affected by *Acremonium* species. Latch et al.[33] reported that live leaf, stem, and root mass were all greater in infected isolines, than in noninfected isolines, of perennial ryegrass from a "Grasslands Nui" population growing in a growth chamber. Other trials on *Lolium* species showed either no effect or an inhibitory effect, of *A. lolii* on grass growth.[30,40] The above trials were conducted in the absence of insect herbivory. Two genotypes of *Festuca pratensis* Huds. produced more shoot mass and tillers, when artificially infected with a strain of *A. uncinatum* Gams, Petrini, and Schmidt, but not when infected with a *Phialophora*-like endophyte.[52]

The effect of the endophyte presence on host growth sometimes interacts with nutrient status of the growing medium. Arachevaleta et al.[1] reported no endophyte effect on shoot yield, at the lowest nitrogen level applied, but yield increased to a greater extent in infected plants, than in noninfected plants, of a single genotype as the nitrogen level increased (Table 2). Cheplick et al.[13] reported, in a series of greenhouse studies, that biomass was usually greater in infected seedlings, than in noninfected seedlings and adults, of tall fescue and perennial ryegrass, especially at high nutrient levels. At intermediate nutrient levels, endophyte presence stimulated growth or had no effect. At the lowest nutrient level, endophyte enhanced the growth of tall fescue adults, but depressed the growth of tall fescue seedlings. In the latter case the authors suggested that the endophyte competed with the host for energy and nutrients at low input levels, thereby reducing the availability of resources for host growth. In support of this hypothesis, Marks and Clay[39] found that infected perennial ryegrass plants grew faster only at high nutrient and high atmospheric carbon dioxide levels.

B. FIELD ENVIRONMENTS

Growth responses of tall fescue to the endophyte, under field conditions, have also been variable, but generally point toward enhanced host growth and vigor, as manifested by greater competitiveness and drought survival of symbiotic, over nonsymbiotic, plants. Hill et al.[26] found that endophyte infection more consistently enhanced shoot mass and tiller number per plant, in field trials than in growth chamber[8] and greenhouse trials,[25] using the same genotypes. They also measured relative crowding coefficients greater than 1.0 for infected, vs. noninfected, tall fescue, based on shoot mass, indicating that endophyte infection

Table 1 Dry matter yield, tiller production, and yield per tiller of greenhouse-grown endophyte-infected (E+) and endophyte-free (E−) tall fescue genotypes[25]

		Tillers		Yield			
		E+	E−	E+	E−	E+	E−
Year	Genotype	(No. plant^{-1})		(g plant^{-1})		(mg tiller^{-1})	
1987	7	75[a]	109	2.67	2.59	35.4[a]	26.1
	9	55[a]	76	2.48	2.60	45.6[a]	29.9
	11	62	69	2.78	2.90	44.4	42.1
	17	46	44	1.91	2.08	45.1	47.8
	27	51[a]	71	2.47	2.61	48.1[a]	38.0
	Mean	58	74	2.46	2.49	43.7	36.8
1988	7	103	98	6.13[a]	4.80	58.9	48.6
	9	64	72	4.45	4.65	67.5	63.1
	11	81	75	6.28[a]	5.48	77.2	68.6
	17	63	65	4.96[a]	4.48	77.4	67.8
	27	91	84	5.84[a]	5.56	63.0	66.6
	Mean	81	79	5.54	5.00	68.8[a]	62.9

[a] Significant difference ($p < 0.05$) between E+ and E− within a variable.

Table 2 Tiller number and shoot dry matter production of endophyte-infected and endophyte-free tall fescue at three nitrogen fertilizer rates (mg pot⁻¹)[1]

Infection Status	Tiller Number Nitrogen Rates			Shoot Yield Nitrogen Rates		
	11	73	220	11	73	220
		(No. pot⁻¹)			(g pot⁻¹)	
Infected	11	22	33	5.18	10.31	13.82
Noninfected	14	24	21	4.81	6.34	8.91
Significance test[a]	ns	ns	*	ns	*	*

[a] ns = Not significant; * = $p < 0.05$.

stimulated plant vigor at the expense of neighboring, noninfected plants. Read and Walker[44] reported higher forage yield and stand density in east Texas with infected "AU Triumph," "Kenhy," and "Kentucky 31" than with noninfected populations of the same cultivars, especially during years of below-normal rainfall. Forage yield was positively correlated with endophyte-infection frequency, and infection frequency increased over time. In southern Georgia, infected Kentucky 31 had a greater yield than did noninfected Kentucky 31, in late summer when it was not irrigated and was under infrequent cutting, but no endophyte effect was detected when it was irrigated.[65]

Lewis and Clements[35] grew infected and noninfected rows of nine perennial ryegrass genotypes at four locations in the U.K. Most genotypes showed no consistent endophyte effect on shoot yield across locations; however, the yield of one genotype was consistently enhanced by the endophyte, while that of another genotype was inhibited by the endophyte. Hume et al.[28] reported no endophyte effect on perennial ryegrass shoot yield in a cool, moist environment in New Zealand. In France, forage yield increases were associated with endophyte infection of perennial ryegrass in a warm, dry summer environment, but not in colder, wetter environments.[23] An enhancing effect of endophyte on ryegrass yield appears not to be as consistent as with tall fescue. This may be due to the milder, maritime climates that perennial ryegrass is usually tested in, compared with the warm, continental environments where tall fescue prevails.

C. HORMONE INVOLVEMENT

Acremonium modification of tiller number, growth, and morphology suggests a phytohormone-mediated response. Indeed, other fungal–plant associations display microbial-derived hormonal responses in the host. For example, root growth and morphology are affected by some auxin-producing mycorrhizae and *Rhizobium* bacteria.[34] Auxin is a key hormone in regulating plant growth, differentiation, and apical dominance and is produced mainly in shoot meristems and expanding leaves. The endophyte is most prevalent near these organs. Porter et al.[42] reported that auxin was produced *in vitro* by the closely related endophyte *Balansia epichloë* (Weese) Diehl. De Battista et al.[16] demonstrated *in vitro* production of auxin by *A. coenophialum* isolated from upright and rhizomatous tall fescue clones. The plants from which the endophytes were isolated produced 24% greater biomass than did their nonsymbiotic isolines. Auxin concentrations in both plant types were unaffected by endophyte infection; however, this does not rule out an involvement of the endophyte in auxin metabolism. Some of the inconsistent endophyte effects on tillering among genotype–endophyte associations[8] may be expressions of differing effects on auxin concentrations or antagonisms by other hormones such as gibberellic acid.

Gibberellin is another class of hormones that are produced by some fungi. Gibberellin may stimulate cell elongation, reduce bud dormancy, or modify host synthesis/activation of polysaccharide hydrolases.[34] The fungal pathogen *Gibberella fujikuroi* (Sawada) Wollenw. (perfect stage of *Fusarium moniliforme* Sheld.) secretes gibberellin, causing cell expansion and erratic stem elongation in rice (*Oryza sativa* L.).[41] De Battista et al.[16] found no gibberellins or cytokinins produced in *in vitro* cultures of two *A. coenophialum* isolates. Again, a possible involvement of symbiotic *Acremonium* in altering plant gibberellin or cytokinin production or deactivation cannot be ruled out.

Abscisic acid is produced by the fungi *Cercospora rosicola* Passevini, *C. cruenta* Sacc., and *Botrytis cinerea* Pers. ex Fr.[66] Bunyard and McInnis[12] reported *in vitro* production of auxin and abscisic acid in *A. coenophialum* broth culture when spiked with the hormone precursors, tryptophan and farnesol or farnesyl phosphate. *Acremonium*-derived abscisic acid may be translocated passively in the transpiration stream from the plant base to leaf stomata, by virtue of the intercellular location of the endophyte in crown and sheath tissues. Abscisic acid can stimulate ethylene production, which may, in turn, affect host growth. There are no reports of endophyte effects on ethylene production.

Kinetin severely reduced *in vitro* growth of *A. coenophialum*; however, the growth of *A. typhinum* Morgan-Jones & Gams, *Epichloë typhina* (Fr.) Tul., and *Atkinsonella hypoxylon* (Peck) Diehl was less inhibited.[24] The authors speculated that relatively high cytokinin levels in leaf blades of symbiotic plants inhibit hyphal extension past the leaf ligule, although no *in planta* tests have confirmed this (K.D. Gwinn, personal communication).

Mechanisms of hormone-mediated plant responses to endophyte presence will be very difficult to elucidate. Such studies will have to deal with cumulative layers of complexity, namely those of host–endophyte genetic interactions and multiple hormone reactions and balances. Furthermore, hormone modes of action are incompletely understood.

III. DROUGHT TOLERANCE

A. FIELD PERFORMANCE

Early accounts of the drought-tolerance-enhancing effect of *A. coenophialum* were based on observations in field trials comparing high and low endophyte-infected stands of tall fescue for grazing[43] and turf characteristics.[21] Read and Camp[43] reported that 94%-infected tall fescue pastures averaged 4% bare ground area, whereas 12%-infected stands averaged 54% bare area, following a drier-than-normal summer in east Texas (Table 3). Forage dry matter yields averaged 55% higher in high-endophyte pastures than in low-endophyte pastures, during nongrazing periods. Funk et al.[21] observed greater stand density after recovery from dry summer conditions in high-endophyte-infected cultivars than in low-endophyte-infected cultivars, of tall fescue grown for turf in New Jersey.

Observations that endophyte presence enhanced tall fescue summer survival were also made in Arkansas,[60] Georgia,[11] Louisiana,[29] and Texas.[44] These states comprise the southern and western extremes of the range of tall fescue in the primary fescue-growing region of the U.S., as defined by high temperature and prolonged soil water deficit during summer. These phenomenological reports suggest that the endophyte has expanded the geographic range of adaptation of tall fescue into drought-stress-prone environments and lowered the economic risk of stand loss.

The role of the endophyte in protecting tall fescue from drought-induced tissue and plant mortality was confirmed in field populations subject to a water-supply gradient in immature stands[57] and in fully established stands.[58] During a dry period in the first summer after autumn sowing of Kentucky 31, nonirrigated infected stands had greater green-leaf mass, less leaf senescence, and lower canopy temperature than did endophyte-free stands (Figure 1); however, there were no differences under high irrigation. In the following year (1988), severe drought in the absence of irrigation caused greater plant mortality in endophyte-free stands than in infected stands, such that end-of-year tiller density was reduced 38% from that of the infected stands (Figure 2). Reduced tiller density in endophyte-free stands continued through most of 1989, a relatively wet year.

These results indicate that endophyte presence elevates the water-deficit threshold for inducing stress in tall fescue, i.e., greater drought resistance, as measured by tiller and plant survival. The fact that tiller density of infected stands also tended to be greater at intermediate and high irrigation levels,[58] though usually not significantly, suggests an additional phenomenon. Infected plants with greater tillering under

Table 3 **Percentage loss of ground cover area during summer of 1984 and mean forage availability (1982–1984) with high (94%) and low (12%) endophyte-infected tall fescue in east Texas**[43]

Pasture No.	Loss of Ground Cover		Forage Availability	
	High	Low	High	Low
	(% of Area)		(Mg ha^{-1})	
1	0	5	—[a]	—
2	11	74	—	—
3	1	84	—	—
Mean	4	54	1.40	0.90

[a] Individual pasture data not available.

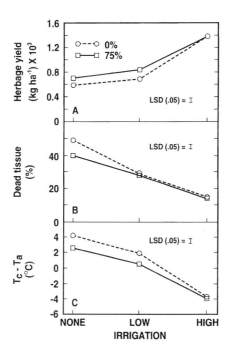

Figure 1 (A) Herbage dry matter yield, (B) percentage dead leaf tissue, and (C) canopy minus air temperature, of tall fescue at 0 and 75% endophyte infection and at three irrigation levels. LSD bars compare endophyte means within irrigation levels. *(Adapted from West, C. P., et al., Plant Soil, 112:3–6, 1988.)*

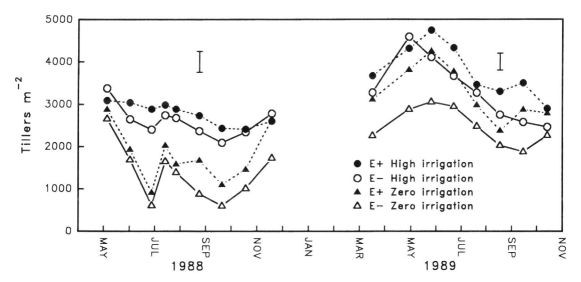

Figure 2 Tiller density in 1988 and 1989 as affected by endophyte infection at high and zero irrigation; means of eight replicates. E+ = endophyte infected (80%); E− = endophyte free (0%). $LSD_{0.05}$ bars for comparing means within dates are 488 in 1988 and 399 in 1989. Endophyte × irrigation interaction was significant ($p < 0.01$) in both years. *(From West, C. P., et al., Agron. J., 85:264–270, 1993.)*

favorable water conditions will require a more prolonged drought in order for tiller density to be depressed to critical levels for whole-plant survival than will noninfected plants.

Endophyte-related drought tolerance is coupled to competitiveness phenomena in that maintenance of dense stands interferes with the establishment and reproductive efficiency of other species and, indeed, with endophyte-free plants of the same species.[15,26] Bare soil areas caused by drought-induced stand

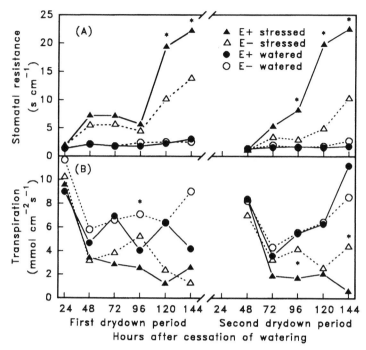

Figure 3 (A) Stomatal resistance and (B) transpiration rate, of endophyte-infected (E+) and endophyte-free (E–) Kentucky-31 tall fescue, with and without water stress during two drying periods. *(From Elmi, A. A., Ph.D. thesis, University of Arkansas, Fayetteville. DAI 53–08B, Diss. Abstr. Order no. 9237355, 1992.)*

reduction allowed substantially greater annual weed invasion in noninfected, than in infected, nonirrigated stands, probably because of reduced competition from tall fescue.[58] Grazing stress[43] and susceptibility to root-feeding nematodes,[56] in nonsymbiotic plants, may intensify the effects of drought stress on reducing plant competitiveness.

Maintenance of high stand density due to the presence of the endophyte would promote the adaptability of tall fescue and perennial ryegrass for turf uses in stress-prone environments.[22] Knox and Karnok[31] observed greater tall fescue root growth and less bare area after relief of water stress in single-genotype stands of endophyte-infected plants than in endophyte-free stands, when maintained as turf. Turf quality index tended to be greater in infected stands, owing mainly to turf density rather than to color.

B. GROWTH AND GAS EXCHANGE

Experiments with endophyte-infected and endophyte-free isogenic populations or clones in controlled environments separate plant physiological responses to the endophyte, from possible pest responses in field trials. In three genotypes derived from a Kentucky 31 population, tiller number was reduced by water stress in nonsymbiotic, but not in symbiotic, isolines; however, the reverse was true for a fourth genotype.[8] In contrast, Maclean et al.[38] observed that the tiller number was greater for water-stressed, noninfected genotypes than for infected genotypes. Higher rates of leaf elongation occurred in infected plants than noninfected plants, during drought stress[38] and after relief of stress.[1,20] Arachevaleta et al.[1] reported that infected plants exhibited leaf rolling sooner than did noninfected plants, of one genotype during the onset of water stress. This response may have been unique to that genotype, because other authors[63] have reported no tendency for the endophyte to stimulate leaf rolling. Indeed, delayed leaf rolling is normally associated with greater drought tolerance in tall fescue.[56,64]

Stomatal closure, induced by water deficit, benefits plants in the short term by reducing water loss via transpiration, with an eventual cost in reduced photosynthesis.[54] Elmi et al.[20] reported that stomatal resistance of leaf blades increased faster, and to a higher degree, in an infected Kentucky 31 population than in the noninfected population, during two water-withholding periods, with no differences in watered controls (Figure 3A). Transpiration rates of stressed, infected plants were lower than those of noninfected plants (Figure 3B). The same results were found with genetically identical infected and noninfected clones.[18]

Reports of the effects of endophyte infection on stomatal conductance and photosynthesis have been mixed. Bates and Joost,[5] Belesky et al.,[6] and Richardson et al.[47] showed that endophyte infection reduced conductance and photosynthesis in some, but not all, genotypes. However, Richardson et al.[50] found that conductance and photosynthesis were increased slightly by the endophyte in one genotype, but unaffected by the endophyte in another. Endophyte enhancement of water conservation via early stomatal closure may be achieved by some indirect stimulation of a phytohormone signal, such as abscisic acid, either from root to shoot, within the shoot,[67] or via fungal secretion of antitranspirant compounds into the apoplasm. The report by Bunyard and McInnis[12] of *in vitro* production of abscisic acid by *A. coenophialum* lends evidence in support of this scenario.

Phenotypes that sensitively regulate transpiration loss early during soil drying may be better able, than less-sensitive phenotypes, to prolong net carbon assimilation or to recover growth after short-term water deficit, because of water conservation in the soil–plant system. Various workers have demonstrated higher instantaneous water-use efficiency (ratio of photosynthesis:transpiration rates on unit-leaf-area basis) in leaves of symbiotic, compared with nonsymbiotic, tall fescue;[5,47] however, this response was not consistent across all soil water levels or all genotypes. Richardson et al.[50] reported no endophyte effect on water-use efficiency in two greenhouse-grown genotypes. More data are needed on the endophyte effects on field water-use efficiency integrated over time and crop canopy. Only then can the costs and benefits of reduced transpiration, in terms of plant competitiveness and survival in water-limiting environments, be adequately assessed.

C. OSMOTIC ADJUSTMENT

Greater survival of endophytic, over endophyte-free, tall fescue during severe drought occurs despite the complete loss of green leaf area and the cessation of growth of endophytic plants.[58] This may be caused by a combination of (1) the higher intrinsic tillering ability of infected plants, thereby postponing the depletion of live tillers; and (2) some direct effect of the endophyte in improving host–water relations in a manner that delays tiller desiccation or enhances desiccation tolerance. Osmotic adjustment is a means of plant adaptation to water stress, in which cell solutes accumulate in response to water deficit. The resulting lower osmotic potential permits the retention of cell turgor pressure, which is necessary for cell growth and metabolic maintenance.[54]

Endophyte presence has been reported to stimulate osmotic adjustment in tall fescue, relative to nonsymbiotic plants, in leaves and basal vegetative growing zones in Kentucky-31 populations[19] and in clonal material[18] (Table 4). The magnitude of osmotic adjustment enhancement was usually greater in the basal 2-cm growing zone than in the leaf blade. White[62] and Richardson et al.[48] have also reported endophyte-enhanced osmotic adjustment in young, emerged leaf blades; however, it is in the meristematic and elongating zone that high osmotic adjustment, and consequent turgor maintenance, may favor the survival of vegetative tillers during drought.[59] Elmi[18] found positive correlations ($r = 0.87**$ and $r = 0.44*$) between magnitudes of growing-zone osmotic adjustment during two drying cycles, and percentage of tillers surviving severe drought in a single genotype. Endophyte-infected plants had higher osmotic adjustment, survival rate, and leaf elongation rate than did noninfected plants (Table 5).

In a study using turf-type tall fescue clones, White et al.[64] observed that superior plant survival during severe drought was associated with turgor maintenance and osmotic adjustment, when averaged across

Table 4 **Osmotic adjustment in leaf tissues of a cultivar and a genotype of endophyte-infected and noninfected tall fescue[18]**

| Endophyte Status | Kentucky 31 | | GA-87-122 | |
| | Leaf Blade | Basal Zone | Leaf Blade | Basal Zone |
	(MPa)			
Infected	0.55[a]	0.52[a]	0.46[a]	0.51[a]
Noninfected	0.22 ns[b]	0.34 ns	0.29 ns	0.13 ns

[a] Osmotic potentials of previously water-stressed tissues are significantly different from controls at $p < 0.05$ and $p < 0.01$; [b] ns = Nonsignificant.

Table 5   **Tiller survival and leaf elongation rates after relief from 3-week drought in response to preconditioning treatment of endophyte-infected (E+) and endophyte-free (E–) tall fescue, genotype GA-87-122**[18]

Water Treatment	Tiller Survival Rate Endophyte			Leaf Elongation Rate Endophyte		
	E+	E– (%)	Mean	E+	E– (mm day^{-1})	Mean
Control	47	28	38	12.3	12.5	12.4
Stress	61	48	54[a]	26.7	15.7	21.2[b]
Mean	54	38[a]		19.5	14.2[a]	

[a] Main effect means different at $p < 0.05$; [b] Water treatment means different at $p < 0.06$ for leaf elongation rate.

Table 6   **Carbohydrate analysis of endophyte-infected (E+) and endophyte-free (E–) tall fescue grown at field capacity (control) or at –1.0 MPa (stress) for 7 days**[49]

Carbohydrate	Endophyte	Leaf Blade			Leaf Sheath		
		Control	Stress	LSD[a] (μmol g^{-1} dry wt)	Control	Stress	LSD
Fructose	E+	89	198	54.5	76	125	35.2
	E–	102	109		127	108	
Glucose	E+	111	244	53.5	106	201	56.9
	E–	105	118		130	130	
Sucrose	E+	65	100	60.7	35	77	20.6
	E–	36	79		31	71	
Mannitol	E+	0	0.3	0.01	2.6	3.8	0.8
	E–	0	0		0	0	
Arabitol	E+	0	0	0.0	0	2.7	1.1
	E–	0	0		0	0	

[a] Fisher's protected LSD ($p < 0.05$) for comparing means within sugar by tissue combinations.

endophyte treatments. There was no endophyte effect on leaf-blade osmotic adjustment or plant survival, in the clones tested; however, drought-stressed, infected plants had a lower bulk modulus of tissue elasticity and a greater turgid weight:dry weight ratio.[63] These results indicate that leaves of infected plants had less rigid cell walls than did leaves of noninfected plants. This means that a given amount of water loss from tissues of infected plants would be accompanied by less reduction in turgor, because of greater volume shrinkage than in tissues of noninfected plants. Differences in the elasticity modulus and turgid weight:dry weight ratio, due to the endophyte, were not, however, large enough to alter leaf water potential at zero turgor. Nevertheless, these results suggest that the endophyte can impact biophysical characteristics of cell walls of leaf blades despite the absence of mycelia in that tissue, again suggesting a hormonally mediated response.

Endophyte enhancement of osmotic adjustment may enable surviving tillers to maintain sufficiently high turgor in the growing zone, to retard desiccation and allow rapid resumption of leaf growth upon relief of stress. The leaf elongation zone of tall fescue is a strong sink for soluble carbohydrates, of which low-molecular-weight fructans make up the bulk of the carbohydrate-derived osmoticum.[53] Richardson et al.[49] reported that water-stressed, symbiotic plants accumulated higher concentrations of glucose and fructose in leaf sheaths and glucose in blades than did nonsymbiotic clones (Table 6). The authors suggested that monosaccharides contribute, at least partially, to the solute pool responsible for increased osmotic adjustment in symbiotic plants.

An alternative or concomitant mechanism by which the endophyte may biochemically mediate tiller survival is by secretion of sugar alcohols (polyols), which may protect host enzymes and membranes from desiccation damage.[4] Richardson et al.[49] detected mannitol and arabitol in sheaths of water-stressed, symbiotic plants, but not in nonsymbiotic plants (Table 6). Since virtually no mannitol or arabitol was found in blades or any leaf tissues of nonsymbiotic plants, these polyols are probably produced only by the fungus. Richardson et al.[46] reported that an *A. coenophialum* isolate had greater pressure potential when grown on media with osmotic potentials of −0.7 and −1.2 MPa than on media with higher or lower osmotic potentials, suggesting that the endophyte itself osmotically adjusts. These results imply that the endophyte may directly mediate biochemical host–response mechanisms to water deficit, despite the extracellular nature of the fungus.

While our understanding of the mechanisms by which the endophyte enhances host survival during drought is incomplete, patterns of responses are emerging. Endophyte presence seems to promote classic responses to water deficit, earlier in a drying trend (as in higher stomatal resistance and reduced transpiration) and to a greater extent (as in osmotic adjustment), relative to noninfected plants, i.e., a combination of drought-stress postponement and drought tolerance at low water potential.[54] This suggests that the endophyte induces a sort of incipient stress that somehow preconditions or sensitizes the host to drought, thereby permitting the plant to exhibit adaptive responses sooner. For example, higher leaf stomatal resistance could arise from endophyte-derived compounds that mimic root-to-shoot signals in plants. Whether this is a drought-induced phenomenon or a preexisting condition is not clear. Early decreases in transpiration and cell elongation may be part of a generalized reduction in leaf growth, which would cause a build-up of soluble carbohydrates near the sink, i.e., leaf sheaths and leaf growing zones. Progressively severe water deficit would favor hydrolysis of storage carbohydrate,[55] thereby providing solutes for osmotic adjustment and turgor maintenance. Additional factors contributing to superior drought survival of endophyte-infected tall fescue in field environments may be enhanced rooting density and depth, which may result from decreased nematode feeding damage.[57]

Inasmuch as *Acremonium*-infected plants are cool-season grasses that normally undergo some degree of dormancy during summer water-deficit periods, it follows that the drought tolerance of crown tissues would be more important than the drought avoidance of leaf blades, as a "strategy" for plant survival. Researchers are seeking a biochemical basis for endophyte-enhanced drought tolerance. For example, fungal-derived sugar alcohols may protect enzymes and membranes from desiccation and serve as antioxidants. On a population level, however, multiple morphological and physiological mechanisms conferring drought tolerance may lend greater flexibility for adapting to stresses, in endophytic populations than in nonendophytic populations.[26] Natural selection for such a successful mutualism probably resulted in multiple adaptation mechanisms for which the benefits of long-term meristem survival exceed short-term costs in energy balance.

IV. NITROGEN METABOLISM

There are two reasons why endophyte effects on nitrogen metabolism are of interest. First, the antiherbivory compounds ergopeptine, pyrrolizidine, and peramine alkaloids contain nitrogen and are only found in endophytic plants. Understanding environmental and management effects on their accumulation will enable one to minimize livestock toxicities or exploit pest deterrent traits. Second, studying endophyte effects on nitrogen uptake and metabolism may aid in understanding endophyte-mediated plant responses to stress. Increased soil nitrogen availability enhanced the concentration of ergopeptines in endophyte-infected tall fescue.[2,3,37,51] Nitrogen fertilization increased total ergopeptine[10] and pyrrolizidine[7] concentrations in pastures, on a unit endophyte-infection-rate basis.

Ergopeptine accumulation in greenhouse-grown tall fescue was substantially greater at a low nitrate supply than at a low ammonium supply; however, no differences occurred between nitrogen sources at a high nitrogen supply.[2] This does not necessarily imply that fescue toxicosis in pastures is of a greater risk with low rates of nitrate than with ammoniacal fertilizers, since in field soils, nitrification of ammonium to nitrate would tend to mask potential nitrogen-source differences. In a companion study,[2] increasing the rate of ammonium nitrate fertilizer increased ergopeptine concentration in drought-stressed plants (−0.50 MPa soil matric potential), but there was no increase in nonstressed plants (−0.03 MPa). Plant growth suppression due to water stress concentrated the alkaloids into smaller plant mass, but ergopeptine yield was around twice as high in stressed plants compared to nonstressed plants, indicating enhanced synthesis. The authors speculated that accumulated soluble carbohydrates in stressed plants served as precursors to alkaloid synthesis. Belesky et al.[9] also showed that drought stress increased ergopeptine and pyrrolizidine synthesis. It is not known whether increased synthesis under water stress of these alkaloids, which are absent in endophyte-free grasses, contributes to enhanced drought resistance of infected grasses.

Lyons et al.[36] provided evidence that endophyte infection alters nitrogen assimilation patterns in tall fescue leaf sheaths, which contain the fungus, and in leaf blades, which are not colonized by the fungus, compared with noninfected plants. The endophyte reduced nitrate concentrations in sheath and blade, whereas tissue ammonium concentrations were stimulated by the endophyte in sheaths only. Endophyte infection increased the concentrations of many free amino acids in the sheath and blade, especially glutamine and asparagine at high nitrogen fertilization. The endophyte also increased glutamine synthetase activity by 32% in leaf blades. The authors speculated that increased nitrogen demand in the sheath, due to fungal activity, would reduce nitrogen availability to the blade. This would necessitate improved ammonium reassimilation in the blade, via glutamine synthetase, to maintain high rates of amino acid synthesis. This may constitute a compensatory mechanism that precludes the endophyte from competing for nitrogen to a point that is deleterious to plant growth.

V. CONCLUSIONS

Plant growth under controlled conditions is generally stimulated by the endophyte, especially with abundant nutrient availability, although individual genotypes vary greatly in this response. Under field conditions, greater growth and persistence due to the presence of the endophyte probably result more from enhanced competitiveness and resistance to drought and herbivory, than from direct stimulation of host growth. Endophyte-enhanced drought tolerance of crown tissues is probably more important than drought avoidance of leaf blades, as a strategy for plant survival. Direct linkage of endophyte-modified plant hormone balance to plant growth, tillering, or stomatal closure has not yet been demonstrated; however, future research will likely uncover a hormonal connection. Endophyte-induced changes in the assimilation and utilization of nitrogen and carbon may contribute to improved host competitiveness and persistence.

There is potential for exploiting the beneficial traits of *Acremonium* endophytes in improving grass drought resistance. The great variation among host genotype–endophyte associations in the degree of growth and drought-resistance enhancement suggests possibilities for genetic selection through breeding for optimum expression of these traits. There is also the prospect for selecting *Acremonium* strains that synthesize little or no ergot-type alkaloids, but that retain the pest- and drought-resistance benefits to the plant. Such strains would have to be evaluated in symbiosis with numerous elite breeding lines and with exposure to pests and drought stress, to select the associations that most consistently express the desirable traits. No single biochemical or genetic marker in the association will probably ever suffice to adequately evaluate new associations for their potential as improved, endophytic cultivars, because of the multifaceted stresses and environmental interactions that define plant population stability.

REFERENCES

1. **Arachevaleta, M., C. W. Bacon, C. S. Hoveland, and D. E. Radcliffe,** Effect of the tall fescue endophyte on plant response to environmental stress, *Agron. J.,* 81:83–90, 1989.
2. **Arechavaleta, M., C. W. Bacon, R. D. Plattner, C. S. Hoveland, and D. E. Radcliffe,** Accumulation of ergopeptide alkaloids in symbiotic tall fescue grown under deficits of soil water and nitrogen fertilizer, *Appl. Environ. Microbiol.,* 58:857–861, 1992.
3. **Azevedo, M. D., R. D. Welty, A. M. Craig, and J. Bartlett,** Ergovaline distribution, total nitrogen and phosphorous content of two endophyte-infected tall fescue clones, In: D. E. Hume, G. C. M. Latch, and H. S. Easton, Eds., *Proc. 2nd Int. Symp.* Acremonium/*Grass Interactions*, AgResearch, Grasslands Research Centre, Palmerston North, New Zealand, 1993, 59.
4. **Bacon, C. W.,** Abiotic stress tolerances (moisture, nutrients) and photosynthesis in endophyte-infected tall fescue, *Agric. Ecosyst. Environ.,* 44:123–141, 1993.
5. **Bates, G. E. and R. E. Joost,** The effect of *Acremonium coenophialum* on the physiological responses of tall fescue to drought stress, In: S. S. Quisenberry and R. E. Joost, Eds., *Proc. Int. Symp.* Acremonium/*Grass Interactions*, Louisiana Agriculture Experiment Station, Baton Rouge, 1990, 121.
6. **Belesky, D. P., O. J. Devine, J. E. Pallas, Jr., and W. C. Stringer,** Photosynthetic activity of tall fescue as influenced by a fungal endophyte, *Photosynthetica,* 21:82–87, 1987a.

7. **Belesky, D. P., J. D. Robbins, J. A. Stuedemann, S. R. Wilkinson, and O. J. Devine,** Fungal endophyte infection-loline derivative alkaloid concentration of grazed tall fescue, *Agron. J.,* 79:217–220, 1987b.

8. **Belesky, D. P., W. C. Stringer, and N. S. Hill,** Influence of endophyte and water regime upon tall fescue accessions. I. Growth characteristics, *Ann. Bot.,* 63:495–503, 1989a.

9. **Belesky, D. P., W. C. Stringer, and R. D. Plattner,** Influence of endophyte and water regime upon tall fescue accessions. II. Pyrrolizidine and ergopeptine alkaloids, *Ann. Bot.,* 64:343–349, 1989b.

10. **Belesky, D. P., J. A. Stuedemann, R. D. Plattner, and S. R. Wilkinson,** Ergopeptine alkaloids in grazed tall fescue, *Agron. J.,* 80:209–212, 1988.

11. **Bouton, J. H. and G. W. Burton,** Effect of endophytic fungal infection on tall fescue performance in the lower South, In: *Agronomy Abstracts,* American Society Agronomy, Madison, WI, 1988, 122.

12. **Bunyard, B. and T. McInnis, Jr.,** Evidence for elevated phytohormone levels in E+ endophyte-infected tall fescue, In: *Proc. Int. Symp.* Acremonium/*Grass Interactions,* Louisiana Agriculture Experiment Station, Baton Rouge, 1990.

13. **Cheplick, G. P., K. Clay, and S. Marks,** Interactions between infection by endophytic fungi and nutrient limitation in the grasses *Lolium perenne* and *Festuca arundinacea, New Phytol.,* 111:89–97, 1989.

14. **Clay, K.,** Effects on fungal endophytes on the seed and seedling biology of *Lolium perenne* and *Festuca arundinacea, Oecologia,* 73:358–362, 1987.

15. **Clay, K.,** Fungal endophytes, grasses and herbivores, In: P. Barbosa, V. A. Kriachik, and C. G. Jones, Eds., *Microbial Mediations of Plant-Herbivore Interactions,* John Wiley & Sons, New York, 1991, 199.

16. **De Battista, J. P., C. W. Bacon, R. Severson, R. D. Plattner, and J. H. Bouton,** Indole acetic acid production by the fungal endophyte of tall fescue, *Agron. J.,* 82:878–880, 1990a.

17. **De Battista, J. P., J. H. Bouton, C. W. Bacon, and M. R. Seigel,** Rhizome and herbage production of endophyte-removed tall fescue clones and populations, *Agron. J.,* 82:651–654, 1990b.

18. **Elmi, A. A.,** Physiological effects of endophyte and nematodes on water relations in tall fescue. Ph.D. thesis, University of Arkansas, Fayetteville. DAI 53–08B, Diss. Abstr. Order no. 9237355, 1992.

19. **Elmi, A. A., C. P. West, and K. E. Turner,** *Acremonium* endophyte enhances osmotic adjustment in tall fescue, *Arkansas Farm Res.,* 38(5):7, 1989.

20. **Elmi, A. A., C. P. West, K. E. Turner, and D. M. Oosterhuis,** *Acremonium coenophialum* effects on tall fescue water relations, In: S. S. Quisenberry and R. E. Joost, Eds., *Proc. Int. Symp.* Acremonium/*Grass Interactions,* Louisiana Agriculture Experiment Station, Baton Rouge, 1990, 137.

21. **Funk, C. R., R. H. Hurley, J. M. Johnson-Cicalese, and D. C. Saha,** Association of endophytic fungi with improved performance and enhanced pest resistance in perennial ryegrass and tall fescue, In: *Agronomy Abstracts,* American Society Agronomy, Madison, WI, 1984, 66.

22. **Funk, C. R., R. H. White, and J. P. Breen,** Importance of *Acremonium* endophyte in turfgrass breeding and management, *Agric. Ecosyst. Environ.,* 44:215–232, 1993.

23. **Grand-Ravel, C., G. Charmet, and F. Balfourier,** A comparison between endophyte infected and uninfected plants, In: D. E. Hume, G. C. M. Latch, and H. S. Easton, Eds., *Proc. 2nd Int. Symp.* Acremonium/*Grass Interactions,* AgResearch, Grasslands Research Centre, Palmerston North, New Zealand, 1993, 204.

24. **Gwinn, K. D. and D. B. Chalkley,** Cytokinin sensitivity of grass endophytes: in vitro growth inhibition, *Phytophathology,* 82:1164, 1992.

25. **Hill, N. S., W. C. Stringer, G. E. Rottinghaus, D. P. Belesky, W. A. Parrott, and D. D. Pope,** Growth, morphological and chemical component responses of tall fescue to *Acremonium coenophialum, Crop Sci.,* 30:156–161, 1990.

26. **Hill, N. S., D. P. Belesky, and W. C. Stringer,** Competitiveness of tall fescue as influenced by *Acremonium coenophialum, Crop Sci.,* 31:185–190, 1991.

27. **Hoveland, C. S.,** Importance and economic significance of the *Acremonium* endophytes to performance of animals and grass plant, *Agric. Ecosyst. Environ.,* 44:3–12, 1993.

28. **Hume, D. E., A. J. Popay, and D. J. Barker,** Effect of *Acremonium* endophyte on growth of ryegrass and tall fescue under varying levels of soil moisture and Argentine stem weevil attack, In: D. E. Hume, G. C. M. Latch, and H. S. Easton, Eds., *Proc. 2nd Int. Symp.* Acremonium/*Grass Interactions,* AgResearch, Grasslands Research Centre, Palmerston North, New Zealand, 1993, 161.

29. **Joost, R. E. and D. F. Coombs,** Importance of *Acremonium* presence and summer management to persistence of tall fescue, In: *Agronomy Abstracts*, American Society Agronomy, Madison, WI, 1988, 130.

30. **Keogh, R. G. and T. Lawrence,** Influence of *Acremonium lolii* presence on emergence and growth of ryegrass seedlings, *N.Z. J. Agric. Res.,* 30:507–510, 1987.

31. **Knox, J. D. and K. J. Karnok,** Root and shoot growth of endophyte infected and endophyte free tall fescue under water stress and non-stress conditions, In: *Agronomy Abstracts*, American Society Agronomy, Madison, WI, 1992, 171.

32. **Latch, G. C. M.,** Physiological interactions of endophytic fungi and their hosts. Biotic stress tolerance imparted to grasses by endophytes, *Agric. Ecosyst. Environ.,* 44:143–156, 1993.

33. **Latch, G. C. M., W. F. Hunt, and D. R. Musgrave,** Endophytic fungi affect growth of perennial ryegrass, *N.Z. J. Agric. Res.,* 28:165–168, 1985.

34. **Leopold, A. C. and P. E. Kriedemann,** *Plant Growth and Development*, 2nd ed., McGraw-Hill, New York, 1975.

35. **Lewis, G. C. and R. O. Clements,** Effect of *Acremonium lolii* on herbage yield of *Lolium perenne* at three sites in the United Kingdom, In: S. S. Quisenberry and R. E. Joost, Eds., Proc. *Proc. Int. Symp.* Acremonium/*Grass Interactions*, Louisiana Agriculture Experiment Station, Baton Rouge, 1990, 160.

36. **Lyons, P. C., J. J. Evans, and C. W. Bacon,** Effects of the fungal endophyte *Acremonium coenophialum* on nitrogen accumulation and metabolism in tall fescue, *Plant Physiol.,* 92:726–732, 1990.

37. **Lyons, P. C., R. D. Plattner, and C. W. Bacon,** Occurrence of peptide and clavine ergot alkaloids in tall fescue grass, *Science,* 232:487–489, 1986.

38. **Maclean, B., C. Matthew, G. C. M. Latch, and D. J. Barker,** The effect of endophyte on drought resistance in tall fescue, In: D. E. Hume, G. C. M. Latch, and H. S. Easton, Eds., *Proc. 2nd Int. Symp.* Acremonium/*Grass Interactions*, AgResearch, Grasslands Research Centre, Palmerston North, New Zealand, 1993, 165.

39. **Marks, S. and K. Clay,** Effects of CO_2 enrichment, nutrient addition, and fungal endophyte-infection on the growth of two grasses, *Oecologia,* 84:207–214, 1990.

40. **Marks, S., K. Clay, and G. P. Cheplick,** Effects of fungal endophytes on interspecific and intraspecific competition in the grasses *Festuca arundinacea* and *Lolium perenne, J. Appl. Ecol.,* 28:194–204, 1991.

41. **Phinney, B. O.,** The history of the gibberellins, In: A. Crozier, Ed., *The Biochemistry and Physiology of the Gibberellins*, Vol. 1, Praeger, New York, 1983, 19.

42. **Porter, J. K., C. W. Bacon, H. G. Cutler, R. F. Arrendale, and J. D. Robbins,** In-vitro auxin production by *Balansia epichloë, Phytochemistry,* 24:1429–1431, 1985.

43. **Read, J. C. and B. J. Camp,** The effect of the fungal endophyte *Acremonium coenophialum* in tall fescue on animal performance, toxicity, and stand maintenance, *Agron. J.,* 78:848–850, 1986.

44. **Read, J. C. and D. W. Walker,** The effect of the fungal endophyte *Acremonium coenophialum* on dry matter production and summer survival of tall fescue, In: S. S. Quisenberry and R. E. Joost, Eds., *Proc. Int. Symp.* Acremonium/*Grass Interactions*, Louisiana Agriculture Experiment Station, Baton Rouge, 1990, 181.

45. **Rice, J. S., B. W. Pinkerton, W. C. Stringer, and D. J. Undersander,** Seed production in tallfescue as affected by fungal endophyte, *Crop Sci.,* 30:1303–1305, 1990.

46. **Richardson, M. D., C. W. Bacon, N. S. Hill, and D. M. Hinton,** Growth and water relations of *Acremonium coenophialum*, In: D. E. Hume, G. C. M. Latch, and H. S. Easton, Eds., *Proc. 2nd Int. Symp.* Acremonium/*Grass Interactions*, AgResearch, Grasslands Research Centre, Palmerston North, New Zealand, 1993a, 181.

47. **Richardson, M. D., C. W. Bacon, and C. S. Hoveland,** The effect of endophyte removal on gas exchange in tall fescue, In: S. S. Quisenberry and J. E. Joost, Eds., *Proc. Int. Symp.* Acremonium/*Grass Interactions*, Louisiana Agriculture Experiment Station, Baton Rouge, 1990, 189.

48. **Richardson, M. D., G. W. Chapman, Jr., C. S. Hoveland, and C. W. Bacon,** Carbohydrates of drought-stressed tall fescue as influenced by the endophyte, In: *Agronomy Abstracts*, American Society Agronomy, Madison, WI, 1991, 133.

49. **Richardson, M. D., G. W. Chapman, Jr., C. S. Hoveland, and C. W. Bacon,** Sugar alcohols in endopyte-infected tall fescue under drought, *Crop Sci.,* 32:1060–1061, 1992.

50. **Richardson, M. D., C. S. Hoveland, and C. W. Bacon,** Photosynthesis and stomatal conductance of symbiotic and nonsymbiotic tall fescue, *Crop Sci.,* 33:145–149, 1993b.

51. **Rottinghaus, G. E., G. B. Garner, C. N. Cornell, and J. L. Ellis,** HPLC method for quantitating ergovaline in endophyte-infested tall fescue: seasonal variation of ergovaline levels in stems with leaf sheaths, leaf blades, and seed heads, *J. Agric. Food Chem.,* 39:112–115, 1991.

52. **Schmidt, D.,** Effects of *Acremonium uncinatum* and a *Phialophora*-like on vigour, insect and disease resistance of meadow fescue, In: D. E. Hume, G. C. M. Latch, and H. S. Easton, Eds., *Proc. 2nd Int. Symp.* Acremonium/*Grass Interactions*, AgResearch, Grasslands Research Centre, Palmerston North, New Zealand, 1993, 185.

53. **Schnyder, H. and C. J. Nelson,** Diurnal growth of tall fescue leaf blades. I. Spatial distribution of growth, deposition of water, and assimilate import in the elongation zone, *Plant Physiol.,* 86:1070–1076, 1988.

54. **Turner, N. C.,** Adaptation to water deficits: a changing perspective, *Aust. J. Plant Physiol.,* 13:175–190, 1986.

55. **Virgona, J. M. and E. W. R. Barlow,** Drought stress induces changes in the nonstructural carbohydrate composition of wheat stems, *Aust. J. Plant Physiol.,* 18:239–247, 1991.

56. **West, C. P. and K. D. Gwinn,** Role of *Acremonium* in drought, pest, and disease tolerances of grasses, In: C. E. Hume, G. C. M. Latch, and H. S. Easton, *Proc. 2nd Int. Symp.* Acremonium/*Grass Interactions: Plenary Papers*, AgResearch, Grasslands Research Centre, Palmerston North, New Zealand, 1993, 131.

57. **West, C. P., E. Izekor, D. M. Oosterhuis, and R. T. Robbins,** The effect of *Acremonium coenophialum* on the growth and nematode infestation of tall fescue, *Plant Soil,* 112:3–6, 1988.

58. **West, C. P., E. Izekor, K. E. Turner, and A. A. Elmi,** Endophyte effects on growth and persistence of tall fescue along a water-supply gradient, *Agron. J.,* 85:264–270, 1993.

59. **West, C. P., D. M. Oosterhuis, and S. D. Wullschleger,** Osmotic adjustment in tissues of tall fescue in response to water deficit, *Environ. Exp. Bot.,* 30:149–156, 1990.

60. **West, C. P., E. L. Piper, G. Duff, and L. B. Daniels,** Endophyte effects on steer gains and stand vigor of Kentucky 31 tall fescue, *Arkansas Farm Res.,* 39(4):9, 1989.

61. **White, J. F., Jr., G. Morgan-Jones, and A. C. Morrow,** Taxonomy, life cycle, reproduction and detection of *Acremonium* endophytes, *Agric. Ecosyst. Environ.,* 44:13–37, 1993.

62. **White, R. H.,** Water relations characteristics of tall fescue as influenced by *Acremonium* endophyte, In: *Agronomy Abstracts*, American Society Agronomy, Madison, WI, 1989, 167.

63. **White, R. H., M. C. Engelke, S. J. Morton, J. M. Johnson-Cicalese, and B. A. Ruemmele,** Acremonium endophyte effects on tall fescue drought tolerance, *Crop Sci.,* 32:1392–1396, 1992a.

64. **White, R. H., M. C. Engelke, S. J. Morton, and B. A. Ruemmele,** Competitive turgor maintenance in tall fescue, *Crop Sci.,* 32:251–256, 1992b.

65. **Wilkinson, S. R.,** Influence of endophytic infection of K31 tall fescue on yield response to irrigation, cutting management and competition with Tifton 44 bermudagrass, In: D. E. Hume, G. C. M. Latch, and H. S. Easton, Eds., *Proc. 2nd Int. Symp.* Acremonium/*Grass Interactions*, AgResearch, Grasslands Research Centre, Palmerston North, New Zealand, 1993, 189.

66. **Zeevaart, J. A. D. and R. A. Creelman,** Metabolism and physiology of abscisic acid, *Ann. Rev. Plant Physiol. Plant Mol. Biol.,* 39:439–473, 1988.

67. **Zhang, J. and W. J. Davies,** Increased synthesis of ABA in partially dehydrated root tips and ABA transport from roots to leaves, *J. Exp. Bot.,* 38:2015–2023, 1987.

Section IV

Chemical Constituents and Toxicity

Chapter 8

Chemical Constituents of Grass Endophytes

James K. Porter

CONTENTS

I. INTRODUCTION

Survival, for most life forms, is dependent on their developing a series of intricate biochemical relationships with their environment. Within these ecological or evolutionary progressions, certain endophytic fungi and their grass hosts have coevolved with independent and interdependent biochemical nexus. Our discussions on grass endophytes will be directed at those fungi that form nonpathogenic, systemic, intercellular relationships within the aerial portion of their grass host.[9,36,108] Bacon and DeBattista[9] have reviewed reports[36,61,111,123] of over 200 grass–endophyte associations, that include 16 tribes within 6 subfamilies of Gramineae (Poaceae), and suggest that the grass–endophyte associations are widespread and that most form biotrophic, mutualistic associations in which the biochemical properties of one are either exploited by or dependent on the other. Current research involving the chemical constituents of grass endophytes developed from the direct association of these fungi with livestock toxicities.[8,10,11,14,42,46,102] Prior thinking on the functional role of the natural toxins isolated from endophyte-infected grasses suggested their production was inconsequential and was the end result of secondary metabolism. More recently, research information on these compounds has defined survival mechanisms, both physiological and defensive, that indicate an ecological justification. Furthermore, biotechnology has emphasized great economic potentials from the endophyte–grass associations, by developing superior lines of turf and forage grasses.[60] Other potential uses are as sources of natural plant growth regulators, e.g., auxin production;[35,85] insecticides; natural herbicides, e.g., production of alleochemicals;[60,80] and unique sources of medicinal agents.[18,115]

This chapter will concentrate on the biological significance of the major chemical classes of compounds that have been isolated and identified from specific grass endophytes and their host associations. Brief emphasis will be placed on the deleterious (i.e., animal toxicities) and beneficial (i.e., grass host disease and insect resistance) aspects of the compounds resulting from the endophyte–grass host relationships. Some discussions will focus on the biological and biosynthetic origins (i.e., plant and/or fungus)

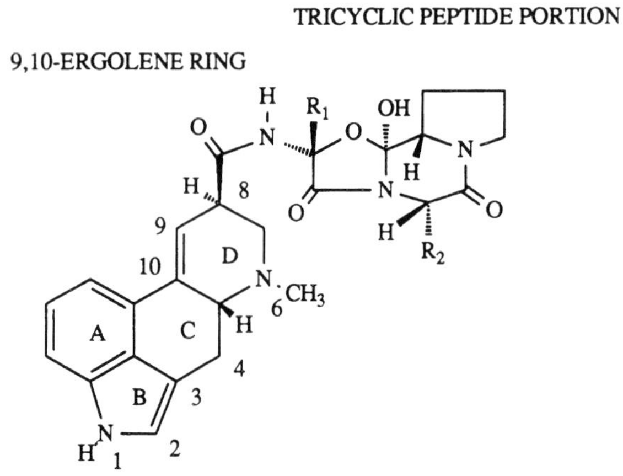

TRICYCLIC PEPTIDE PORTION

9,10-ERGOLENE RING

Figure 1 Basic ergopeptine alkaloids.

of these compounds. More specifically, this review will focus on those endophytes of the genus *Acremonium* and the endophytes of the tribe Balansiae.[7,9,36,61,70,108,112]

II. ENDOPHYTES (CLAVICIPITACEAE), GRASS (GRAMINEAE) HOST, AND MAJOR CLASSES OF COMPOUNDS

A. *ACREMONIUM* ENDOPHYTES

The endophytes *A. coenophialum* Morgan-Jones and Gams, from tall fescue (*Festuca arundinacea* Schreb.), and *A. lolii* Latch, Christensen, and Samuels, from perennial ryegrass (*Lolium perenne* L.), have been the major focus of research associated with toxic syndromes in livestock grazed on infected pastures. The economic impact on animal agriculture, and the descriptions of animal toxicities as a result of endophyte-infected grasses, have been reported.[42,46,54,96,118,120,122] Ergopeptine alkaloids, primarily ergovaline (Figure 1; Table 1), and the unique cyclic indole-isoprenoid lolitrems, primarily lolitrem B (Figure 2), have been associated with fescue toxicity and ryegrass staggers, respectively, in livestock.[10,46,64,128] The ergopeptine and lolitrem alkaloids have been isolated *in vitro* and *in planta*, from *A. coenophialum*- and *A. lolii*-infected fescue and perennial ryegrasses, respectively.

Fescue toxicity and ryegrass staggers in livestock are analogous to the ergot syndromes produced when livestock ingest sclerotia of *Claviceps purpurea* (Fries) Tulasne and *C. paspali* Stevens and Hall, respectively.[102] The association of these "ergot-like" syndromes in livestock on endophyte-infected grasses has been reviewed.[10,102] The endophytes of ryegrass and tall fescue also produce the clavine (Figure 3), ergopeptine (Figure 1) and the indole-isopreniod lolitrem (Figure 2) alkaloids[46,64,68,69,88,89,91,128] (see below).

The natural ergot alkaloids may be divided into five major classes: the clavine, lysergic acid, simple lysergic acid amides, ergopeptine, and the ergopeptam alkaloids (Figures 1, 3–5). The ergopeptine class may be further subdivided into three major groups, based on the amino acids comprising the tricyclic peptide portion of these compounds, and thus on the substituents at R_1 [i.e., $-CH_3$ = ergotamine group; $-CH_2CH_3$ = ergoxine group; $-CH(CH_3)_2$ = ergotoxine group] with $R_2 = -CH_2Ph$; –isoBu; –secBu; –isoPr, or –

Table 1 **Ergopeptine (or ergot cyclol) alkaloids**

Ergotamine Group ($R_1 = CH_3$)	R_2
Ergotamine[a]	–CH₂Ph
Ergosine[a]	–*i*–Bu
Beta-ergosine	–*sec*–Bu
Ergovaline[a]	–*i*–Pr
Ergobine	–Et
Ergoxine group ($R_1 = -C_2H_5$)	R_2
Ergostine	–CH₂Ph
Ergoptine[a]	–*i*–Bu
Beta-ergoptine	–*sec*–Bu
Ergonine	–*i*–Pr
Ergobutine	–Et
Ergotoxine group ($R_1 = -i$–Pr)	R_2
Ergocristine	–CH₂Ph
Alpha-ergocryptine	–*i*–Bu
Beta-ergocryptine	–*sec*–Bu
Ergocornine[a]	–*i*–Pr
Ergobutyrine	–Et

[a] Found in both endophyte-infected grass and cultures.

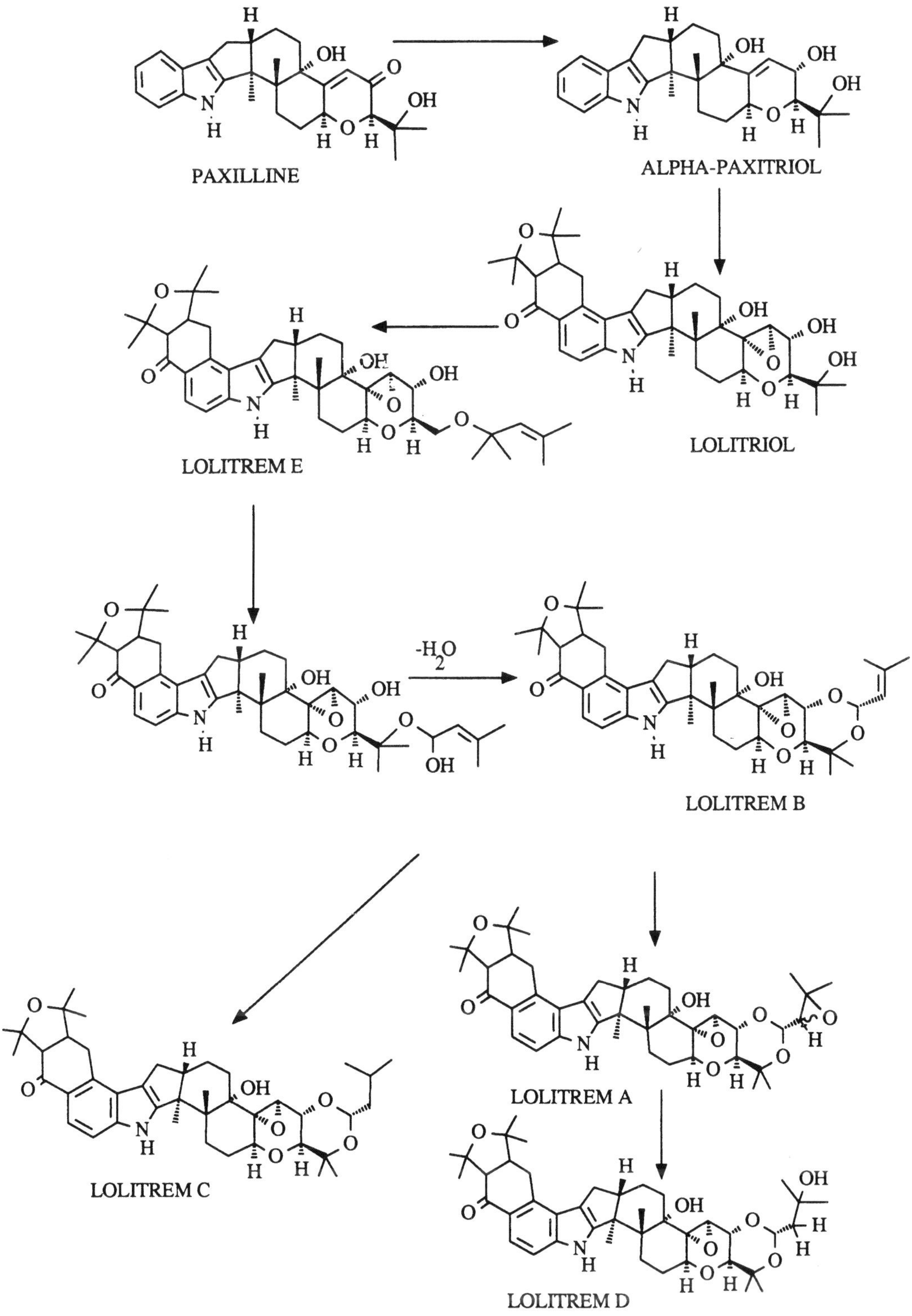

Figure 2 Lolitrems.

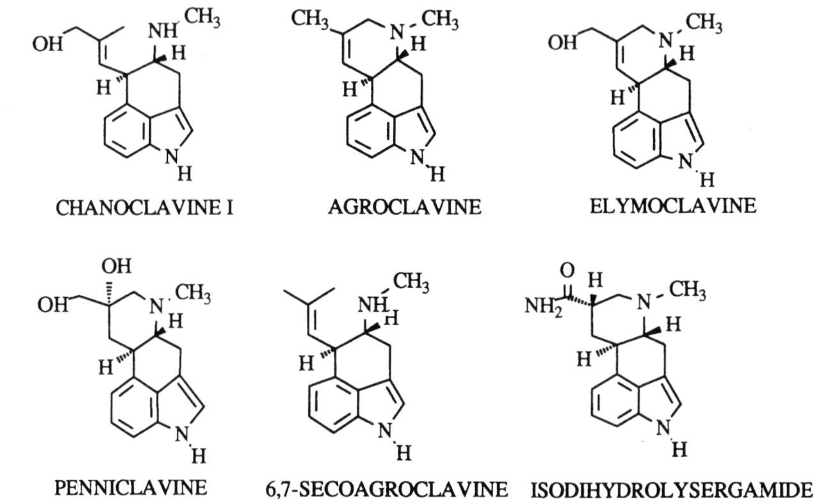

Figure 3 Clavine alkaloids.

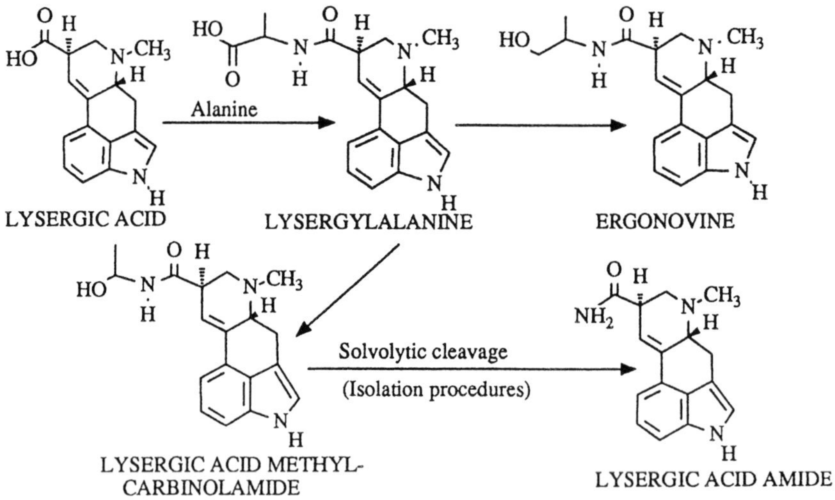

CHANOCLAVINE I AGROCLAVINE ELYMOCLAVINE

PENNICLAVINE 6,7-SECOAGROCLAVINE ISODIHYDROLYSERGAMIDE

LYSERGIC ACID LYSERGYLALANINE ERGONOVINE

Alanine

Solvolytic cleavage
(Isolation procedures)

LYSERGIC ACID METHYL-
CARBINOLAMIDE

LYSERGIC ACID AMIDE

Figure 4 Lysergic acid and amides (structures and schematic).

Et (Figure 1, Table 1).[19,79] The ergopeptam class (Figure 5) is defined with an open ring in the peptide portion of the molecule.[44,78] They occur in minor quantities from cultures of *Claviceps*, and have yet to be identified from the *Acremonium* endophytes. References to the ergot alkaloids, throughout this manuscript, will follow the nomenclature and stereochemistry outlined in Berde and Schild[18] and other references and reviews.[45,49,79,84]

Other compounds associated with *A. coenophialum*-infected tall fescue are the pyrrolizidine alkaloids, *N*-acetylloline and *N*-formylloline (Figure 6),[15,24,25,112] and the pyrrolopyrazine alkaloid, peramine (Figure 7).[112,113] Currently, the potential for fescue toxicity is based on concentrations of ergovaline and the loline alkaloids, *N*-acetyl- and *N*-formyllolines, and the level of *A. coenophialum* infection in the grass. Minor ergopeptine alkaloids detected by tandem mass spectrometry[83] from *A. coenophialum*-infected fescue seed include ergosine with trace amounts of ergotamine, ergonine, ergocrystine, ergocryptine, ergocornine, and the C-8 epimers ergosinine and ergotaminine (Figure 1; Table 1).[128] Investigations of the infected grass revealed ergovaline,[64,104,112] ergosine, and ergonine, in the blades and sheath, with trace amounts of ergoptine and ergocornine (Figure 1; Table 1).[64] Ergovaline was the predominant ergopeptine in all samples, and the minor alkaloids occurred in the same proportions. Prior investigations revealed ergovaline as the major ergopeptine alkaloid produced in laboratory cultures of this fungus, with minor amounts of

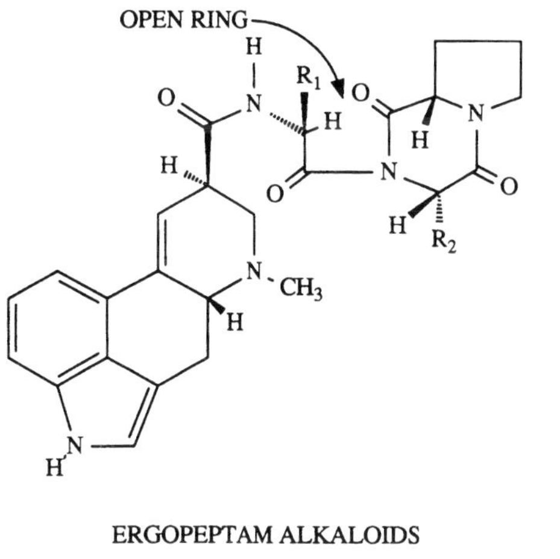

OPEN RING

ERGOPEPTAM ALKALOIDS

Figure 5 Ergopeptam alkaloids (Ergovalam:
R1 = -Me; R2 = -iso-Pr).

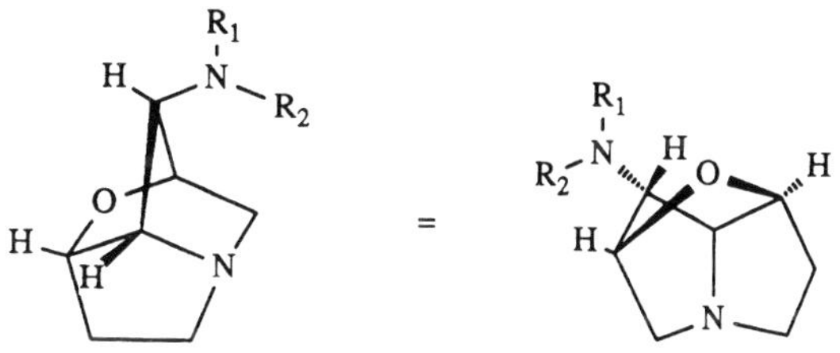

	R_1	R_2
LOLINE	CH_3	H
N-ACETYLLOLINE	CH_3	$COCH_3$
N-FORMYLLOLINE	CH_3	CHO
N-ACETYLNORLOLINE	H	$COCH_3$
N-METHYLLOLINE	CH_3	CH_3

Figure 6 Loline alkaloids.

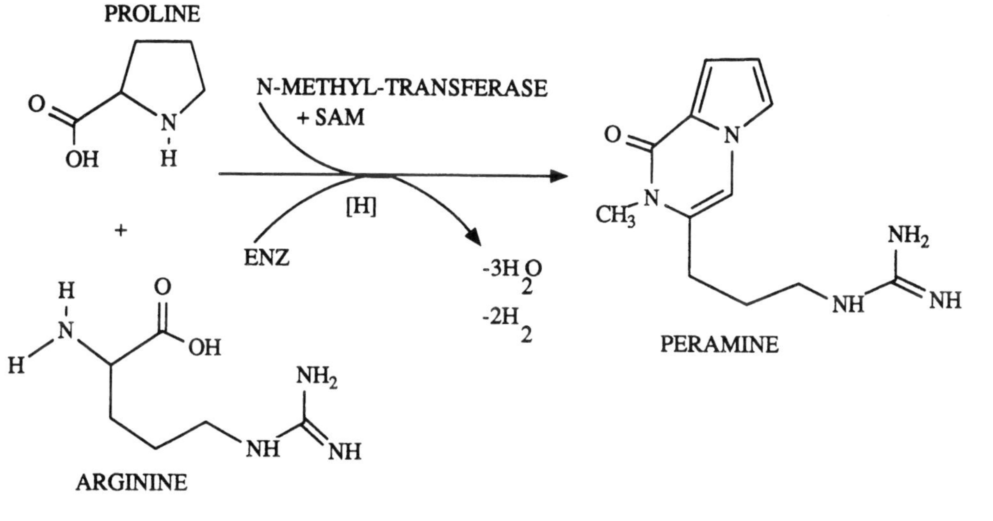

PROLINE

N-METHYL-TRANSFERASE
+ SAM

[H]

ENZ

$-3H_2O$

$-2H_2$

PERAMINE

ARGININE

Figure 7 Peramine (structure and bioschematic).

ergosine and their C-8 epimers ergovalinine and ergosinine.[89,91] Recently, ergovaline and ergotamine were found in 120-day-old grasses fertilized with liquid fertilizer,[5] a fact that may be related to senescent material (C.W. Bacon, USDA/ARS, Athens, GA, private communication). Since the C-8 epimeric ergopeptines (i.e., defined with the suffixal "-inine" instead of "-ine") are artifacts from extraction and isolation procedures, they are not considered true natural products and therefore will not be mentioned further in these discussions.

The clavine class of ergot alkaloids (Figure 3) also have been isolated from both laboratory cultures of *A. coenophialum*[89,91] and from infected tall fescue.[64] These alkaloids include chanoclavine 1, agroclavine, elymoclavine, 6,7-secoagroclavine, and penniclavine.[64,89,91] Lysergic acid amide (or ergine, Figure 4) has been found in *A. coenophialum*-infected fescue.[81,110] However, this compound most probably is produced during the extraction procedures (i.e., solvolytic cleavage) via lysergic acid methylcarbinolamide (or another simple lysergic acid amide derivative) (Figure 4).[2,45,50,51] Ergonovine (Figure 4), another simple lysergic acid amide, has been found in *A. coenophialum*-infected fescue seeds.[81,129] This compound may be from *Claviceps*-contamination of the fescue seeds,[129] and subsequently, *Claviceps* may be the source of ergonovine reported from this and other studies. Then, too, age and handling of samples may be contributing factors. Further studies are required before we can conclude, unequivocally, that *A. coenophialum*-infected seeds and grass contain this alkaloid.

Ergovaline, ergotamine, and peramine (Figures 1 and 7; Table 1) also have been isolated from *A. lolii*-infected perennial ryegrass.[106,107] The potential for the neurotoxic ryegrass staggers syndrome in livestock is defined by the lolitrem concentrations, primarily lolitrem B and paxilline (Figure 2),[41,68] and the level of *A. lolii* infection in the host grass. The ergot alkaloids and/or the minor lolitriol (Figure 2) in *A. lolii*-infected ryegrass may be more related to decreased animal performances (i.e., reduced weight gains, reproduction problems, heat stress) or the "ill thrift" condition observed in livestock.[40,69]

The loline alkaloids in *A. coenophialum*-infected fescue may be produced by the grass in response to the endophyte, or vice versa.[112] To date, the loline alkaloids have not been isolated from *A. lolii*-infected ryegrass. Peramine (Figure 7), however, is produced by the endophytes in both *A. coenophialum*-infected fescue and *A. lolii*-infected ryegrass.[106,112] The loline alkaloids and peramine in infected fescue and peramine in infected ryegrass are considered responsible for the specific mutualistic defensive associations related to insect deterrences,[112] rather than ruminant toxicities. Therefore, peramine and the loline alkaloids are considered important, desirable metabolites to be maintained, because of their agronomic benefits.

The ability of these endophyte-infected grasses to produce ergovaline, lolitrems, peramine, and the lolines, either singularly or in combination, have been reported. The tall fescue endophyte *A. coenophialum* and the perennial ryegrass endophyte *A. lolii* were associated with the same three alkaloids, in their naturally infected host grasses as well as when they were artificially introduced into other grasses.[112] Ergovaline, peramine, and the lolines were detected in tall fescue (KY 31) naturally infected with *A. coenophialum* and also when this fungus was artificially introduced into ryegrass (*L. perenne*, cultivar Ruanui) and another cultivar of *F. arundinacea* (cultivar G1-320, Johnstone). Lolitrem B, peramine, and ergovaline were detected in ryegrass (cultivars Nui and Repel) naturally infected with *A. lolii* and when artificially introduced into fescue (*F. arundinacea*, cultivar G1-320, Johnstone). Other *Acremonium* species were not characterized for their production of alkaloids, although peramine and ergovaline were the most frequent combination. Lolines were detected only in *A. coenophialum*-infected grasses and in combination with either ergovaline or peramine.

Petroski et al.[82] reported that *Stipa robusta* infected with an unknown species of *Acremonium* contained *N*-formylloline (Figure 6) along with a few simple lysergic acid amides. However, peramine was not looked for in these samples. More recently, Penn et al.[77] reported that isolates of *A. coenophialum* from tall fescue in culture produced paxilline and lolitrem B. Furthermore, *A. uncinatum* Gams, Petrini, and Schmidt and other species of the section *Albo-lanosa* produced paxilline, lolitriol, and lolitrem B (Figure 2).

These reports establish that the ability to synthesize both the tricyclic ergopeptine and tremorgenic lolitrems resides with these endophytes, chemically grouping them together with *Claviceps*. Prior to the isolation and identification of ergovaline and ergosine from cultures of the endophyte of tall fescue, (i.e., *A. coenophialum*),[88,89,91] only *Claviceps* was listed as the genus capable of producing the ergopeptine alkaloids. Furthermore, compounds similar in structure and tremorgenic activity to the lolitrems were previously reported from *C. paspali* and from species of *Penicillium* and *Aspergillus* (see below).[28-30]

B. *EPICHLOË TYPHINA*

The teleomorphic state of *A. typhinum* Morgan-Jones & Gams is *Epichloë typhina*,[70] the choke fungus of several grasses.[8,36,112] Certain isolates of *E. typhina* have been studied in relation to their ability to produce ergovaline, lolitrems, peramine, and the loline alkaloids. Siegel et al.[112] established that *F. longifolia* infected by *E. typhina* contained ergovaline, peramine, and lolitrem B. Furthermore, they indicate that *Elymus canadensis* L., *Sitanion longifolium* Smith, and *Agrostis hiemalis* (Walt.) B.S.P. infected by *E. typhina* contained ergovaline and/or peramine. However, *E. typhina*-infected *Poa ampla* Merr. contained peramine, with lolines only being detected in combination with peramine and/or ergovaline.

C. *BALANSIA* SPECIES

"Ergot-like" syndromes in livestock grazing on pasture grasses have been related to endophytic fungi of the genus *Balansia* (tribe Balansiae).[10-13,102,126] Bacon et al.[10,11] related cattle symptoms analogous to fescue toxicity and ryegrass staggers to weed grasses infected with *Balansia* spp. Clavine and lysergic

Table 2 **Ergot alkaloids identified from *Balansia* species**

Balansia epichloë[13,90,91]	*Balansia strangulans*[12,91]
Chanoclavine I	Chanoclavine I
Isochanoclavine I	6,7-Secoagroclavine
Agroclavine	*Balansia claviceps*[12,90]
Elymoclavine	Chanoclavine I
6,7-Secoagroclavine	Ergonovine
Penniclavine	Ergonovinine
Ergonovine	*Balansia obtecta*[98]
Ergonovinine	Ergobalansine
Balansia henningsiana[12,13,91]	Ergobalansinine
Chanoclavine I	*Balansia cyperi*[98]
Dihydroelymoclavine	Ergobalansine
Isodihydrolysergamide[a]	Ergobalansinine
Ergonovine	
Ergonovinine	
Epichloë typhina[112]	
Ergovaline	
Ergovalinine	

[a] Tentative identification.

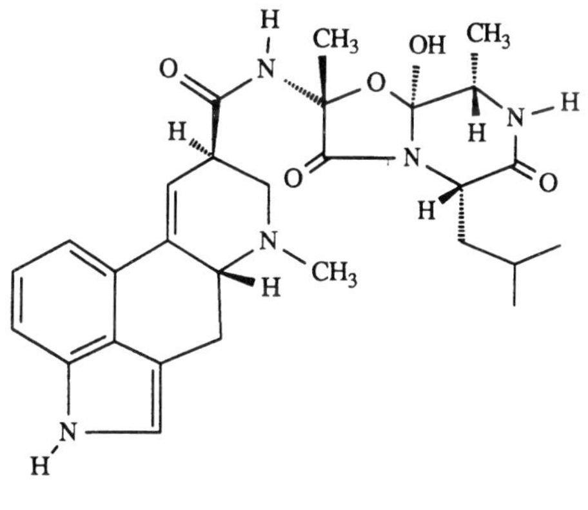

Figure 8 Ergobalansine.

acid amide alkaloids (Figures 3 and 4; Table 2) were isolated from *B. epichloë* (Weese) Diehl from smut grass [*Sporobolus poiretii*, (Roem. and Schult.) Hitchc.], *B. strangulans* from *Panicum hians*, *B. henningsiana* from broomsedge (*Andropogon virginicus*), and *B. claviceps* from *Chasmanthium laxum*.[6,12,13,88,90,91] Recently, a new ergopeptine alkaloid, ergobalansine (Figure 8), was isolated from *B. obtecta* and *B. cyperi*.[98] Ergobalansine lacks the proline ring moiety in the tricyclic peptide portion of the basic ergopeptine alkaloids (see Figure 1). The biological and toxicological significance of this new ergopeptine alkaloid is unknown.

Other compounds isolated from cultures of *B. epichloë* include 3-indoleacetic acid, 3-indole acetamide, 3-indole ethanol, and methyl-3-indolecarboxylate (Figure 9).[85] Indoleacetic acid also has been isolated from *A. coenophialum*-infected fescue and from cultures of this fungus.[35] Epimeric *erythro*- and *threo*-3-indolyglycerols [1-(3-indolyl)-propane-1,2,3-triol], along with 3-indolebutanetriol [4-(3-indolyl)-butane-1,2,3-triol] and 3-(3,3′-diindolyl)-propane-1,2-diol, also have been reported from cultures of *B. epichloë* (Figure 10).[93,94]

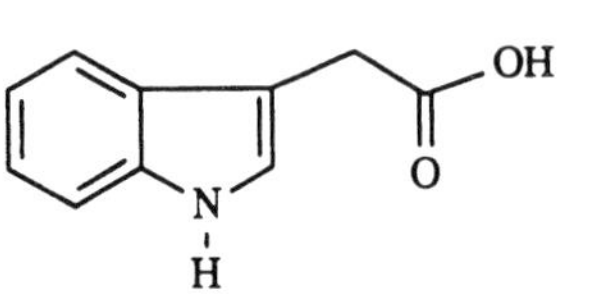

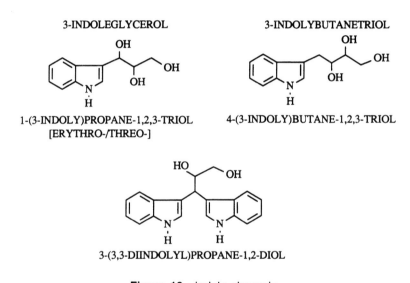

3-INDOLEACETIC ACID

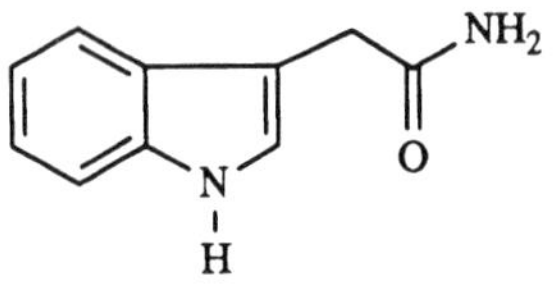

3-INDOLEACETAMIDE

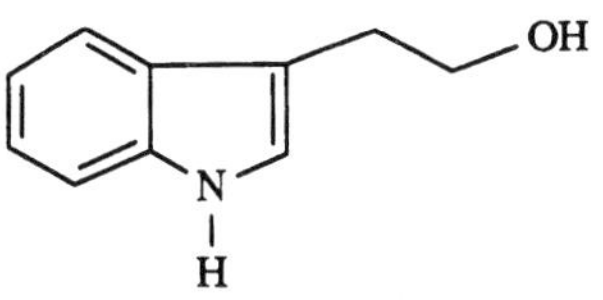

3-INDOLE ETHANOL

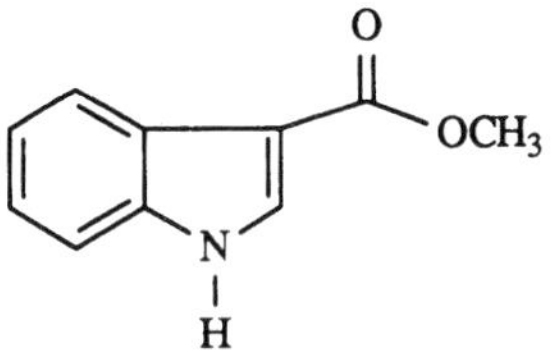

METHYL-3-INDOLECARBOXYLATE

Figure 9　3-Indoleacetic acid, 3-indoleacetamide, 3-indoleethanol, and methyl-3-indolecarboxylate.

3-INDOLEGLYCEROL

3-INDOLYBUTANETRIOL

1-(3-INDOLY)PROPANE-1,2,3-TRIOL
[ERYTHRO-/THREO-]

4-(3-INDOLY)BUTANE-1,2,3-TRIOL

3-(3,3-DIINDOLYL)PROPANE-1,2-DIOL

Figure 10　Indole glycerols.

III. COMPOUNDS ASSOCIATED WITH GRASS ENDOPHYTES: BOTANICAL AND BIOSYNTHETIC ORIGINS

A. ERGOT ALKALOIDS: CLAVINE, LYSERGIC ACID, LYSERGIC ACID AMIDES, AND ERGOPEPTINES

The biosynthesis of ergot alkaloids by *Claviceps* has been reviewed in detail[18,45,49,115] and, in all probability, should follow the same general basic pathway in *Acremonium* and *Balansia* species. Briefly, tryptophan and mevalonic acid (Figure 11) form dimethylallytryptophan, which, through a series of steps, produces chanoclavine 1, the precursor to agroclavine (Figures 3 and 11). Hydroxylation of agroclavine to elymoclavine, and subsequent oxidation and isomerization to d-lysergic acid, is the key intermediate to the formation of the ergopeptines and simple lysergic acid amide alkaloids. The tricyclic peptide portion of the ergopeptine alkaloids (Figure 1) is assembled linearly from proline and two other variable amino acids, presumably by a multienzyme complex. In the case of ergovaline, alanine (L-*alpha*-OH-alanine or another closely related compound)[2,18,45] and valine are the other two amino acids necessary for its biosynthesis. Ergovaline appears to be the major ergopeptine alkaloid produced by *E. typhina* and most *Acremonium* species, both *in vivo*[91,112] and *in planta*.[64,112,128] Although produced in minor concentrations by some isolates of *Claviceps*, ergovaline synthesis and accumulation may be unique to this group of endophytes. Porter et al.[86] studied and compared the ergopeptine alkaloids from *Claviceps*-infected fescue, wheat, and barley with those from *A. coenophialum*-infected tall fescue and suggested that the varying concentrations (or ratios) of the ergopeptine alkaloids (i.e., ergovaline/ergosine, etc.) may define which fungus (i.e., *Acremonium* vs. *Claviceps*) may be responsible for animal toxicities.

The hypothetical biosynthesis of the simple lysergic acid amides ergonovine and lysergic acid methylcarbinolamide may result via lysergic acid and alanine (or a closely related compound)[2,45,50] to give lysergylalanine (Figure 4). Although in question, lysergylalanine then may be converted to ergonovine via reduction of the carboxylic acid moiety or to lysergic acid methylcarbinolamide (lysergyl-*alpha*-hydroxyethylamide) via hydroxylation and decarboxylation.[2,45] Lysergic acid methylcarbinolamide decomposes to lysergic acid amide, following extraction and isolation (Figure 4).[2,45,50,51] Both lysergic acid amide and the methylcarbinolamide have varying degrees of biological activity[18,72] and most probably contribute to animal toxicity. Shelby[110] compared concentrations of ergovaline and lysergic acid amide in infected and noninfected fescue grass and seeds and suggested monitoring both compounds as an indicator of the fescue endophyte.

Since the biosynthesis for the ergopeptine alkaloids proceeds through the clavine alkaloids that have been reported in both cultures and the infected grass, future investigations will no doubt uncover other analogs similar to those produced by species of *Claviceps*.

The clavine and lysergic acid amide derivatives, up to and including ergonovine, were isolated from the *Balansia* species, suggesting that these fungi may be used as sources for these medicinal agents. The isolation and identification of ergobalansine from *B. obtecta* and *B. cyperi*[98] suggest a unique synthesis within these species. Interestingly, proline is absent in ergobalansine, whereas all the tricyclic ergopeptine alkaloids identified from *E. typhina* and its anamorphs, *Acremonium,* to date contain the proline ring moiety. In the biosynthesis of ergobalansine, it would appear that alanine and *alpha*-OH-alanine (or two molecules of alanine) and one molecule of leucine, instead of proline, combine to make up the cyclic peptide portion of this compound. No doubt, future investigations of *Balansia* will uncover other analogues in this class. Additionally, proline is the hypothetical precursor to peramine; whether this is related to *Acremonium*'s ability to produce ergovaline and/or peramine (see below) should be considered.

There is a preponderance of evidence in the literature that other factors must be considered in the biosynthesis of these compounds by the endophytes. The endophyte–host grass associations, insect herbivory, and livestock grazing pressure, along with fertilization and environmental factors, all influence alkaloid production.[4,5,16,17,24,25,53] Hill et al.[53] suggested the plant plays a major role in controlling the expression of the genetic potential of endophytes for ergopeptine production. Furthermore, if the biosynthesis of the ergot alkaloids parallels *Claviceps*, synthesis of the peptide moiety is controlled by the relative concentrations of free amino acids[44] within the apoplasm. Lyons et al.[63] have reported increases

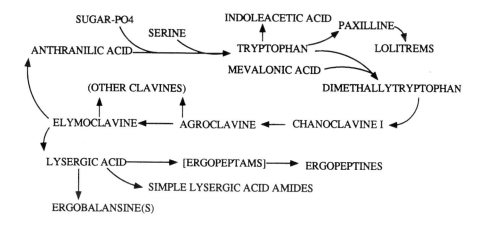

Figure 11 General relation between tryptophan, ergot alkaloid, and lolitrem biosynthesis.

in total soluble amino acids, along with increases in arginine and glutamine synthetase activity, in *A. coenophialum*-infected fescue. Glutamine is involved in transamination reactions, as well as reserve pools of ammonia and glutamic acid. Glutamic acid is the precursor for the biosynthesis of proline and serves as the beginning synthesis of arginine[121,131] (see below for correlation with peramine biosynthesis). In this regard, the first step is the production of ornithine, the proposed precursor involved with the biosynthesis of the loline alkaloids.[103] Subsequently, intermediate precursors, enzyme availability, amino acid composition, and/or sources of these during stress on both endophyte and host should be considered.

Nitrogen influences alkaloid production in culture and on *Acremonium*-infected pastures.[16,17,63,64,104] With regard to animal responses to fescue toxicity and pasture fertilization, there also may be a "nitrogen relationship."[95,118] Furthermore, high concentrations of inorganic phosphate in saprophytic cultures of *Claviceps* sp. inhibit alkaloid biosynthesis and also increase alkaloid-degradative enzymes.[43,101] Whether varying concentrations of soil inorganic nitrogen and phosphorus may be used to influence or prevent ergot alkaloid production and subsequently limit toxic effects in livestock on endophyte-infected pastures awaits future investigations.

B. SIMPLE INDOLE ALKALOIDS: INDOLEGLYCEROLS, INDOLEACETIC ACID

Tryptophan biosynthesis by the endophytes may involve unique mechanisms associated with their host plants and vice versa. Generally, in most organisms, tryptophan biosynthesis proceeds through a series of steps that combine anthranilic acid and 5-phosphoribosyl-1-pyrophosphate to produce indole-3-glycerol phosphate. Indole-3-glycerol phosphate then reacts with serine to produce tryptophan.[56,121,131] Another unique but novel consideration is the pathway in which anthranilic acid combines with fructose-6-phosphate, through a series of steps, to produce indole-3-tetritol phosphate, which then reacts with serine to produce tryptophan.[124] The isolation and identification of 3-indoleglycerol and 3-indolebutanetriol (Figure 10) from cultures of *B. epichloë*[93,94] suggest a similar process is used by this endophyte. De Battista et al.[35] and Richardson et al.[100] suggested that drought tolerance, carbohydrate metabolism (i.e., sugar alcohols), and certain auxin production (i.e., indole acetic acid) (Figure 9) may be related to the endophyte–host associations in *A. coenophialum*-infected fescue. Auxin production is the end result of tryptophan metabolism,[131] and other indole related compounds. Indole acetic acid, indole acetamide, indole ethanol, and methyl-3-indolecarboxylate (Figure 9) have been reported in cultures of *B. epichloë*.[85] Recently, the biosynthesis of indole acetic acid from tryptophan, through indole acetamide, has been reported.[58] The major auxin produced in culture of *B. epichloë* was indoleacetic acid. This also was the major auxin produced both in cultures and in fescue infected with *A. coenophialum*.[35] The identification of epimeric *erythro*- and *threo*-3-indoleglycerols from cultures of *B. epichloë*[93] may suggest novel plant growth regulatory activity, along with the control of tryptophan (and subsequently alkaloid production) in the plant, by the endophyte.

Based on the inhibition of ergot alkaloid biosynthesis (via inorganic phosphate) and the simultaneous stimulation of the alkaloid degradation enzymes in saprophytic cultures of a *Claviceps* species, Robbers[101] suggested the concentration of these compounds is a dynamic state and not simply an end product of metabolism. Flieger et al.[43] proposed possible mechanisms of clavine alkaloid turnover (i.e., elymoclavine) that involve feedback synthesis of anthranilic acid, tryptophan, dimethylallyltryptophan, and subsequently chanoclavine I, etc. (Figure 11). Therefore, it is not inconceivable that interdependent relationships exist between plant and fungal growth, as well as tryptophan and alkaloid biosynthesis.

C. LOLITREM ALKALOIDS: PAXILLINE, PAXITRIOL(S), LOLITRIOL, LOLITREMS

Lolitrem biosynthesis in *A. lolii*-infected perennial ryegrass seeds and cultures of the fungus has been suggested to occur through paxilline,[68,69,122] which is produced via tryptophan and a series of reactions with various modifications of geraniol.[1,29] This may serve as a guide to lolitrem production by *A. lolii* and *A. coenophialum*.

The synthesis of *alpha*-paxitriol and lolitriol (Figure 2) and the subsequent demonstration of lolitriol in extracts of *A. lolii*-infected ryegrass leaf and seeds[68,69] provided the foundations for the proposed biosynthesis of lolitrem B. Most recently, structures for the minor lolitrems, A, C, D, and E (Figure 2), in *A. lolii*-infected seeds were determined.[68] Miles et al.[68,69] proposed that the biosynthesis of lolitrem B proceeds via paxilline, *alpha*-paxitriol, and lolitriol, to lolitrem E. Lolitrem E may then undergo allylic oxidation followed by cyclization and condensation, respectively, to yield lolitrem B. Oxidation of lolitrem B would yield lolitrem A, and subsequent hydrolysis would yield lolitrem D. Reduction of lolitrem B would yield lolitrem C. Future studies with labeled compounds will no doubt provide more detailed information on the biosynthesis of these neurotoxic alkaloids.

D. PYRROLIZIDINE (LOLINE) ALKALOIDS: *N*-ACETYLLOLINE, *N*-FORMYLLOLINE

Loline was originally isolated from toxic tall fescue as festucine.[130] Robbins et al.[103] first isolated and identified *N*-acetylloline, *N*-formylloline, and *N*-acetylnorloline (or demethyl-*N*-acetylloline) from fescue seeds and reported that these alkaloids decreased feed intake and weight gains in rats. Bush et al.[24,25] reported that *N*-acetylloline and *N*-formylloline were directly related with *A. coenophialum*-infected tall fescue. Subsequent studies correlated these alkaloids with certain insect deterrence properties of the infected grass.[39,57,112] The chemistry of the lolines isolated from *A. coenophialum*-infected fescue and other synthetic analogues has been reviewed.[97] Although in question,[97] Robbins et al.[103] suggest loline biosynthesis proceeds through *N*-acetylnorloline (demethyl-*N*-acetylloline) via norloline and invokes the alkaloid retronecine, which is formed by the dimerization of two molecules of ornithine.[20,55]

Belesky et al.[16] reported that the concentration of loline alkaloids (*N*-acetyl-, *N*-formyl-) in infected fescue was increased by nitrogen fertilization and varied according to the season. These authors suggested production of the lolines may be influenced by many biotic and abiotic factors, including grazing animals, plant response, endophyte growth and metabolism, and environmental factors. Greenhouse investigations[23,24] have shown lolines increase with plant age and in regrowth tissues following clipping. The lolines may be produced by the plant, in response to the fungus and/or as a defense mechanism in response to insect herbivory, which may involve both the endophyte and its host. Labeling studies will be needed to substantiate both sources and the actual pathway for these alkaloids. Interestingly, Siegel et al.[112] reported the presence of the loline alkaloids in conjunction with peramine and ergovaline in tall fescue naturally infected with *A. coenophialum*, and when the tall fescue endophyte was artificially introduced into other cultivars. The lolines have been shown to occur with either peramine or ergovaline, in *Poa autumnalus* (Muhl.) Ell. and *F. gigantea* (L.) Vill. naturally infected with *A. coenophialum* and another *Acremonium* sp.

If the biosynthesis of these alkaloids is a defense mechanism involving both endophyte and its host, it is doubtful that these mechanisms are unique to *A. coenophialum*-infected fescue. Certain insects (see below) have developed the ability to use plant pyrrolizidine alkaloids in their own defense mechanisms against other predator insects.[52,109] Therefore, it is not inconceivable that the endophyte(s)–plant host associations have developed similar lines of defense. The endophyte–grass associations may contain these and other avenues of defense, and ultimately they may relate to the biosynthesis or production of these alkaloids.

E. PYRROLOPYRAZINE: PERAMINE

Peramine[106] is produced by *A. coenophialum*, *A. lolii*, other *Acremonium* species, as well as in *A. starrii*-infected *Bromus anomalus* Rupt. It is also found in several *E. typhina*-infected grasses.[112] The biosynthesis of peramine may occur through the cyclization and condensation reactions of proline with arginine (Figure 7),[106] with the final *N*-methylation via an *N*-methyltransferase and *S*-adenosylmethionine. If these two amino acids accumulate linearly on an enzyme surface (or a multienzyme complex similar to that described for the tricylic peptide portion of the ergopeptine alkaloids), reduction of the carbonyl moiety from arginine, with elimination of H_2O and $2H_2$ from the proline moiety, would lead to peramine. Specific C^{14} labeling studies will be needed to confirm this proposal (see above for hypothetical correlations of proline and arginine biosynthesis in *A. coenophialum*-infected fescue).

F. MISCELLANEOUS: ERGOSTATETRAENEONE, ERGOSTEROL, ERGOSTEROL-PEROXIDE, HARMAN, NORHARMAN, AND HALOSTACHINE

Several compounds not directly related to endophyte–grass associations have been detected in grasses on which livestock exhibited "ergot-like syndromes" or in grasses naturally infected with endophytes. These compounds are grouped here and briefly, especially since the biological effects are unknown.

Ergosta-4,6,8(14),22-tetraen-3-one (trivial name ergostatetraeneone) and ergosta-4,6,8(14)-triene-3-one (Figure 12) were isolated from endophyte-infected toxic tall fescue, *B. epichloë*-infected smut grass (*S. poiretii*) and cultures of this fungus, Bermuda grass (*Cynodon dactylon* L.) on which cattle were showing "Bermuda grass tremors" (a neurotoxic syndrome similar to "ryegrass staggers"), and also cultures of *Claviceps* spp. that were isolated from toxic bermuda grass.[87,92] Production of the

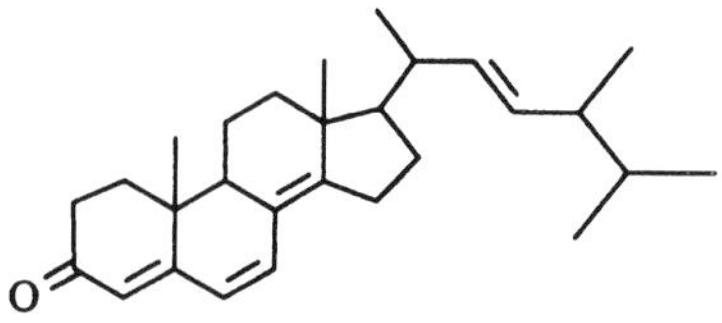

ERGOSTA-4,6,8,(14),22-TETRAENE-3-ONE

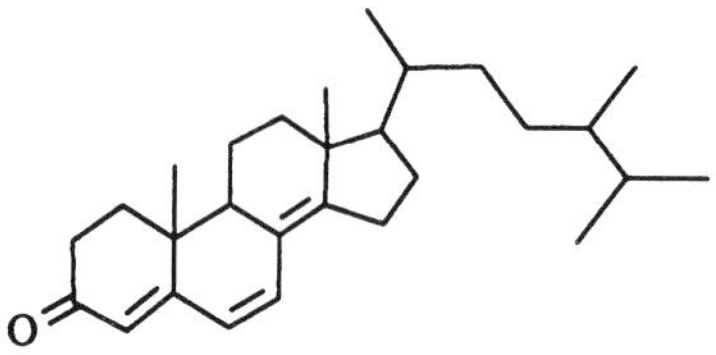

ERGOSTA-4,6,8(14)-TRIENE-3-ONE

Figure 12 Tetraene-3-one and triene-3-one.

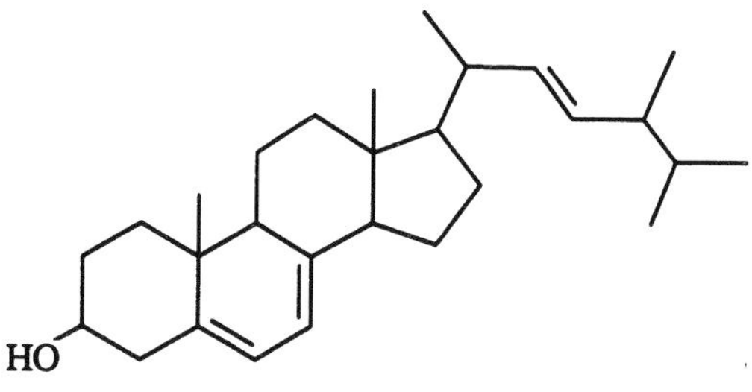

ERGOSTEROL

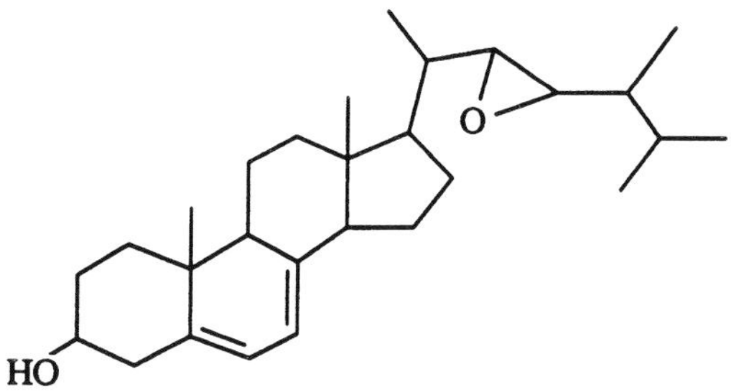

ERGOSTEROL PEROXIDE

Figure 13 Ergosterol and ergosterol peroxide.

ergostatetraeneone along with ergosterol and ergosterol peroxide (Figure 13) was later associated with toxic *A. coenophialum*-infected fescue.[34] Subsequent studies have shown ergosterol (and/or the oxidation of ergosterol) to be either an indicator of growth in certain species of fungi or a signal for mycotoxin production.[75] Although ergostatetraeneone, ergosterol, and ergosterol peroxide are produced by a number of fungi[34,92] and found in plant tissues, varying concentrations of these compounds in endophyte-infected grasses may be used to indicate growth of the endophytes, toxicity, or production of other compounds.

Phytoalexins are antibiotics produced by a plant, in response to a disease-producing agent, i.e., a fungus. Ergosterol sensitivity has been employed as a bioassay for screening compounds for polyene-like antibiotic activity.[3] Latch[60] has reviewed *Acremonium* spp. and *E. typhina*-infected grasses and their resistance to certain fungal and viral diseases. The data suggest that endophytes do possess antibiotic-like activity, although specific chemicals have not been identified, nor have modes of action been established. However, a highly speculative and overly simplified description of activity may be based on the selective binding of an antibiotic to fungal cell-membrane ergosterol, which alters membrane permeability and subsequently leads to membrane potential collapse, lysis, and cell death. Mammalian cells, containing cholesterol instead of ergosterol, are less susceptible to these types of polyenes. Thus, screening for fungicidal agents, with membrane-ergosterol activity, may provide compounds with therapeutic utility in the treatment of various mycoses.[3]

Harman, norharman, and halostachine (Figure 14) have been found in tall fescue toxic to livestock.[33,125] Possible mechanisms associated with these compounds, and their toxicity, have been reviewed.[33,95,96,120] More recently, norharman and halostachine were associated with insect protection and feeding deterrence mechanisms in other forage species.[109] Whether these three alkaloids are related to endophyte infection, and possibly as other defense mechanisms, awaits future investigations.

Bond et al.[21] reported that the tall fescue cultivar fescue G1-307, was toxic to livestock, contained significantly higher concentrations of *gamma*-aminobutyric acid (GABA) than did a nontoxic cultivar

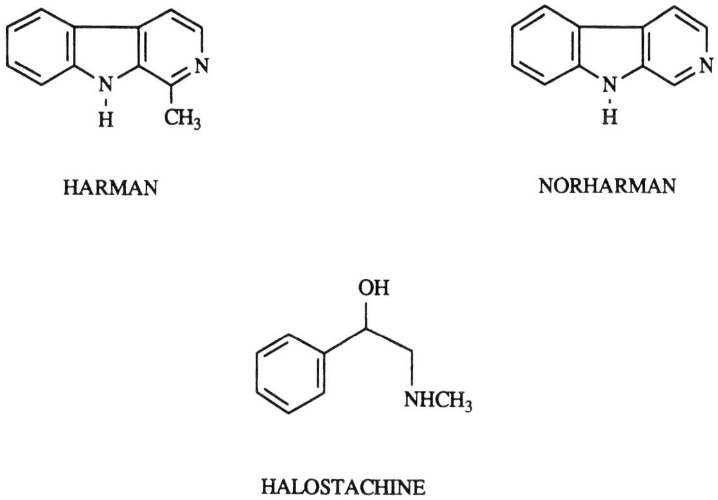

HARMAN

NORHARMAN

HALOSTACHINE

Figure 14 Harman, norman, and halostachine.

fescue G1-306. GABA is considered because some of its biological activities have been mentioned in animal performance problems on endophyte-infected grasses[95] (C. T. Dougherty, Lexington, KY, private communication, see below). (For additional compounds from grasses associated with toxicities that have a more historical significance, see Yates.[125,126])

IV. COMPOUND MODE OF ACTION AND BIOTECHNOLOGICAL RELEVANCE

A. MAMMALIAN: LIVESTOCK, WILDLIFE, HUMANS

Mechanisms of action and the mode of toxicity associated with *A. coenophialum*-infected fescue were recently reviewed.[95,96,120] Reduced serum prolactin in cattle, sheep, and horses on *A. coenophialum*-infected fescue support the basic dopaminergic mechanisms of the ergopeptine alkaloids.[18] Most importantly, the suppression of prolactin by the endophyte has been ameliorated by administration of a dopamine antagonist, metoclopramide,[62] and monoclonal antibodies specifically against the ergoline ring nucleus have been used to reverse the prolactin suppression in steers, by the endophyte.[119] In addition, the vasoconstriction activity of the ergopeptine alkaloids may contribute to reduced animal performances (i.e., weight gains, milk production), reproduction problems, and possibly gangrene of the foot and tail.[37,96,120] Garner et al.[48] reported that cattle, at low ambient temperatures (−5°C), appear to lose their natural vasodilitary process, when receiving toxin(s) from endophyte-infected fescue. Rhodes et al.[99] reported restrictive blood flow to peripheral and core body tissues in sheep and cattle, which is also consistent with the vasoconstrictive effects of the ergot alkaloids,[18] and Dyer[37] reported ergovaline is a potent vasoconstrictor of isolated bovine uterine and umbilcal ateries. Then too, Oliver et al.[72] observed the analogous vasoactivity of lysergic acid amide on isolated bovine vasculature.

Heat stress in cattle exacerbates the effects of the ergot alkaloids and contributes significantly to toxicity.[74] Cornell et al.[31] have established that 50 ng of ergovaline per gram of infected grass is sufficient to produce some of the symptoms of fescue toxicity in cattle under heat stress conditions. Similar thermoregulatory imbalances have been reported in rats dosed with ergovaline isolated from *A. coenophialum*-infected fescue.[116] (For further discussions concerning effects of heat and neuroendocrine mechanisms of toxicity, see an earlier review.[96])

Oliver et al.[73] reported that *N*-acetylloline has vasoconstrictive activity *in vitro* that, subsequently, may augment the vasoactivity of the ergot alkaloids. The pyrrolizidine alkaloids isolated from fescue decreased feed intake and subsequent weight gains in rats,[103] but they do not appear to affect prolactin.[117]

Reduced melatonin has been reported in cattle ingesting endophyte-infected fescue.[95] This may reflect the ability of the ergot alkaloids to inhibit the norepinephrine-induced synthesis of serotonin-*N*-acetyltransferase in the pineal gland. Serotonin-*N*-acetyltransferase is responsible for the conversion of serotonin to *N*-acetylserotonin, the precursor to melatonin. The dopaminergic activity of the ergopeptines also may influence melatonin concentrations. These and other possible mechanisms associated with reduced melatonin have been discussed in detail.[95,96] Imbalances in circulating prolactin and melatonin may have a pronounced effect on endogenous biological rhythms in both photo- and nonphotobreeders, by altering the day-length signal and, subsequently, the physiological and morphological responses, e.g., seasonal changes in hair coat and weight gains.

Recently, Moubarak et al.[71] reported that ergovaline resulted in a dose-dependent inhibition of rat brain synaptosomal Na$^+$/K$^+$ ATPase activity. The ATPase enzyme system plays a key role in the maintenance of neuronal membrane potential, thus suggesting this inhibition may contribute to fescue toxicity.

The effects of GABA are extremely diverse and complex,[38,59] and some of the effects have been suggested as affecting satiety in livestock on infected grass (C. T. Dougherty, private communication). Both the ergopeptine alkaloids and the *beta*-carboline alkaloids (harman; norharman) have "GABA-like" activities.[95] Additionally, GABA decreases both prolactin from the pituitary,[59] and melatonin in the bovine pineal gland.[26,38] This suggests yet another mechanism by which endophyte-infected grasses may affect livestock toxicity.

The tremorgenic, neurotoxic mechanism associated with the lolitrems in ryegrass staggers is currently inconclusive. Several reviews have compared the clinical signs of ryegrass staggers with other similar tremorgenic diseases in animals, caused by C. paspali, Penicillium, and Aspergillus spp.[29,30,65,66] Selective tremorgenic compounds produced by these fungi, i.e., paspalinine, aflatrem, veruculogen, and paxilline, which are structurally similar to the lolitrems, have been studied. Gant et al.[47] reported that these neurotoxins, including paxilline, inhibited GABA-induced Cl$^-$ influx and t-butylcyclophosphorothionate binding in rat brain membranes. These authors suggested that the tremorgenic mycotoxins inhibit GABA receptor function, by binding close to the receptor Cl$^-$ channel, suggesting that the structurally similar lolitrems also may alter GABA receptor function. Additionally, Selala et al.[114] have reported an *in vitro* cholinergic effect with these tremorgens, on the isolated guinea pig ileum.

There is little doubt that the combined effects of the classes of ergot alkaloids (clavines, simple lysergic acid amides, and ergopeptine), lolines, lolitrems, and possibly peramine (along with the other miscellaneous alkaloids) act with some degree of synergism in both animal and insect toxicity.[127] Individually, these compounds also may represent an initial insult to the biological system, i.e., at subclinical concentrations over time, which predisposes animals, or makes them more susceptible, to other "less toxic" compounds.

The ability of the endophytes to produce either one or both of the ergopeptine and lolitrem alkaloids may help explain the varying syndromes in animals grazed on infected grasses and the difficulty in accessing animal toxicities. Additionally, the genetic capability of producing either the ergopeptine and/or the lolitrem alkaloids is possible with grass endophytes. Therefore, the genetic manipulation of endophytes, in order to produce pasture grasses for grazing livestock free of one specific toxin, should be observed with extreme discipline.

Care should be taken when supplementing both livestock and ruminant wildlife with winter hay infected with the endophytes. Replanting feed strips (for wildlife) with endophyte-infected grasses may have adverse effects on wildlife reproduction, growth, and maturation, similar to those in livestock.[96] Therefore, controlling wildlife pests (i.e., small mammals, deer, etc.) with endophyte-infected grasses should be approached with caution. Before these practices are fully endorsed, controlled experiments should be conducted to determine the long-term manifestations of such practices on wildlife populations. The effects both native and introduced endophyte-infected grasses may have on both ruminant and nonruminant wildlife populations are currently unknown. Moreover, extreme care should be taken not to release endophyte–grass symbionts that are extremely aggressive in growth, drought tolerance, and insect resistance and that also are toxic to livestock and wildlife.

The effects of ergot poisoning in humans, e.g., gangrene of the peripheral extremities, convulsions, spontaneous abortions, and death, have been recognized since antiquity.[18,22] As an ancillary effect, certain ergot alkaloids and their semisynthetic analogues have therapeutic applications in the treatment of migraines, other vascular headaches, and senile cerebral insufficiency.[18,115] The uterotonic and vasoactive effects of ergonovine and ergotamine have been used to stimulate the gravid uterus, induce labor, and inhibit postpartum bleeding and prevent uterine atony after childbirth. The dopaminergic activities of these alkaloids show potential in the treatment of parkinsonism, acromegaly, amenorrhea, prolactin-dependent galactorrhea, breast cancer, and possibly prostate cancer.[18,115] The hallucinogenic properties of the semisynthetic lysergic acid diethylamide (LSD) are well documented.[18,115]

With regards to human health, the accumulation of ergot alkaloids or the neurotropic lolitrems in animals grazed on fungal infected grasses have not been investigated. Long-term ingestion of these compounds from contaminated meat, milk, and/or dairy products could pose a significant public health concern.[120] Given the known effects of the ergot alkaloids, along with their embryotoxic activity and effects on the gravid uterus,[18] pregnant women may be at greater risk than the general population. In this regard, important interactions between compounds may result in increased toxicity. (For further discussions in this area see Thompson and Porter.[120])

B. INSECT TOXICITY

Several reviews and reports have been published on the endophyte–grass associations, insect activity, and the compounds responsible,[27,32,39,57,76,105,106,109,112,113,127] and more detailed coverage will be presented in other chapters. Endophyte–grass associations that produce the largest number of toxic compounds will have the greatest spectrum of biological activity[112] and will have the most effective genetic capability at diverse survival mechanisms. Currently, peramine and the lolines appear to have a more direct effect on insect activity. However, a combination of certain alkaloids also appear to have synergistic activity.[39,127] Whether any correlation exists between the production of the insect deterrence peramine and other similar pyrrolopyrazines (or keto- and diketopiperazines) awaits future investigations.

Although pyrrolizidine alkaloids are usually avoided by insect herbivores, a number of insects have evolved adaptations, not only to cope with these compounds, i.e., metabolize them, but also to utilize them for their own protection. Several insects can store pyrrolizidine alkaloids obtained from consuming the plants, which are then used for protection from predators.[52] In addition to the use of the pyrrolizidines for defense, males of various Lepidoptera synthesize sex pheromones from these plant-derived alkaloids.[52] Furthermore, some insects seem to be drawn to certain endophyte-infected grasses, or endophyte-infected grasses attract certain insects.

C. PLANT GROWTH REGULATION: AUXIN AND ALLEOCHEMICAL PRODUCTION

Plant auxins regulate or modify plant growth, while alleochemicals are toxic substances released by plants, which cause repression or destruction of other nearby plants. Auxin and alleochemical production seems to reside within the endophyte–grass associations. Indole acetic acid and derivatives are normal metabolites of the uninfected grasses and also are produced by *Acremonium* and *Balansia* species.[35,85] Plants infected by several species of *Balansia* are often stunted, faciated with deformation of the flag leaf and inflorescence.[36] The identification of epimeric indoleglycerols from cultures of *B. epichloë*[93] may suggest that these compounds are responsible for mechanisms for controlling altered growth of the grass. Biosynthetic interrelationships exist between the simple indoleglycerols, tryptophan, indole auxins, and ergot alkaloids, for these symbionts.[6,12,85] Similar to some root mycorrhizal symbioses,[67] deviations in the host–plant carbohydrate pools also may result from auxin production by the endophyte(s).[100] Studies with loline and certain synthetic analogues have shown that N-formylloline has significant phytotoxicity to annual ryegrass (*L. multiflorum*),[80] which implicates the lolines with the mechanisms to improve competition with other grass species. The quantities of loline alkaloids found in *A. coenophialum*-infected fescues and poas[112] suggest this as a rationale for these compounds. Therefore, the potential for natural plant growth regulators, insecticides, and herbicides all resides with symbiotic grasses.

V. CONCLUSIONS

The endophyte-infected grass has a symbiosis that is mammalian, insect, and nematode resistant; it is also drought tolerant and has improved competition and persistence. Thus, these benefits preclude the immediate eradication of the fungi to prevent animal toxicities. Viable options to animal toxicities are treatment with compounds that inhibit the activities of the toxic alkaloids, development of vaccines to immunize animals against toxicity, and selective animal breeding for those less susceptible to the toxic alkaloids. Another option is to genetically manipulate the grass host or fungus to produce endophyte-infected grass cultivars with stress resistance and insect deterrences, without affecting livestock and wildlife health. Natural product chemistry, e.g., large-scale fermentation, extraction, isolation, identification, and production, along with pharmacologic and toxicologic testing and screening, should provide unique sources of natural plant growth regulators, insecticides, herbicides, and medicinal agents, for both livestock and humans. For over 200 years, *Claviceps* and research on the ergot alkaloids have been described as a virtual endless treasure trove of medicinal agents.[18,22] The next 200 years will no doubt involve research dedicated to the unlimited treasure troves related to these endophyte–grass associations.

REFERENCES

1. **Acklin, W., F. Weibel, and D. Arigoni,** Zur biosynthese von paspalin und verwandten metaboliten aus *Claviceps paspali, Chimia,* 31:63, 1977.
2. **Agurell, S.,** Biosynthetic studies of ergot alkaloids and related indoles, *Acta Pharm. Suecica,* 3:71–100, 1966.
3. **Antonio, J. and T. F. Molinski,** Screening of marine invertebrates for the presence of ergosterol-sensitive antifungal compounds, *J. Nat. Prod.,* 56:54–61, 1993.
4. **Arachevaleta, M., C. W. Bacon, C. S. Hoveland, and D. E. Radcliffe,** Effects of tall fescue on plant response to environmental stress, *Agron. J.,* 81:83–90, 1989.
5. **Arachavaleta, M., C. W. Bacon, R. D. Plattner, C. S. Hoveland, and D. E. Radcliffe,** Accumulation of ergopeptide alkaloids in symbiotic tall fescue grown under deficits of soil water and nitrogen fertilizer, *Appl. Environ. Microbiol.,* 58:857–861, 1992.
6. **Bacon, C. W.,** A chemically defined medium for the growth and synthesis of ergot alkaloids by species of *Balansia, Mycologia,* 77:418–423, 1985.
7. **Bacon, C. W.,** Procedures for isolating the endophyte from tall fescue and screening isolates for ergot alkaloids, *Appl. Environ. Microbiol.,* 54:2615–2618, 1988.
8. **Bacon, C. W.,** Grass-endophyte research, 1977–present, an ecological miasma clarified through history, In: D. E. Hume, G. M. C. Latch, and H. S. Easton, Eds., *Proc. 2nd Int. Symp.* Acremonium/ *Grass Interactions,* AgResearch, Grasslands Research Centre, Palmerston North, New Zealand, 1993.
9. **Bacon, C. W. and J. De Battista,** Endophytic fungi of grasses, In: D. K. Arora, B. Rai, K. G. Mukerji, and G. R. Knudsen, Eds., *Handbook of Applied Mycology. Soils and Plants,* Vol. 1, Marcel Dekker, New York, 1991, 231.
10. **Bacon, C. W., P. C. Lyons, J. K. Porter, and J. D. Robbins,** Ergot toxicities from endophyte-infected grasses: a review, *Agron. J.,* 78:106–116, 1986.
11. **Bacon, C. W., J. K. Porter, and J. D. Robbins,** Toxicity and occurrence of *Balansia* on grasses from toxic fescue pastures, *Appl. Microbiol.,* 29:553–556, 1975.
12. **Bacon, C. W., J. K. Porter, and J. D. Robbins,** Laboratory production of ergot alkaloids by species of *Balansia, J. Gen. Microbiol.,* 113:119–126, 1979.
13. **Bacon, C. W., J. K. Porter, and J. D. Robbins,** Ergot alkaloid biosynthesis by isolates of *Balansia epichloe* and *Balansia henningsiana, Can. J. Bot.,* 59:2534–2538, 1981.
14. **Bacon C. W., J. K. Porter, J. D. Robbins, and E. S. Luttrell,** *Epichloe typhina* from toxic tall fescue, *Appl. Environ. Microbiol.,* 34:576–581, 1977.
15. **Bacon, C. W. and M. R. Siegel,** Endophyte parasitism of tall fescue, *J. Prod. Agric.,* 1:45–55, 1988.
16. **Belesky, D. P., J. D. Robbins, J. A. Stuedemann, S. R. Wilkinson, and O. J. Devine,** Fungal endophyte infection-loline derivative alkaloid concentration of grazed tall fescue, *Agron. J.,* 79:217–220, 1987.
17. **Belesky, D. P., J. A. Stuedemann, R. D. Plattner, and S. R. Wilkinson,** Ergopeptine alkaloids in grazed tall fescue, *Agron. J.,* 80:209–212, 1988.
18. **Berde B. and H. O. Schild,** *Ergot Alkaloids and Related Compounds. Handbook Experimental Pharmacol.,* Vol 49, Springer-Verlag, New York, 1978.
19. **Bianchi, M. L., N. C. Perellino, B. Gioia, and A. Minghetti,** Production by *Claviceps purpurea* of two new peptide ergot alkaloids belonging to a new series containing *alpha*-aminobutyric acid, *J. Nat Prod.,* 45:191–196, 1982.
20. **Bottomley, W. and T. A. Geissman,** Pyrrolizidine alkaloids. the biosynthesis of retronecine, *Phytochemistry,* 3:357–360, 1964.
21. **Bond, J., J. B. Powell, D. J. Undersander, P. W. Moe, H. F. Tyrrell, and R. R. Oltjen,** Forage composition and growth and physiological characteristics of cattle grazing several varieties of tall fescue during the summer, *J. Anim. Sci.,* 59:584–593, 1984.
22. **Bove, F. J.,** *The Story of Ergot,* S. Karger, New York, 1970.
23. **Bush, L. P. and P. B. Burrus,** Tall fescue forage and quality and agronomic performance as affected by the endophyte, *J. Prod. Agric.,* 1:55–60, 1988.
24. **Bush, L. P., P. L. Cornelius, R. C. Buckner, D. R. Varney, R. A. Chapman, P. B. Burrus, II, C. W. Kennedy, T. A. Jones, and M. J. Saunders,** Association of N-acetylloline deravitives and N-formylloline derivative with *Epichloe typhina* in tall fescue, *Crop. Sci.,* 22:941–943, 1982.

25. **Bush, L. P., F. F. Fannin, M. R. Siegel, D. L. Dahlman, and H. R. Burton,** Chemistry of compounds associated with endophyte-grass interactions: saturated pyrrolizidine alkaloids, In: R. Joost and S. Quisenberry, Eds., Acremonium/*Grass Interactions*, Vol. 44, *Agriculture, Ecosystems, and Environment,* Elsevier, Amsterdam, 1993, 81.

26. **Chan, A. and M. Ebadi,** The kinetics of norepinephrine induced stimulation of serotonin N-acetyltransferase in bovine pineal gland, *Neuroendocrinology,* 31:244–251, 1980.

27. **Clay, K. and G. P. Cheplick,** Effect of ergot alkaloids from fungal endophyte-infected grasses on fall armyworm (*Spodoptera frugiperda*), *J. Chem. Ecol.,* 15:169–182, 1989.

28. **Cole R. J. and R. H. Cox,** *Handbook of Toxic Fungal Metabolites,* Academic Press, New York, 1981.

29. **Cole, R. J. and J. W. Dorner,** Role of fungal tremorgens in animal disease, In: P. S. Steyn and R. Vleggaar, Eds., *Mycotoxins and Phycotoxins,* Elsevier, Amsterdam, 1986, 501.

30. **Cole, R. J., J. W. Dorner, J. A. Lansden, R. H. Cox, C. Pape, B. Cunfer, S. S. Nicholson, and D. M. Bedell,** Paspalum staggers: isolation and identification of tremorgenic metabolites from sclerotia of *Claviceps paspali, J. Agric. Food Chem.,* 25:1197–1203, 1977.

31. **Cornell, C. N., J. V. Lueker, G. B. Garner, and J. L. Ellis,** Establishing ergovaline levels for fescue toxicosis, with and without endoparasites, under controlled climatic conditions, In: S. S. Quisenberry and R. E. Joost, Eds., *Proc. Int. Symp.* Acremonium/*Grass Interactions*, Louisiana Agriculture Experiment Station, Baton Rouge, 1990, 75.

32. **Dahlman, D. L., H. Eichenseer, and M. R. Siegel,** Chemical perspectives on endophyte-grass interactions and their implications to insect herbivory, In: P. Barbosa, V. Krischik, and C. Jones, Eds., *Microbial Mediation of Plant-Herbivore Interactions*, John Wiley & Sons, New York, 1991, 227.

33. **Davis, C. B., B. J. Camp, and J. C. Read,** The vasoactive potential of halostachine, an alkaloid of tall fescue (*Festuca arundinacea*, Schreb.) in cattle, *J. Vet. Anim. Toxicol.,* 25:408–411, 1983.

34. **Davis, N. D., R. J. Cole, J. W. Dorner, J. D. Weete, P. A. Backman, E. M. Clark, C. C. King, S. P. Schmidt, and U. L. Diener,** Steroid metabolites of *Acremonium coenophialum*, an endophyte of tall fescue, *J. Agric. Food Chem.,* 34:105–108, 1986.

35. **De Battista, J. P., C. W. Bacon, R. F. Severson, R. D. Plattner, and J. H. Bouton,** Indole acetic acid production by the fungal endophyte of tall fescue, *Agron. J.,* 82:878–880, 1990.

36. **Diehl, W. W.,** *Balansia* and the Balansiae in America, U.S. Department of Agriculture, Monograph No. 4, U.S. Gov. Printing Office, Washington D.C., 1950.

37. **Dyer, D. C.,** Evidence that ergovaline acts on serotonergic receptors, *Life Sci.,* 53:223–228, 1993.

38. **Ebadi, M., A. Chan, M. Itoh, H. M. Hammad, S. Swanson, and P. Govitrapong,** The role of *gamma*-aminobutyric acid in the regulation of melatonin synthesis in bovine pineal gland, In: I. Hanin, Ed., *Dynamics of Neurotransmitter Function*, Raven Press, New York, 1984, 177.

39. **Eichenseer, H., D. D. Dahlman, and L. P. Bush,** Influence of endophyte infection, plant age and harvest interval on *Rhopalosiphum padi* survival and its relation to quantity of N-formyl and N-acetyl loline in tall fescue, *Entomol. Exp. Appl.,* 60:29–38, 1991.

40. **Fletcher, L. R.,** Heat stress in lambs grazing ryegrass with different endophytes, In: D. E. Hume, G. C. M. Latch, and H. S. Easton, Eds., *Proc. 2nd Int. Symp.* Acremonium/*Grass Interactions*, AgResearch, Grasslands Research Centre, Palmerston North, New Zealand, 1993, 114.

41. **Fletcher, L. R., I. Garthwaite, and N. R. Towers,** Ryegrass staggers in the absence of lolitrem B, In: D. E. Hume, G. C. M. Latch, and H. S. Easton, Eds., *Proc. 2nd Int. Symp.* Acremonium/*Grass Interactions*, AgResearch, Grasslands Research Centre, Palmerston North, New Zealand, 1993, 119.

42. **Fletcher, L. R. and I. C. Harvey,** An association of a *Lolium* endophyte with ryegrass staggers, *N.Z. Vet. J.,* 29:185–186, 1981.

43. **Flieger, M., P. Sedmera, J. Novak, L. Cvak, J. Zapletal, and J. Stuchlik,** Degradation products of ergot alkaloids, *J. Nat. Prod.,* 54:390–395, 1991.

44. **Flieger, M., P. Sedmera, J. Vokoun, Z. Rehacek, J. Stuchlik, Z. Malinka, L. Cvak, and P. Harazim,** New alkaloids from saprophytic culture of *Claviceps purpurea, J. Nat. Prod.,* 47:970–976, 1984.

45. **Floss, H. G.,** Biosynthesis of ergot alkaloids and related compounds, *Tetrahedron,* 32:873–912, 1976.

46. **Gallagher, R. T., A. D. Hawkes, P. S. Steyn, and R. Vleggaar,** Tremorgenic neurotoxins from perennial ryegrass causing ryegrass staggers disorders of livestock: structure and elucidation of lolitrem B, *J. Chem. Soc. Chem. Commun.,* 1984:614–616, 1984.

47. **Gant, D. B., R. J. Cole, J. J. Valdes, M. E. Eldefrawi, and A. T. Eldefrawi,** Action of tremorgenic mycotoxins on GABA receptor, *Life Sci.,* 41:2207–2214, 1987.

48. **Garner, G. B., D. E. Spiers, C. N. Cornell, and G. E. Rottinghaus,** Bovine thermal response to cold stress with and without *Acremonium coenophialum* infested tall fescue seed in the ration measured in a climate controlled chamber, In: D. E. Hume, G. C. M. Latch, and H. S. Easton, Eds., *Proc. 2nd Int. Symp.* Acremonium/*Grass Interactions*, AgResearch, Grasslands Research Centre, Palmerston North, New Zealand, 1993, 128.

49. **Garner, G. B., G. E. Rottinghaus, C. N. Cornell, and H. Testereci,** Chemistry of compounds associated with endophyte/grass interaction: ergovaline and ergopeptine related alkaloids, In: R. Joost and S. Quisenberry, Eds., Acremonium/*Grass Interactions*, Vol. 44, *Agriculture, Ecosystems, and Environment*, Elsevier, Amsterdam, 1993, 65.

50. **Groger, D., D. Erge, and H. G. Floss,** Uber die herfkunft der seitkette im d-lysergsaure-methlcarbinolamide, *Z. Naturforsch. [B],* 23B:177–180, 1968.

51. **Groger, D. and V. E. Tyler,** Alkaloid production by *Claviceps paspali* in submerged culture, *Lloydia,* 26:174–191, 1963.

52. **Hartman, T., A. Biller, L. Witte, L. Ernst, and M. Boppre,** Transformation of plant pyrrolizidine alkaloids into novel insect alkaloids by the Arctiid moths (Lepidoptera), *Biochem. Systematics Ecol.,* 18:549–554, 1990.

53. **Hill, N. S., W. A. Parrott, and D. D. Pope,** Ergopeptine alkaloid production by endophytes in a common tall fescue genotype, *Crop Sci.,* 31:1545–1547, 1991.

54. **Hoveland, C. S.,** Importance and economic significance of the *Acremonium* endophytes to performance of animals and grass plant, In: R. Joost and S. Quisenberry, Eds., Acremonium/*Grass Interactions*, Vol. 44, *Agriculture, Ecosystems, and Environment,* Elsevier, Amsterdam, 1993, 3.

55. **Hughes, C. A., R. Letcher, and F. L. Warren,** The senecio alkaloids. XVI. The biosynthesis of the necine bases from carbon-14 precursors, *J. Chem. Soc. (London),* 1964:4974–4978, 1964.

56. **Hutter, R. and P. Niederberger,** Tryptophan biosynthetic genes in eukaryotic microorganisms, *Ann. Rev. Microbiol.,* 40:55–77, 1986.

57. **Johnson, M. C., D. L. Dahlman, M. R. Siegel, L. P. Bush, G. C. M. Latch, D. A. Potter, and D. R. Varney,** Insect feeding deterrents in endophyte-infected tall fescue, *Appl. Environ. Microbiol.,* 49:568–571, 1985.

58. **Kawaguchi, M., S. Fujioka, A. Saurai, Y. T. Yamaki, and K. Syono,** Presence of a pathway for the biosynthesis of auxin via indole-3-acetamide in *Trifoliata* orange, *Plant Cell Physiol.,* 34:121–128, 1993.

59. **Lamberts, S. W. J. and R. M. Macleod,** Regulation of prolactin secretion at the level of the lactotroph, *Physiol. Rev.,* 70:279–318, 1990.

60. **Latch, G. C. M.,** Physiological interactions of endophytic fungi and their host. Biotic stress tolerance imparted to grasses by endophytes, In: R. Joost and S. Quisenberry, Eds., Acremonium/*Grass Interactions*, Vol. 44, *Agriculture, Ecosystems, and Environment,* Elsevier, Amsterdam, 1993, 143.

61. **Latch, G. C. M., M. J. Christensen, and G. J. Samuels,** Five endophytes of *Lolium* and *Festuca* in New Zealand, *Mycotaxon,* 20:535–550, 1984.

62. **Lipham, L. B., F. N. Thompson, Jr., J. A. Stuedemann, and L. Sartin,** Effects of metoclopramide on steers grazing endophyte-infected fescue, *J. Anim. Sci.,* 67:1090–1097, 1989.

63. **Lyons, P. C., J. J. Evans, and C. W. Bacon,** Effects of fungal endophyte *Acremonium coenophialum* on nitrogen accumulation and metabolism in tall fescue, *Plant. Physiol.,* 92:726–732, 1990.

64. **Lyons, P. C., R. D. Plattner, and C. W. Bacon,** Occurrence of peptide and clavine ergot alkaloids in in tall fescue grass, *Science,* 232:487–489, 1986.

65. **Mantle, P. G.,** Ergotism in cattle, In: T. D. Wyllie and L. G. Morehouse, Eds., *Mycotoxic Fungi, Mycotoxins, Mycotoxicosis*, Vol. 2, Marcel Dekker, New York, 1978, 145.

66. **Mantle, P. G., P. H. Mortimer, and E. P. White,** Mycotoxic tremorgens of *Claviceps paspali* and *Penicillium cyclopium*: a comparative study of effects on sheep and cattle in relation to natural staggers syndromes, *Res. Vet. Sci.,* 24:49–56, 1977.

67. **Meyer, F. H.,** Physiology of mycorrhiza, *Ann. Rev. Plant Physiol.,* 25:567–586, 1974.

68. **Miles, C. O., S. C. Munday, A. L. Wilkins, R. M. Ede, A. D. Hawkes, P. P. Embling, and N. R. Towers,** Large scale isolation of lolitrem B, structure determination of some minor lolitrems, and tremorgenic activities of lolitrem B and paxilline in sheep, In: D. E. Hume, G. C. M. Latch, and H. S. Easton, Eds., *Proc. 2nd Int. Symp.* Acremonium/*Grass Interactions*, AgResearch, Grasslands Research Centre, Palmerston North, New Zealand, 1993, 85.

69. **Miles, C. O., A. L. Wilkins, R. T. Gallagher, A. D. Hawkes, S. C. Munday, and N. Towers,** Synthesis and tremorgenicity of paxitrols and lolitriol: possible biosynthetic precursors of lolitrem B, *J. Agric. Food Chem.*, 40:234–238, 1992.

70. **Morgan-Jones, G. and W. Gams,** Notes on hypomycetes. XLI. An endophyte of *Festuca arundinacea* and the anamorph of *Epichloe typhina*, new taxa in one of two new sections of *Acremonium*, *Mycotaxon*, 15:311–318, 1982.

71. **Moubarak, A. S., E. L. Piper, C. P. West, and Z. B. Johnson,** Interaction of purified ergovaline from endophyte-infected tall fescue with synaptosomal ATP*ase* enzyme system, *J. Agric. Food Chem.*, 41:407–409, 1993.

72. **Oliver, J. W., L. K. Abney, J. R. Strickland, and R. D. Linnabary,** Vasoconstriction in bovine vasculature induced by the tall fescue alkaloid lysergamide, *J. Anim. Sci.*, 71:2708, 1993.

73. **Oliver, J. W., R. G. Powell, L. H. Abney, R. D. Linnabary, and R. J. Petroski,** N-acetyllolline-induced vasoconstriction of the lateral saphenous vein (cranial branch) of cattle, In: S. S. Quisenberry and R. E. Joost, Eds., *Proc. Int. Symp.* Acremonium/*Grass Interactions*, Louisiana Agriculture Experiment Station, Baton Rouge, 1990, 239.

74. **Osborn, T. G., S. P. Schmidt, D. N. Marple, C. H. Rahe, and J. R. Steenstra,** Effects of consuming fungus-infected and fungus-free tall fescue and ergotamine tartrate on selective physiological variables of cattle in environmentally controlled conditions, *J. Anim. Sci.*, 1992:2501–2509, 1992.

75. **Passi, S., C. Dehuca, A. A. Fabbri, and C. Fanelli,** Ergosterol oxidation: signal either for fungal growth and aflatoxin production in *Aspergillus parasiticus*, In: Proc. VIII Int. IUPAC Symp. Mycotoxins and Phycotoxins, National Advisors for Culture and Arts (INAH), Mexico City, 1992, 50.

76. **Patterson, C. G., D. A. Potter, and F. F. Fannin,** Feeding deterrency of alkaloids from endophyte infected grasses to Japanese beetle grubs, *Entomol. Exp. Appl.*, 61:285–289, 1991.

77. **Penn, J., I. Garthwaite, M. J. Christensen, C. M. Johnson, and N. R. Towers,** The importance of paxilline in screening for potentially tremorgenic *Acremonium* isolates, In: D. E. Hume, G. C. M. Latch, and H. S. Easton, Eds., *Proc. 2nd Int. Symp.* Acremonium/*Grass Interactions*, AgResearch, Grasslands Research Centre, Palmerston North, New Zealand, 1993, 88.

78. **Perellino, N. C., J. Malyszko, M. Ballabio, B. Gioia, and A. Minghetti,** Directed biosynthesis of unnatural ergot peptide alkaloids, *J. Nat. Prod.*, 55:424–427, 1992.

79. **Perellino, N. C., J. Malyszko, M. Ballabio, M. Gioia, and A. Minghetti,** Identification of ergobine, a new natural peptide ergot alkaloid, *J. Nat. Prod.*, 56:489–493, 1993.

80. **Petroski, R. J., D. L. Dornbos, Jr., and R. G. Powell,** Germination and growth inhibition of annual ryegrass (*Lolium multiflorum* L.) and alfalfa (*Medicago sativa* L.) by loline alkaloids and synthetic N-acetylloline derivatives, *J. Agric. Food Chem.*, 38:1716–1718, 1990.

81. **Petroski, R. J. and R. J. Powell,** Preparative separation of complex alkaloid mixture by high-speed countercurrent chromatography, In: P. A. Hedin, Ed., *Naturally Occurring Pest Bioregulators*, Am. Chem. Soc. Sym. Series 449, Washington, D.C., 1991, 426.

82. **Petroski, R. J., R. G. Powell, and K. Clay,** Alkaloids of *Stipa robusta* (sleepy grass) infected with an *Acremonium* endophyte, *Nat. Toxins*, 1:84–88, 1992.

83. **Plattner, R. D., S. G. Yates, and J. K. Porter,** Quadrupole mass spectrometry/mass spectrometry of the ergot cyclol alkaloids, *J. Agric. Food Chem.*, 31:785–789, 1983.

84. **Porter, J. K. and D. Betowski,** Chemical ionization mass spectrometry of the ergot cyclol alkaloids, *J. Agric. Food Chem.*, 29:650–653, 1981.

85. **Porter, J. K., C. W. Bacon, H. G. Cutler, R. F. Arrendale, and J. D. Robbins,** *In vitro* auxin production by *Balansia epichloe*, *Phytochemistry*, 24:1429–1431, 1985.

86. **Porter, J. K., C. W. Bacon, R. D. Plattner, and R. F. Arrendale,** Ergot peptide alkaloid spectra of *Claviceps*-infected tall fescue, wheat and barley, *J. Agric. Food Chem.*, 35:359–361, 1987.

87. **Porter, J. K., C. W. Bacon, and J. D. Robbins,** Major alkaloids of a *Claviceps* isolated from toxic bermuda grass, *J. Agric. Food Chem.*, 22:838–841, 1974.

88. **Porter, J. K., C. W. Bacon, and J. D. Robbins,** Clavicipitaceae: *Claviceps* related fungi and their production of ergot alkaloids, *Lloydia*, 41:654–655, 1978.

89. **Porter, J. K., C. W. Bacon, and J. D. Robbins,** Ergosine, ergosinine and chanoclavine I from *Epichloe typhina*, *J. Agric. Food Chem.*, 27:595–598, 1979.

90. **Porter, J. K., C. W. Bacon, and J. D. Robbins,** Lysergic acid amide derivatives from *Balansia epichloe* and *Balansia claviceps* (Clavicipitaceae), *J. Nat. Prod.*, 42:309–314, 1979.

91. **Porter, J. K., C. W. Bacon, J. D. Robbins, and D. Betowski,** Ergot alkaloid identification in Clavicipitaceae systemic fungi of pasture grasses, *J. Agric. Food Chem.*, 29:653–657, 1981.

92. **Porter, J. K., C. W. Bacon, J. D. Robbins, and H. C. Higman,** A field indicator in plants associated with ergot-type toxicities in cattle, *J. Agric. Food Chem.,* 23:771–775, 1975.

93. **Porter, J. K., J. D. Robbins, C. W. Bacon, D. S. Himmelsbach, and A. F. Haeberer,** Determination of epimeric 1-(3-indolyl)-propane-1,2,3-triol isolated from *Balansia epichloe, Lloydia,* 41:43–49, 1978.

94. **Porter, J. K., C. W. Bacon, J. D. Robbins, D. S. Himmelsbach, and H. C. Higman,** Indole alkaloids from *Balansia epichole* (Weese), *J. Agric. Food Chem.,* 25:88–93, 1977.

95. **Porter, J. K., J. A. Stuedemann, F. N. Thompson, Jr., B. A. Buchanan, and H. A. Tucker,** Melatonin and pineal neurochemicals in steers grazed on endophyte-infected tall fescue: effects of metoclopramide, *J. Anim. Sci.,* 71:1526–1531, 1993.

96. **Porter, J. K. and F. N. Thompson, Jr.,** Effects of fescue toxicosis on reproduction in livestock, *J. Anim. Sci.,* 70:1594–1603, 1992.

97. **Powell, R. G. and R. J. Petroski,** The loline group of pyrrolizidine alkaloids, In: S. W. Pelletier, Ed., *The Alkaloids: Chemical and Biological Perspectives,* Vol. 8, Springer-Verlag, New York, 1992, 320.

98. **Powell, R. G., R. D. Plattner, S. G. Yates, K. Clay, and A. Leuchtmann,** Ergobalansine, a new ergot-type peptide alkaloid isolated from *Cenchrus echinatus* (sandbur grass) infected with *Balansia obtecta,* and produced in liquid cultures of *B. obtecta* and *Balansia cyperi, J. Nat. Prod.,* 53:1272–1279, 1990.

99. **Rhodes, M. T., J. A. Paterson, M. S. Kerley, H. E. Garner, and M. H. Laughlin,** Reduced blood flow to peripheral and core body tissues in sheep and cattle induced by endophyte-infected tall fescue, *J. Anim. Sci.,* 69:2033–2043, 1991.

100. **Richardson, M. D., G. W. Chapman, Jr., C. S. Hoveland, and C. W. Bacon,** Sugar alcohols in endophyte-infected tall fescue under drought, *Crop Sci.,* 32:1060–1061, 1992.

101. **Robbers, J. E.,** The fermentative production of ergot alkaloids, In: A. Mizrahi and A. L. van Wezel, Eds., *Advances in Biotechnological Processes,* Vol. 3, A. R. Liss, New York, 1984, 197.

102. **Robbins, J. D., J. K. Porter, and C. W. Bacon,** Occurrence and clinical manifestation of ergot and fescue toxicoses, In: J. L. Richards and J. R. Thurston, Eds., *Diagnosis of Mycotoxicoses,* Martinus Nijhoff, Dordrecht, Netherlands, 1986, 61.

103. **Robbins, J. D., J. G. Sweeny, S. R. Wilkinson, and D. Burdick,** Volatile alkaloids of Kentucky 31 tall fescue seed (*Festuca arundinancea* Screb.), *J. Agric. Food Chem.,* 20:1040–1043, 1972.

104. **Rottinghaus, G. E., G. B. Garner, C. N. Cornell, and J. L. Ellis,** An HPLC method for quantitating ergovaline in endophyte-infected tall fescue: seasonal variation of ergovaline levels in stems with leaf sheaths, leaf blades and seed heads, *J. Agric. Food Chem.,* 39:112–115, 1991.

105. **Rowan, D. D. and G. L. Gaynor,** Isolation of feeding deterrants against Argentine stem weevil from ryegrass infected with the endophyte *Acremonium loliae, J. Chem. Ecol.,* 12:647–658, 1986.

106. **Rowan, D. D., M. B. Hunt, and D. L. Gaynor,** Peramine, a novel insect feeding deterrent from rye grass infected with *Acremonium loliae, J. Chem. Soc. Chem. Commun.,* 1986:935–936, 1986.

107. **Rowan, D. D. and G. J. Shaw,** Dectection of ergopeptine alkaloids in endophyte-infected perennial ryegrass by tandem mass spectrometry, *N.Z. Vet. J.,* 35:197–198, 1987.

108. **Sampson, K.,** The systemic infection of grasses by *Epichloe typhina* (Pers.) Tul., *Trans. Br. Mycol. Soc.,* 18:30–47, 1933.

109. **Saunders, J. A., N. R. O'Neill, and J. T. Romeo,** Alkaloid chemistry and feeding specificity of insect herbivores, In: S. W. Pelletier, Ed., *Alkaloids: Chemical and Biological Perspectives,* Vol. 8, Springer-Verlag, New York, 1992, 151.

110. **Shelby, R. A.,** Improved HPLC method for the detection of ergot alkaloids in endophyte-infected tall fescue, In: *Proc. Tall Fescue Toxicoses Workshop,* Southern Extension and Research Activity Information Exchange-8, Memphis, 1992, 3.

111. **Shelby, R. A. and L. W. Dalrymple,** Incidence and distribution of the tall fescue endophyte in the United States, *Plant Dis.,* 71:783–786, 1987.

112. **Siegel, M. R., G. C. M. Latch, L. B. Bush, F. F. Fannin, D. D. Rowan, B. A. Tapper, C. W. Bacon, and M. C. Johnson,** Fungal endophyte-infected grasses: alkaloid accumulation and aphid response, *J. Chem. Ecol.,* 16:3301–3315, 1990.

113. **Siegel M. R. and C. L. Schardl,** Fungal endophytes of grasses: detrimental and beneficial associations, In: J. H. Andrews and S. S. Hirano, Eds., *Microbial Ecology of Leaves,* Springer-Verlag, New York, 1990, 198.

114. **Selala, M. I., G. M. Laekeman, B. Loenders, A. Musuku, A. G. Herman, and P. Schepens,** In vitro effects of tremorgenic mycotoxins, *J. Nat. Prod.,* 54:207–212, 1991.

115. **Socic, H. and V. Gaberc-Porekar,** Biosynthesis and physiology of ergot alkaloids, In: D. K. Arora, R. P. Elander, and K. G. Mukerji, Eds., *Handbook of Applied Mycology, Fungal Biotechnology,* Vol. 4, Marcel Dekker, New York, 1992, 475.

116. **Spiers, D. E.,** Fescue toxicosis: the environmental connection, *J. Anim. Sci.,* 71(Suppl. 1):116 (Abstr. 106), 1993.

117. **Strickland, J. D., D. L. Cross, T. C. Jenkins, R. J. Petroski, and R. G. Powell,** The effect of alkaloid and seed extracts of endophyte-infected tall fescue on prolactin and secretion in an *in vitro* rat pituitary perfusion system, *J. Anim. Sci.,* 70:2779–2786, 1992.

118. **Stuedemann J. A. and C. S. Hoveland,** Fescue toxicity: history and impact on animal agriculture, *J. Prod. Agric.,* 1:39–44, 1988.

119. **Thompson, F. N., N. S. Hill, D. L. Dawe, and J. A. Stuedemann,** In: D. E. Hume, G. C. M. Latch, and H. S. Easton, Eds., *Proc. 2nd Int. Symp.* Acremonium/*Grass Interactions*, AgResearch, Grasslands Research Centre, Palmerston North, New Zealand, 1993, 135.

120. **Thompson, F. N. and J. K. Porter,** Tall fescue toxicoxes in cattle: could there be a public health problem, *Vet. Human Toxicol.,* 32:51–57, 1991.

121. **Umbarger, H. E.,** Amino acid biosynthesis and its regulation, *Ann. Rev. Biochem.,* 47:533–606, 1978.

122. **Weedon, C. M. and P. G. Mantle,** Paxilline biosynthesis by *Acremonium loliae*; a step toward defining the origin of lolitrems neurotoxins, *Phytochemistry,* 26:969–971, 1987.

123. **White, J. F., Jr.,** Widespread distribution of endophytes in poaceae, *Plant Dis.,* 71:340–342, 1987.

124. **White, A., P. Handler, and E. L. Smith,** *Principles of Biochemistry,* McGraw-Hill, New York, 1964.

125. **Yates, S. G.,** Toxicity of tall fescue: a review, *Econ. Bot.,* 16:295–303, 1962.

126. **Yates, S. G.,** Tall fescue toxins, In: M. Recheigl, Ed., *Handbook of Naturally Occurring Food Toxicants*, CRC Press, Boca Raton, FL, 1983, 249.

127. **Yates, S. G., J. C. Fenster, and R. J. Bartlet,** Assay of tall fescue seed extracts, fractions, and alkaloids using the large milkweed bug, *J. Agric. Food Chem.,* 37:354–357, 1989.

128. **Yates, S. G., R. D. Plattner, and G. B. Garner,** Detection of ergopeptine alkaloids in endophyte-infected, toxic K-31 fescue by mass spectrometry/mass spectrometry, *J. Agric. Food Chem.,* 33:719–722, 1985.

129. **Yates, S. G. and R. G. Powell,** Analysis of ergopeptine alkaloids in endophyte-infected tall fescue, *J. Agric. Food Chem.,* 36:337–340, 1988.

130. **Yates, S. G. and H. L. Tookey,** Festucine, an alkaloid from tall fescue (*Festuca arundinacea,* Schreb.): chemistry of the functional groups, *Aust. J. Chem.,* 18:53–60, 1965.

131. **Zubay, G.,** *Biochemistry*, 2nd ed., Macmillan, New York, 1988.

Chapter 9

Vaccines and Pharmacological Agents to Alleviate Fescue Toxicosis

Frederick N. Thompson and George B. Garner

CONTENTS

I. INTRODUCTION

Tall fescue (*Festuca arundinacea* Schreb) has had a long and colorful history before, during, and after the acceptance of it as a desirable forage species.[14] This grass was separated from meadow fescue (*F. elatior* L.) and made a distinct species in 1771. As defined, tall fescue consisted of plants with very broad leaves and tall robust growth habits. As a distinct species, Europeans considered tall fescue too coarse and unpalatable to be a first-class pasture grass. In the U.S. this grass was officially introduced from Sweden in 1901, and in 1931 E. N. Fergus of the University of Kentucky collected seed from tall fescue from the Suiter Farm in Menife County, KY. Twelve years later the Kentucky Agricultural Experiment Station and Drs. Fergus and R. Buckner released the cultivar "KY-31." Six years later, in 1949, Fescue foot, the crippling disease in cattle, was first described by Cunningham in New Zealand.[8] This disease was described in Missouri in 1952.[14] Thus, immediately after its introduction, tall fescue was observed to be toxic to cattle. These symptoms in cattle all resemble ergotism produced by species of *Claviceps*; however, taxa of this genus were absent, and the disease was considered idiopathic, limited in occurrence, and of little economic significance.

Fescue foot, fescue toxicosis, agalactia, and fat necrosis are four different aspects of a toxicosis observed in cattle grazing endophyte-infected tall fescue (see Garner[14] for precise definitions for other diseases within this syndrome). Of these cattle maladies, fescue toxicity is the most recognized. Signs of fescue toxicity are more prevalent in hot weather than cold weather, whereas fescue foot occurs primarily under cool-to-cold conditions. Both fescue foot and fescue toxicosis are, contrary to the early consideration, extremely economically significant.[39] Fescue toxicosis is associated with a reduction in body weight gains and calving rates.

It took approximately three decades after the initial 1949 report[8] of toxicity of this grass before a fungal endophyte and its ergot alkaloids were associated with this grass[4,21] (see Chapter 8 for appropriate references and compounds isolated from this grass). These findings suggested that tall fescue and other grasses could be rendered nontoxic by the removal of the fungus from the grasses. While axiomatic, this proved not to be the case, as evidenced by the subsequent finding of enhanced fitness characteristics of endophyte-infected tall fescue (for example, see Chapter 5). Thus, a solution to the toxicity aspect of endophyte-infected grasses remains an enigma. Nevertheless, it is the tacit assumption that the toxicity of this grass is due to these ergot alkaloids, primarily ergovaline, and it is on this assumption that experimental protocols have been designed to reduce or prevent fescue toxicity in cattle. This chapter is concerned with two very current and specific approaches designed primarily for ameliorating fescue toxicosis, although similar approaches should apply to ryegrass toxicity and other endophyte-induced toxicities.

II. VACCINES

A. HAPTENS

Vaccination and antibody production may be an appropriate strategy to reduce losses to toxic fescue. Assuming that the ergot alkaloids found in endophyte-infected fescue are causative, then antibodies against these agents should bind the alkaloids and, hopefully, result in a loss of activity. However, a molecule is antigenic only if its size is greater than 1000 Da. Thus, ergot alkaloids are not antigenic, since their molecular size is insufficient. Nevertheless, there are several procedures available that can make a small-molecular-weight substance antigenic (see Robbins[35] and references cited therein). One procedure involves the conjugation of the small molecule, i.e., an ergot, to a molecule larger than 1000 Da. The smaller molecule, in this specific instance an ergot alkaloid, is referred to as the hapten (from the Greek to fasten), and it must be covalently linked to a protein or a polysaccharide. The larger molecule is referred to as a carrier; in addition to the above substances, other large particles, such as bacteriophages, erythrocytes, and synthetic macromolecules, may also be used. Using the hapten process, a number of fungal antigens have been prepared, these include aflatoxin B_1, sterigmatocystin, and ochratoxin A.[35]

The degree of substitution, or the number of hapten molecules per carrier molecule, is important to the immunogenic outcome of this process. Additional factors that may be important for success include the choice of the ergot alkaloid and the construction of the immunogen, which should maintain the integrity of the ergot alkaloid structure as well as the magnitude, duration, and quality of the immune response. The degree to which a hapten binds to the carrier is a chemical function; often times a hapten may form different determinants, resulting in variation in antiserum, which recognizes several compounds within a class. This lack of specificity, while not desired in general, may be advantageous in the case of tall fescue toxicity, because of the numbers of specific ergot alkaloids produced.[30,42]

Haptens-carrier conjugates have been constructed[19] by utilizing ergot alkaloids that basically contain only the ergoline ring, such as ergonovine and lysergol. In order to increase the number of reactive groups that permit covalent binding to the carrier functional groups, the ergot alkaloid is then joined to two coupling agents, glutamate and succinate, using the mixed anhydride method.[3] The resulting ergonovine-glutaric-succinate haptens are then covalently conjugated to human serum albumin, using the carbodiimide reaction.[35]

The hapten-carrier conjugate is purified and administered subcutaneously, intradermally, or intramuscularly. Since intake of low molecular weight alkaloids via grazing will not stimulate the immune system, the delivery of the antigen conjugate in small amounts over time may be necessary to constantly challenge the immune system for continued antibody production. Thus, delivery via injection of microspheres[1] may be the vehicle of choice. The ultimate successful criterion for a hapten-carrier conjugate prepared as a vaccine would be an immunized animal. A successful immunized animal can be judged by reproductive performance and weight gain.

Although production of antiergot alkaloid antibodies has been achieved (Figure 1),[19] prevention of the toxicosis, in this manner, has not been demonstrated. However, the successful use of the hapten-type immunological techniques for preventing toxicoses from ingested material has precedent. Vaccines have been protective against the mycotoxicoses, lupinosis,[33] and trichothecene D-2.[6] Sheep vaccinated against digitoxin conjugated to bovine serum albumin were largely protected against cardiac glycosides or cardiac glycoside-containing plants.[24] Ten of 14 immunized sheep survived challenge with digitoxin. However, in other trials sheep vaccinated against digitoxin were not protected against other cardiac glycosides. Passive immunization of mice, using rabbit antiricin antibodies, protected the mice against ricin toxicity.[20] Passive immunization also is used to protect animals and humans from the externally generated botulism toxin.[25]

In contrast to the immunological successes against mycotoxins, signs associated with zearalenone toxicity intensified in gilts following vaccination against zearalenone.[31] The height of the vaginal epithelium in these gilts was the measured response. Vaccination had no effect on vaginal epithelium height; however, following challenge with zearalenone, the vaginal epithelium was increased in the vaccinated animals compared to controls. In ewes, zearalenone reduced the incidence of estrus, the ovulation rate, and the lambing rate. Ewes vaccinated against zearalenone had an even further reduction

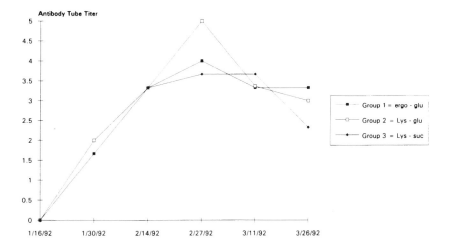

Figure 1 Mean antibody tube titer from Angus heifers after receiving immunogens created by conjugating ergonovine to human serum albumin (HSA) via glutaric anhydride (ergo-glu), lysergol to HSA via glutaric anhydride (lys-glu), or lysergol to HSA via succinic anhydride (lys-suc). Heifers (n = 3 per group) were treated with 1.0 mg of immunogen each on 1/16/92, 1/30/92, and 2/14/92. Each unit of antibody tube titer represents a 2X serial dilution of serum, beginning at a 1:10 dilution at zero (0).

in reproductive performance.[38] While there is no explanation for these results, immunization may have decreased zearalenone clearance. Similarly, immunization of ewes against sporidesmin, another mycotoxin and the etiological agent for facial eczema, failed to protect the animals from challenge.[13] In these studies the agents may have been initially bound by the immunoglobulins, but dissociation occurred prior to metabolism of the antibody ligand, allowing the alkaloid to mediate a biological effect. To overcome such an effect, the affinity of immunoglobulin for the ligand must be appropriate.

B. ANTI-IDIOTYPES

An anti-idiotype vaccine may be necessary to overcome the effects of endophyte-infected fescue, by virtue of being able to generate high-affinity antisera. In order to produce such a vaccine, anti-idiotype antibodies are produced in laboratory species, in response to vaccination with a monoclonal antibody created against the targeted antigen.[18] The species to be protected is then vaccinated using the anti-idiotype antibody. The hypervariable region of the antibody contains site-specific regions or idiotopes, and antibodies that bind idiotopes are known as anti-idiotypes. Anti-idiotype antibodies fit with the antibody combining site for the antigen and are, consequently, internal images of the antigen.[15] Nonproteinaceous low-molecular-weight antigens, as is the case with the ergot alkaloids, frequently may not result in the production of a high-affinity antibody response.[7] However, a specific anti-idiotype vaccine did confer immunity against the trichothecene mycotoxin T-2.

Additional problems may still occur with this procedure. Even with this advancement, a vaccine may not be protective if the binding affinity of the tissue receptor for the toxin is greater than the affinity of the antibody is for the toxin. Antigen/antibody binding affinities (Ka) between 10^{-6} to 10^{-10} liters/mole (L/M) may not be sufficient.

Serum prolactin is decreased in cattle grazing endophyte-infected fescue,[41] and ergot alkaloids are potent inhibitors of prolactin secretion.[16] While prolactin itself may not be an important determinant for growth, a decrease in prolactin secretion is indicative of effects of the endophyte. Serum prolactin increased in steers grazing endophyte-infected fescue following intravenous administration of a mouse monoclonal antibody directed against lysergol (Figure 2).[40] This is definitive evidence that ergot alkaloids found in endophyte-infected fescue are causative[30] and suggests that in the presence of sufficient circulating antibodies, other symptomatologies of fescue toxicosis may be reduced or prevented.

The monoclonal antibody used in the aforementioned experiment cross-reacted with other ergot alkaloids (Table 1), including ergovaline.[19] The IC 50 and cross-reactivity of this antisera compares favorably with the polyvalent antisera produced in rabbits, in response to ergonovine conjugated to Keyhole Limpet hemocyanin, reported by Reddick et al.[34] In both cases the generated immune globulins

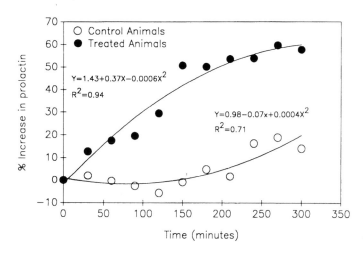

Figure 2 Mean percent serum prolactin increase in Angus steers, beginning just prior to, and for 300 min following, infusion of lysergic moiety-specific monoclonal antibody infusion (treated animals) or bovine serum albumin (control animals).

Table 1 **Cross reactivity of lysergic acid derivatives to monoclonal antibody 15F3.E5**

Lysergic Acid Derivative	Molar Conc. for 50% Maximum Abs.
Lysergol	3.93×10^{-15}
Lysergic acid	3.73×10^{-11}
Ergovaline	1.00×10^{-11}
Ergonovine	1.69×10^{-9}
Ergotamine tartrate	7.69×10^{-9}
2 Bromo-a-ergocryptine	$>1.33 \times 10^{-6}$
Dihydroergocornine	$>1.68 \times 10^{-6}$
Dihydroergocrystine	$>1.78 \times 10^{-6}$

possessed substantial cross-reactivity to ergovaline, the suggested causative agent.[30,42] The Ka and binding capacity of the immune globulins are further indications of antibody quality.[17] These entities are calculated using Scatchard plots via radioactive-ligand-binding studies. If the binding capacity and Ka of the antibody, along with the amount of circulating ergot alkaloids, were known, one could postulate the effectiveness of an immunological solution.

Antisera against ergot alkaloids has been produced in laboratory species in order to generate reagents for radioimmune assay purposes.[12] Antilysergic acid sera was generated in rabbits, using lysergic acid first joined to poly-L-lysine and then conjugated to Keyhole limpet hemocyanin as the antigen.[29] Results indicate that the C and D rings of the ergolene nucleus were the immunodeterminant portions of the molecule, since cross-reaction with indole amines was slight. Similarly, antisera to be used in a radioimmunoassay for dihydroergotamine was produced by immunizing rabbits, using 9,10-dihydroergotamine joined to carboxymethyl and conjugated to bovine serum albumin.[36] In contrast to these efforts to generate specific antisera, it would appear desirable to generate antisera that will bind a variety of ergot alkaloids, in order to overcome the toxicosis.

III. PHARMACOLOGICAL AGENTS

A. DOPAMINERGIC AND ADRENERGIC RECEPTOR ANTAGONISTS
1. Metoclopramide

The symptomatology of fescue foot in cattle resembled ergot poisoning, to Cunningham,[8] without evidence of an ergot alkaloid intake. The discovery of ergovaline in tall fescue pasture samples,[42] coupled with the finding that prolactin was depressed in cattle ingesting endophyte-infected tall fescue forage,[22] stimulated serious studies into the possible causative effects of ergot alkaloids for fescue toxicosis. The relationship between prolactin suppression and ergot alkaloid had been previously documented, including the mechanism via D-2 receptors. Thus, the search for D-2 receptor antagonists was undertaken.

Metoclopramide, a D-2 receptor antagonist, was administered to steers grazing on endophyte-infected tall fescue. The steers grazed the pastures 4 weeks prior to metoclopramide (15 mg/kg) administration. Then the drug was given orally three times weekly for 10 weeks.[27] (A minimal effective dose was not determined.) This resulted in increased body weight gains, greater time spent grazing, higher serum prolactin, and improved hair-coat appearance, compared to controls. Whether the steers were grazing highly infected fescue or fescue with lower levels of endophyte infection, metoclopramide-treated steers had greater increased weight gains on fescue with higher endophyte infection, compared to fescue with lower endophyte infections. This indicates a specific drug effect. The administration of metoclopramide to sheep ingesting endophyte-infected tall fescue seed rations improved dry-matter consumption, but did not reverse the depressed serum prolactin.[2]

2. Phenothiazine and Derivatives

The signs of fescue toxicity in equine species are distinctly different from those in cattle. Reproduction in the mare is primarily affected by tall fescue toxins. A reduction in the digestibility of endophyte-infected hay occurs along with reduced growth.[11] Signs of toxicity in mares are dystocia, agalactia, and prolonged gestation. Associated with the toxicity are a reduction in serum prolactin and progesterone.[32]

In a search for a treatment to reverse the toxicity in pregnant mares grazing highly infected tall fescue, two experiments were conducted giving dopamine antagonists to pony mares. In the first experiment, 16 nonpregnant mares were assigned to one of four treatments consisting of oral administration of perphenazine (0.5 and 1.0 mg/kg body wt), phenothiazine (10 mg/kg body wt), and a control group.[28] Blood samples collected 3 and 6 h after drug administration indicated that perphenazine increased plasma prolactin. Serum prolactin returned to normal by 11 h after drug administration. Phenothiazine had no effect on serum prolactin. Perphenazine at the 1.0 mg/kg level was discontinued due to signs of hyperesthesia. In the second experiment, oral administration of bromocriptine, a synthetic ergot alkaloid, simulated consumption of toxic fescue, in pony mares.[23] Twice-daily oral administration of perphenazine (0.375 mg/kg) partially restored serum prolactin and progesterone and prevented the adverse effects of bromocriptine (0.8 mg/kg$^{.75}$ given i.m. daily).

3. Domperidone

Domperidone, another D-2 receptor antagonist, was administered first orally (1.1 to 2.0 mg/kg) to pregnant mares, beginning day 10 prepartum, grazing on tall endophyte-infected fescue pastures. An increase in serum prolactin and progesterone resulted, with foaling at the expected time (D. L. Cross, personal communication). In a second experiment, domperidone (0.2 mg/kg i.m. daily) administration beginning day 7 prepartum had similar effects. Therefore, three dopamine antagonists have been identified that have beneficial effects against this toxicosis.

4. Prazosin

Increased core body temperatures, decreased peripheral blood flow, and decreased food intake are associated with fescue toxicosis. Adrenergic receptors participate in the regulation of these processes.[37] Prazosin, an alpha-1 adrenergic antagonist, was examined (1.0 mg/kg i.p. daily) for its effectiveness in reducing skin temperature and increasing food intake, in rats consuming endophyte-infected tall fescue seeds.[26] This drug reduced body temperature and increased food intake. Therefore, this drug potentially may be effective against tall fescue toxicosis in cattle.

B. OTHER CHEMICAL AGENTS

The anthelmintic drug ivermectin has reduced the impact of fescue toxicosis in cattle. In the initial observation, weight gain and improved hair coat were noted in the treated animals compared to controls (D. Wallace, personal communication, 1985). Between 1987 and 1992 at the University of Missouri Southwest Center, Mt. Vernon, yearling steers grazed on endophyte-infected tall fescue pasture for 84 to 169 days, with and without ivermectin given both as a sustained release bolus or by injecting a dose used for internal nematode control biweekly.[9,10] With the exception of 1988, treated animals gained more weight than did controls, even in the presence of additional nutritional supplementation. In grazing trials conducted in Alabama that used endophyte-infected pastures, ivermectin-treated steers also had numerically greater weight gains than did controls.[5]

In the ivermectin studies, visual signs of fescue toxicosis — abnormal hair coats, heat sensitivity, and grazing period shifts — were not eliminated, but were reduced. Regardless of the drug to be applied toward alleviating the adverse effects of fescue toxicoses, there are practical considerations to be considered due to the grazing animal's behavior and lack of confinement. The drug must be efficacious: the route of administration must be compatible with the normal management of the animal and meet regulations relative to safety and cost.

REFERENCES

1. **Aguado, M. T. and P. M. Lambert,** Controlled-release vaccines-biodegradable polylactide/polyglycolide (PL/PG) microspheres as antigen vehicles, *Immunobiology,* 184:113–125, 1992.
2. **Aldrich, C. G., M. T. Rhodes, J. L. Mines, M. S. Kerley, and J. A. Paterson,** The effects of endophyte-infected tall fescue consumption and use of a dopamine antagonist on intake, digestibility, body temperature and blood constituents in sheep, *J. Anim. Sci.,* 71:158–163, 1993.
3. **Anderson, G. W., J. E. Zimmerman, and F. M. Callahan,** The use of esters of N-hydroxysuccinimide in peptide synthesis, *J. Am. Chem. Soc.,* 85:3039, 1963.
4. **Bacon, C. W., J. K. Porter, J. D. Robbins, and E. S. Luttrell,** *Epichloe typhina* from toxic tall fescue grasses, *Appl. Environ. Microbiol.,* 34:567–581, 1977.
5. **Bransby, D. I., J. Holliman, and J. T. Eason,** Ivermectin could partially block fescue toxicosis, In: W. Faw and G. A. Pederson, Eds., Proc. Am. Forage and Grassland Council Meet., Des Moines, IA, 1993, 81.
6. **Chanh, T. C., G. Rappocciolo, and J. F. Hewetson,** Monoclonal anti-idiotype induces protection against the cytotoxicity of the trichothecene mycotoxin T-2, *J. Immunol.,* 144:4721–4728, 1990.
7. **Chanh, T. C., E. S. Siwak, and J. F. Hewetson,** Anti-idiotype-based vaccines against biological toxins, *Toxicol. Appl. Pharmacol.,* 108:183–193, 1991.
8. **Cunningham, I. J.,** A note on the cause of tall fescue lameness in cattle, *Australian Vet. J.,* 25:27–28, 1949.
9. **Crawford, R. S.,** Ivermectin: the real silver bullet?, In: C. Roberts, Ed., *Fescue Toxicosis and Management,* In-Service Education Workshop, Columbia, MO, March 1993, University of Missouri Extension, Columbia, 1993, 23.
10. **Crawford, R. S. and G. B. Garner,** Evaluation of corn gluten feed and ivermectin to ameliorate the negative impact of endophyte-infected tall fescue on cattle performance, In: Proc. Am. Forage and Grassland Conference, April 1–4, 1991, Columbia, MO, 1991, 255.
11. **Cross, D. L.,** Equine fescue toxicosis: symptoms and solutions, *J. Anim. Sci.,* 71(Suppl. 1):200, 1993.
12. **Eckert, H., F. R. Kiechel, J. Rosenthaler, R. Schmitt, and E. Schieier,** Biopharmaceutical aspects, In: B. Berde and H. O. Schild, Eds., *Ergot Alkaloids and Related Compounds,* Springer-Verlag, New York, 1978, 719.
13. **Fairclough, R. J., J. W. Ronaldson, W. W. Jonas, P. H. Mortimer, and A. G. Eramuson,** Failure of immunization against sporidesmin or a structurally related compound to protect ewes against facial eczema, *N.Z. Vet. J.,* 32:101–104, 1984.
14. **Garner, G. B.,** Fescue toxicosis: a general description, In: C. Roberts, Ed., *Fescue Toxicosis and Management,* In-Service Education Workshop, University of Missouri Extension, Columbia, 1993, 23.
15. **Gaulton, G. N. and M. I. Greene,** Idiotypic mimicry of biological receptors, *Ann. Rev. Immunol.,* 4:253–280, 1986.
16. **Goldstein, M., J. Y. Lew, A. Sauter, and A. Lieberman,** The affinities of ergot compounds for dopamine agonist and dopamine antagonist receptor sites, In: *Advances in Biochemical Psychopharmacology,* Vol. 23, Raven Press, New York, 1980, 75.
17. **Guillaume, V., B. Conte-Devolx, E. Magnan, F. Boudouresque, M. Grino, M. Cataldi, L. Muret, A. Priou, J. C. Figaroli, and C. Oliver,** Effect of chronic active immunization with antiarginine vasopressin on pituitary-adrenal function of sheep, *Endocrinology,* 130:3007–3014, 1992.
18. **Halliwell, R. E. W. and N. T. Gorman,** The immunoglobulins: structure, genetics, and function, In: *Veterinary Clinical Immunology,* W.B. Saunders, Philadelphia, 1989, 19.
19. **Hill, N. S., F. N. Thompson, D. L. Dawe, and J. A. Stuedemann,** Antibody binding of circulating ergopeptine alkaloids in cattle grazing tall fescue, *Am. J. Vet. Res.,* in press.
20. **Houston, L. L.,** Protection of mice from ricin poisoning by treatment directed against ricin, *Clin. Toxicol.,* 19:385–389, 1982.
21. **Hoveland, C. S., S. P. Schmidt, C. C. King, Jr., J. W. Odom, E. M. Clark, J. A. McGuire, L. A. Smith, H. W. Grimes, and J. L. Holliman,** Association of *Epichloe typhina* fungus and steer performance on tall fescue pasture, *Agron. J.,* 72:1064–1065, 1980.

22. **Hurley, W. L., E. M. Convey, K. Leung, L. A. Edgerton, and R. W. Hemken,** Bovine prolactin TSH, T_4 and T_3 concentrations as affected by tall fescue summer toxicosis and temperature, *J. Anim. Sci.,* 51:374–379, 1981.

23. **Ireland, F. A., W. E. Loch, K. Worthy, and R. V. Anthony,** Effects of bromocriptine and perphenazine on prolactin and progesterone concentrations in pregnant pony mares during late gestation, *J. Reprod. Fert.,* 92:179–186, 1991.

24. **Kellerman, T. S., J. A. W. Coetzer, and T. W. Naude,** Heart, In: T. S. Kellerman, J. A. W. Coetzer, and T. W. Naude, Eds., *Plant Poisoning and Mycotoxicoses of Livestock in South Africa,* Oxford University Press, Cape Town, 1988, 83.

25. **Lagrange, P. H. and A. Capron,** Immune responses directed against infections and parasitic agents, In: J. F. Bach, Ed., *Immunology,* John Wiley & Sons, New York, 1978, 410.

26. **Larson, B. T., S. B. Holste, M. D. Samford, M. S. Kerley, J. T. Turner, and J. A. Patterson,** Prazosin reduces body temperature and increases food intake of rats consuming endophyte-infected tall fescue without changing brain monoamine receptor density, *J. Anim. Sci.,* 71(Suppl. 1):442, 1993.

27. **Lipham, L. B., F. N. Thompson, J. A. Stuedemann, and J. L. Sartin,** Effects of metoclopramide on steers grazing endophyte-infected fescue, *J. Anim. Sci.,* 67:1090, 1989.

28. **Loch, W., K. Worthy, and F. Ireland,** The effect of phenothiazine on plasma prolactin levels in non-pregnant mares, *Equine Vet. J.,* 22:30, 1990.

29. **Lopatin, D. E. and E. W. Voss, Jr.,** Anti-lysergyl antibody measurement of binding parameters in IgG fractions, *Imunochemistry,* 11:285–293, 1974.

30. **Lyons, P. C., R. D. Plattner, and C. W. Bacon,** Occurrence of peptide and clavine ergot alkaloids in tall fescue grass, *Science,* 232:487–489, 1986.

31. **MacDonald, O. A., A. J. Thulin, W. C. Weldon, J. J. Pestka, and R. L. Fogwell,** Effects of immunizing gilts against zearalenone on height of vaginal epithelium and urinary excretion of zearalenone, *J. Anim. Sci.,* 68:3713–3718, 1990.

32. **Monroe, J. L., D. L. Cross, L. W. Hudson, D. M. Hendricks, S. W. Kennedy, and W. C. Bridges, Jr.,** Effect of selenium and endophyte-contaminated fescue on performance and reproduction in mares, *Equine Vet. Sci.,* 8:148–152, 1988.

33. **Ralph, W.,** Lupinosis: tests and a vaccine on the way, *Rural Res.,* 146:4–7, 1990.

34. **Reddick, B. B., K. D. Gwinn, B. J. Savory, and M. H. Collins-Shepard,** Development of an immunoassay for detection of ergoline alkaloids in tall fescue, *Tennessee Farm Home Sci.,* 160:78–81, 1991.

35. **Robbins, R. J.,** The measurement of low-molecular-weight, non-immunogenic compounds by immunoassay, In: H. F. Linskens and J. F. Jackson, Eds., *Immunology in Plant Sciences,* Springer-Verlag, New York, 1986.

36. **Rosenthaler, J. and H. Munzer,** 9,10-dihydroergotamine: production of antibodies and radioimmunoassay, *Experientia,* 32:234–236, 1976.

37. **Samford, M. D., B. T. Larson, J. C. Forcherio, M. S. Kerley, J. T. Turner, and J. A. Paterson,** Characterization of the effects of endophyte-infected tall fescue consumption on changes in adrenergic and dopaminergic receptor in selected body tissues of the bovine, *J. Anim. Sci.,* 71(Suppl. 1):201, 1993.

38. **Smith, J. F., M. E. diMenna, and N. R. Towers,** *Zearalenone and Its Effects on Sheep,* Vol. 3, Proc. 12th Int. Cong. Anim. Reprod. Aug. 23–27, 1992, The Hague, The Netherlands, 1219.

39. **Stuedemann, J. A. and C. S. Hoveland,** Fescue endophyte: history and impact on animal agriculture, *J. Prod. Agric.,* 1:38–44, 1988.

40. **Thompson, F. N., N. S. Hill, D. L. Dawe, and J. A. Stuedemann,** The effects of passive immunization against lysergic acid derivatives on serum prolactin in steers grazing endophyte-infected tall fescue, In: D. E. Hume, G. C. M. Latch, and H. S. Easton, Eds., *Proc. 2nd Int. Symp.* Acremonium/*Grass Interactions: Plenary Papers,* AgResearch, Grasslands Research Centre, Palmerston North, New Zealand, 1993.

41. **Thompson, F. N., J. A. Stuedemann, J. L. Sartin, D. P. Belesky, and O. J. Devine,** Selected hormonal changes with summer fescue toxicosis, *J. Anim. Sci.,* 65:727–733, 1987.

42. **Yates, S. G., R. D. Plattner, and G. B. Garner,** Detection of ergopeptine alkaloids in endophyte infected, toxic KY-31 tall fescue by mass spectrometry/mass spectrometry, *J. Agric. Food Chem.,* 33:719–722, 1985.

Section V

Biotechnological Uses of Endophytes:
Grass Improvement Based on
Molecular Techniques

Chapter 10

Importance of Endophytes in Forage Grasses, a Statement of Problems and Selection of Endophytes

Malcolm R. Siegel and Lowell P. Bush

CONTENTS

I. INTRODUCTION

The beneficial and detrimental characteristics of grasses infected by parasitic mycosymbionts have received considerable attention.[4,18,42,84,86] While the original impetus for research was the mammalian toxicoses associated with pasture grasses, other important effects of infection have been noted. Mycosymbionts are unique in that they produce or induce the plant to produce secondary metabolites that confer an array of benefits on their host. The ecological benefits accrued by the host greatly affect persistence, ensuring enhanced survival of infected plants. The benefits include antiherbivore activity (grazing mammals and insects), nematode resistance, disease resistance, improved growth, and drought

0-8493-6276-8/94/$0.00+$.50
© 1994 by CRC Press, Inc.

tolerance. Human perception of the benefits and detriments of infected grasses adds another dimension to grass–mycosymbiont associations (symbiota).

Persistence of perennial grasses is required for their agricultural usefulness. Persistence of infected grasses is often coupled to toxicity to grazing animals. However, while mammalian toxicosis is an ecological benefit for the host, those concerned with animal husbandry view this toxicity as a detriment to production. Symptoms of infection by *Epichloë typhina* (Fr.) Tul. and the anamorphic *Acremonium* Link: endophytes range from sterilization of the host reproductive system (choke disease) to a nonpathogenic endophytic existence.[96] Consequently, the symbiota exist in a continuum that can range from antagonistic to strictly mutualistic, depending on the genetic constitution of the fungus and grass, and on environmental factors.[19,86] When the mycosymbionts are endophytic or partially ectophytic, they are seed disseminated. Therefore, they are, in essence, maternally inherited components of symbiotic entities. Nevertheless, the mycosymbionts can be cultured, manipulated (via natural selection or genetic engineering), and reintroduced into their host species or related grass species, to produce novel associations for biological research and cultivar development.

In this chapter we will discuss the problems associated with endophyte-infected *Festuca* L. and *Lolium* L. pasture grass species; the desired beneficial characteristics of infected grasses; and the selection, methods of introduction of desirable endophytes into grasses, and characterization of the new associations. In addition, the progress and problems associated with manipulation of nonpathogenic endophytes will be discussed with reference to cultivar improvement.

II. PROBLEMS ASSOCIATED WLTH UTILIZATION OF ENDOPHYTE-INFECTED FORAGE

A. DISTRIBUTION OF ENDOPHYTES AND TOXIC ALKALOIDS IN FORAGE GRASSES

The number and patterns of distribution of cool-season forage grasses infected with endophytes continues to increase as detection methods improve and a greater number of plants are examined for the presence of fungal endophytes.[55] However, in many of these associations, comparative biological and biochemical investigations have not been done. Because of their agricultural impact, the *Festuca arundinacea* Schreb. (tall fescue)/*Acremonium coenophialum* Morgan-Jones & Gams and *Lolium perenne* L. (perennial ryegrass)/*A. lolii* Latch, Christensen & Samuels symbiota have received the greatest amount of research attention, but other mycosymbionts of the tribe Balansieae (*Balansia* Speg., *Atkinsonella* Diehl, and *Myriogenospora* Atk.) that infect predominantly warm-season host species have also contributed to our knowledge base.[55]

1. Alkaloid Profiles

There have been many reports of biochemical differences between endophyte-infected plants and noninfected plants, but toxic compounds have been largely confined to the ergopeptine alkaloids and their precursors, indole diterpenoids (paxilline and lolitrems), peramine, and the lolines (pyrrolizidine alkaloids). All these alkaloids are of fungal origin except the lolines, which have been found only in infected plants. Within a symbiotum there are ecological site-specific associations.[70] Ecological site-specific production of toxic alkaloids have not been shown, but qualitative and quantitative differences among different symbiota have been reported.[14,40,87] In tall fescue–*A. coenophialum* associations, the alkaloids most often found are ergopeptines, lolines, and peramine.[87] Penn et al.[68] found paxilline, also a tremorgen and an intermediate in lolitrem B biosynthesis, and lolitrem B, in agar cultures of *A. coenophialum*. Lolitrems have been detected in seeds of tall fescue–*A. coenophialum* symbiota (C. O. Miles, 1993, personal communication). Herbage from tall fescue containing artificially introduced *A. lolii* did not contain lolines, but ergopeptine alkaloids and lolitrem B were detected in amounts equal to, or greater than, that found in *A. lolii*-infected perennial ryegrass.[87] Christensen et al.[14] reported the presence of lolitrem B in tall fescue infected with an *Acremonium* endophyte, but one not classified as *A. coenophialum*. This association did not produce lolines and thus was like a tall fescue–*A. lolii* symbiotum. Greatest loline accumulation has been found in *F. pratensis* Hudson (meadow fescue) infected with *A. uncinatum* Gams, Petrini, & Schmidt.[12] Differences in alkaloid accumulation among symbiota may relate directly to either vigor of endophyte growth in the plant (compatibility of symbionts), a regulatory effect of a plant genome on the endophyte, or to the specific capacity of the endophyte genome to produce an alkaloid.

The concept that specific symbiota control different expressions of alkaloid content has been well established. Hill et al.[40] found that when a low ergovaline genotype was cured of its endophyte and artificially infected with the high-ergovaline-producing mycosymbiont, a low ergovaline symbiotum was produced. This observation indicates that the plant genotype plays a significant role in the accumulation of ergovaline in tall fescue–*A. coenophialum* symbiotum. Both the genotype of the host and the endophyte significantly altered the accumulation of ergovaline and lolitrem B in *A. lolii*-infected perennial ryegrass.[26] When *A. lolii* was introduced into tall fescue, lolitrem B was present. Lolines were present when *A. coenophialum* was introduced into perennial ryegrass.[87]

2. Distribution Within Plants

A. lolii and *A. coenophialum* are not distributed uniformly throughout an infected host. The endophyte is present in higher concentrations in the leaf sheath than in the leaf blade.[12,48] Alkaloid data do not necessarily mimic endophyte concentration.[5,11,12,48,78] These observations may be explained by not knowing the site(s) of biosynthesis within the symbiota; whether translocation occurs, especially for the loline group of alkaloids; and whether sites of accumulation can be different from sites of biosynthesis. Keogh and Tapper[48] and Ball et al.[5] reported that lolitrem B concentration decreased from tip to ligule within the leaf blade of perennial ryegrass and was highest in the oldest leaf sheath; however, peramine concentration was greatest in the younger leaves, and little differences between leaf blade and leaf sheath were found. Loline alkaloids decreased in senescing leaf blades of vegetative tall fescue–*A. coenophialum*,[29] and the pseudostem contained the highest loline concentration.[12] Ergovaline concentration in tall fescue–*A. coenophialum* was higher in panicles than in stems or leaves.[78] In vegetative plants, the crown tissue contained the highest concentration of ergovaline, followed in descending order by pseudostem, leaf blade, and roots.[3] Data indicate that ergovaline concentrations followed endophyte hyphal density in tissue, more closely than the other alkaloids.

B. MAMMALIAN ACTIVITY

Mammalian activity of forage grasses infected with *Acremonium* spp. has been described in detail by Porter in Chapter 10 and in other reviews.[32,73,74,83] Symptoms of mammalian toxicity are well documented, but our understanding of the mechanism of action is incomplete.

1. Fescue Toxicosis

Fescue toxicity, which is exacerbated at high ambient air temperatures, is most prevalent in southeastern U.S., where most of the tall fescue pastures are infested with *A. coenophialum*. While lolines are present in the greatest concentration, reports of their mammalian toxicity is limited to vasoconstriction of the lateral saphenous vein in cattle,[65] and the proposed cosubstrate for rumen thiaminase and subsequent thiamine deficiency.[27] The ergopeptine alkaloids, primarily ergovaline, are considered the primary toxic entity affecting reproduction, milk production, blood prolactin- and melatonin-mediated events[90] (G. Garner, 1991, personal communication). Concentrations of ergovaline are greater in the spring and fall in the U.S.,[7] periods when animal growth symptoms are less than during the summer, but they are the critical times for breeding and birthing. *A. coenophialum*-infected tall fescue also contains peramine, but no mammalian toxicity has been attributed to this compound. Lolitrems have not been detected in tall fescue–*A. coenophialum* forage. However, as stated earlier, precursors and lolitrem B have been found in cultures of *A. coenophialum*,[68] in tall fescue infected with an unidentified *Acremonium* endophyte,[14] and in KY 31 tall fescue–*A. coenophialum* seed (O. C. Miles, 1993, personal communication). The potential for these tremorgenic compounds to occur in tall fescue–*A. coenophialum* symbiota could be high, and consequently, additional monitoring is warranted.

2. Ryegrass Staggers

Ryegrass staggers is an important neuromuscular disease of livestock grazing *A. lolii*-infected perennial ryegrass in New Zealand.[32] Lolitrem B has been identified as the primary causal agent of ryegrass staggers, but several other lolitrems and paxilline may be present, some of which are tremorgenic.[62] Ergopeptine alkaloids (e.g., ergovaline and ergotamine) have been found in infected perennial ryegrass, but only recently has mammalian toxicity in ryegrass been ascribed to these alkaloids in symbiotum that do not produce lolitrems.[26,30]

3. Other Toxicoses

A byproduct of the grass seed industry in northwest U.S. is the grass straw. Much of the seed production comes from *A. lolii*- and *A. coenophialum*-infected perennial ryegrass and tall fescue. Consequently, the straw contains the same toxins, but at lower concentrations, as the herbage of the endophyte-infected grasses. Significant livestock toxicities have been reported from consumption of the straw as a winter feed or as a feed supplement. Ergovaline concentration in straw[91] was sufficient to cause fescue toxicosis, and the concentration of lolitrem B was sufficient to cause ryegrass staggers (A. M. Craig, 1993, personal communication).

C. IN-SOIL EFFECTS

Allelopathic properties of tall fescue residues have been reported,[58] and alkaloids have been postulated as causal agents. Reduced establishment of companion and subsequent crops may occur[12,58,69,71] by inhibition of seed germination. Chu-Chou et al.[15] also found propagule population densities of mycorrhizal species in field plots of tall fescue without *A. coenophialum* to be twofold greater than in plots with *A. coenophialum*-infected tall fescue. Sporulation of these mycorrhizal isolates was reduced by the roots of *Acremonium*-infected tall fescue plants. Guo et al.[35] also reported that *A. coenophialum* in tall fescue did not alter root infection by mycorrhizal fungi. However, more roots from noninfected plants were colonized than from infected plants. Sporulation was correlated with the extent of colonization.

III. DESIRED CHARACTERISTICS OF INFECTED GRASSES

Desirable properties of endophyte-infected forage grasses are the same as for endophyte-free herbage. These include tolerance to pests, tolerance to overgrazing by animals, improved plant growth, and lack of toxicity to consuming mammalian herbivores. Removal of the endophyte from tall fescue readily meets the last objective, of lack of toxicity to the consuming livestock,[6] but persistence of grasses, when biotic and abiotic stresses occur, is not assured.[9,32,75] To date, plant breeders have improved tall fescue and perennial ryegrass, in the presence of their respective indigenous endophytes, with the main goal being to obtain all the desirable properties in infected cultivars. Based upon data available,[14,26,40,87] both the plants and the endophytes control the expression of the accumulation of toxic substances in the symbiota. However, to date, most research activity has been directed to manipulations of the endophyte, by artificial introduction of naturally occurring isolates into grasses, to study host compatibility, alkaloid profiles, and characterization of desirable traits.

A. REDUCED TOXICOSIS

Livestock toxicosis may be alleviated by appropriate pasture and animal management practices that reduce the intake of toxic substances, by reduction or elimination of the toxic compounds in the herbage and total diet, and by reduced susceptibility of the grazing livestock.[47,78,84,100] Host introduction of naturally occurring or genetically engineered endophyte isolates that do not produce the toxic alkaloids may accomplish the primary goal of reduction or elimination of the toxic substances while maintaining the desirable properties of infected grasses.

B. IMPROVED GROWTH

The very early reports of the effect of endophytes on plant growth were contradictory and often not repeatable. Beginning with the reports of Latch et al.[54] and Read and Camp,[76] on perennial ryegrass and tall fescue infected with *A. lolii* and *A. coenophialum*, respectively, a much greater understanding of the interactions among plants, endophyte infection, and environment has accumulated.

1. Establishment

Keogh and Lawerence[49] and Clay[21] found no difference in seedling densities of *A. lolii*-infected or noninfected perennial ryegrass when germination and emergence was evaluated at near-optimum conditions. No differences in germination rate, percentage of germination, or seedling growth vigor test were found following an accelerated aging treatment or a hydration-dehydration treatment of *A. coenophialum*-infected or noninfected tall fescue (C. J. Kim, 1983, personal communication). However, in field trials more seedlings were established from infected perennial ryegrass seed[89] and greater first-year dry-matter accumulation was measured in infected tall fescue seedlings than in endophyte-free seedlings.[67] Collins et al.[23] showed that in certain environments, *A. coenophialum*-infected tall fescue had a greater number of seedlings per area, but first-harvest dry matter was not significantly higher in infected plants compared to noninfected plants.

2. Biomass

A. lolii-infected perennial ryegrass grown in a controlled environment produced more total dry matter than did noninfected plants.[54] The greatest relative difference between endophyte treatments was in the pseudostem, and the greatest absolute weight difference was in leaf production. Read and Camp[76] showed that in Texas, herbage growth was reduced to a greater extent in noninfected, than in *A. coenophialum*-infected, tall fescue. KY 31 tall fescue infected with *A. coenophialum* had 8 to 25% greater yield than noninfected KY 31, in a clipping management experiment in Georgia.[99] The benefit in yield occurred mostly in late summer, but the author stated that endophyte-free tall fescue did produce sufficient herbage to justify its use in the southern Piedmont of the U.S. Bouton et al.[9] suggested that without summer water stress, *A. coenophialum*-infected tall fescue was not superior to noninfected plants with regard to yield and stand persistence. Field results from Kentucky have shown that endophyte infection frequency was not related to yield, in hay-pasture management. In a seed-production management system, infection frequencies of 7 to 75% did not affect seed yield, autumn accumulated regrowth, or stand density.[88] Few data on root growth are available, but Latch et al.[54] reported 44% more root dry weight accumulation in *A. lolii*-infected perennial ryegrass grown in near-optimum conditions in controlled environments. Biomass production could be the result of the photosynthetic rate and/or area. Belesky et al.[8] reported a small, but significant, increase in leaf photosynthesis in *A. coenophialum*-infected tall fescue. However, the increased dry matter accumulation of *A. coenophialum*-infected tall fescue is primarily based upon increased tiller number and canopy photosynthesis.[11] Leaves of pot-grown tall fescue infected with *A. coenophialum* are thicker and narrower than leaves from endophyte-free plants.[2] Leaf extension rates in perennial ryegrass were not enhanced by *A. lolii* infection and could not account for the enhanced growth observed in some experiments.[28] Enhanced growth has also been observed in meadow fescue where seedling weight and tiller number were increased when plants were infected with *A. uncinatum*.[82] In tall fescue–*A. coenophialum* symbiotum, tiller density was higher in endophyte-infected, than in endophyte-free, swards, but herbage yields were not different, because dry weight per tiller was greater in the endophyte-free plants.[92]

In summary, *Acremonium* endophytes are not beneficial to shoot production of fescue or ryegrass, under minimum stress. However, under abiotic or biotic stress, the presence of the endophyte is beneficial to dry-matter accumulation.

C. DROUGHT RESISTANCE

Drought resistance in tall fescue may be due to drought tolerance or drought avoidance. Drought tolerance includes adjustment in osmotic potential to allow maintenance of turgor and physiological function at low water potentials. Tall fescue infected with *A. coenophialum* had greater osmotic adjustment than endo-phyte-free plants.[60,93] The resulting lower osmotic potential apparently results in greater plant growth at lower soil-water potentials. Osmotic adjustment also occurs in *A. coenophialum* grown on media with different water potentials.[77] Growth of the fungus was maximal at relatively high water potentials (−0.4 MPa); however, the fungus grew at more than 50% of the maximum rate at −2.32 MPa, a water potential the fungus would be subjected to *in situ*. However, White et al.[98] did not measure osmotic adjustment differences between infected and noninfected tall fescue selections and ascribed the difference in plant survival and production to drought avoidance. They suggested that due to pest resistance, deeper rooting and greater rooting density occurred in infected plants, and this resulted in more available soil water. Either mechanism could explain the general observation that under water stress, the endophyte increases the survivability of tiller and plants, and, consequently, increases biomass.

D. NUTRITIVE VALUE

A. coenophialum-infected and noninfected tall fescue had similar ADF, NDF, crude protein, and digest-ibility values,[13,33] but total soluble-amino-acid concentration was increased in endophyte-infected plants compared to noninfected plants.[59] However, a small (3%) decrease of IVDMD of endophyte-infected plants, compared to endophyte-free herbage, has been reported.[2] All the results available indicate that very little nutritive-value difference exists between different endophyte-status plants.

E. INSECTICIDAL AND NEMATICIDAL ACTIVITY

All the alkaloids previously described either deter or are lethal to insects.[20,25,44,75] Many of the observations of deterrence or insecticidal activity have been made with unnatural insect pests of these grasses.

However, much of the research serves as a model for development of new symbiota that will have insect resistance or tolerance to natural pests (e.g., wheat).[61] Many of the insects used to measure negative insect response — *Schizaphis grarninum, Diuraphis noxia, Rhopalosiphurn padi, R. maidis,* and *Sitobion avenae* — are important pests to cereal grasses worldwide. One desirable scenario would be to introduce a mycosymbiont into cereals, which would cause these negative insect responses. However, the toxic alkaloids accumulate in the seeds of the forage grasses, a condition unacceptable for food grains. Natural insect populations are also altered by the presence of the endophyte.[51,63,75] Populations of Russian wheat aphid (*D. noxia*) were lower in *A. lolii*-infected perennial ryegrass plants compared to noninfected plants, suggesting that biocontrol may be achieved with the appropriate endophyte or metabolite, in wheat.[22]

Populations of four soil bome nematodes were reduced in soils in which *A. coenophialum*-infected tall fescue had been grown.[66,94] Reduced populations could be the result of their inability to reproduce on *A. coenophialum*-infected tall fescue[50] or to produce fewer and smaller egg masses containing fewer eggs. The chemical agents responsible for reproductive differences of nematodes in soils with *A. coenophialum*-infected tall fescue are not known, but speculation centers on those alkaloids found in the roots and shoot residue. The relationship among the different types of biological toxicities is not understood, but it seems desirable not to sacrifice insect and nematode toxicity in order to decrease the mammalian toxicity of endophyte-infected forage. The possibility of separating the different toxicities cannot be addressed until we have a better understanding of the causative agents in each situation.

IV. SELECTION OF DESIRABLE ENDOPHYTES

The criteria for selection of the nonpathogenic endophytes for manipulation and introduction into grasses, for cultivar improvement, is dependent on numerous interrelated factors. These include host–fungus compatibility (host selectivity), the profile of desirable characteristics in the symbiotum from which the candidate endophyte originated, and the method of selection: naturally selected vs. genetically engineered biotypes. What has emerged from research strongly suggests that differences in taxonomy and biodiversity exits among *Epichloë* teleomorphs and *Acremonium* anamorphs, and the successful introduction of these mycosymbionts into the same and different species is not guaranteed, in some cases, it is not even desired. A thorough understanding of the taxonomy, biology, ecology, and chemistry associated with this diverse group of fungi is necessary for developing a rational approach to the manipulation of endophytes for cultivar improvement.

A. SELECTION OF NATURALLY OCCURRING ENDOPHYTES

As previously discussed, symbiota show large variations in alkaloid types and levels, and responses to drought persistence, and pests. Differences in phylogenetic relatedness occur among endophytes, as well as variation in cultural, morphological, and physiological characteristics between isolates.[14,80] These *in planta* and in-culture responses have been used as the rationale for the selection of naturally occurring endophytes for crop improvement. Because of the importance of the animal toxicoses and characterization of the responsible alkaloids, there has been an impetus to find and introduce endophytes that do not produce the alkaloids into grass cultivars. Endophytes from symbiota exhibiting other desirable characteristics may also be selected for manipulation once the traits are more fully characterized. Answers to the following questions are required before a rational approach to successful endophyte introduction can be achieved. What are the origins of the candidate endophytes? What are the roles of host and fungal genomes in host compatibility and in symbiota expression? What methods are needed for successful introduction?

Introduction of mycosymbionts, by artificial inoculation[45,53,55,64,86] or by maternal line breeding,[34] into host and nonhost grasses offers some insights into host specificity and compatibility requirements. These studies indicate that endophytes that naturally infect their respective host grasses (indigenous endophytes) are the primary candidates for manipulation and introduction. This implies a high compatibility between host and fungus, but does not guarantee complete intraspecific host–fungus compatibility. Less certain is the success of interspecific and intergeneric transfer of endophytes.

Taxonomic and mating studies[55,95,97] and molecular phylogenetic analysis using isozyme[14,57] and rRNA gene-sequence analyses[1,80,81] have yielded valuable information concerning the genetics and host specificity of the *Epichloë* and *Acremonium* symbionts. These studies indicate that the genus *Epichloë* is composed of more than one species, some of which have infected sympatric hosts, while others appear to exhibit a high degree of host specificity; different grass species were often infected with different mating populations of *Epichloë* species; *Acremonium* endophytes are the anamorphs of the different *Epichloë* species; more than one mycosymbiont species can naturally infect a given grass species; and phylogenetic analysis of the fungi may, in some instances, help indicate whether successful interspecific and intergeneric infection of grasses will occur. Leuchtmann[55] has summarized the data on the compatibility of *Epichloë* and *Acremonium* endophytes with nonhost *Festuca* and *Lolium* species, as assessed by artificial inoculation of coleoptile meristems. The artificial introductions resulted in no infections, transient infections (infection is lost within a year), and persistent infections (low and high seed dissemination and/or ectophytic expression). It was concluded that while nonpathogenic endophytes and *E. typhina* exhibit a high degree of specificity for their original hosts, some species and biotypes had narrow host ranges that were limited to a few congenic host species. The host compatibility relationships of transferred endophytes were not necessarily reciprocal, and variation in the host range may be related to the degree of host similarity.[57]

Molecular phylogeny studies using rRNA sequence analysis have indicated the presence of distinct clades containing both *Epichloë* and *Acremonium* species.[80,81] One of the clades contains four *Acremonium* types (three species and one biotype) and one *E. typhina,* infecting three *Festuca* species. The *Acremonium* endophytes have all been introduced, as seeds disseminated persistent infections, into nonhost grasses[86] (Siegel, unpublished data). As previously mentioned, *A. coenophialum* from tall fescue and *A. lolii* from perennial ryegrass were artificially introduced into their reciprocal hosts. Also, *A. starrii* from *F. arizonica* infected tall fescue. An ascospore isolate from a mating *E. typhina* population on *F. rubra* was introduced into tall fescue, but long-term persistence has not been determined. It was not possible to infect tall fescue with an *E. typhina* isolate from perennial ryegrass or with an *Acremonium* isolate from *Poa autumnalis.* Based on rRNA sequences these mycosymbionts are in different clades from the tall fescue and perennial ryegrass endophytes. The *Acremonium* (*Poa*) isolate could not even be transferred, by maternal line breeding, to an F1 hybrid offspring of a cross between infected *P. autumnalis* and noninfected *P. pratensis*, indicating species specificity (C. R. Funk, 1990, personal communication). These results suggest that, related groups or clades of endophytes exist with narrow host specificity, even though some *Epichloë* and *Acremonium* spp. are highly host specific. Not all nonpathogenic endophytes infecting tall fescue and perennial ryegrass can be designated as *A. coenophialum* or *A.lolii* types.

An in-depth study using isozyme analysis, morphological characters, sensitivity to the fungicide benomyl *in vitro*, and the synthesis of alkaloids in symbiota examined the relationship between *Acremonium* endophytes of tall fescue, meadow fescue, and perennial ryegrass.[14] Isozyme analysis of the numerous isolates identified six taxonomic groupings: three groupings occurred in tall fescue, two in perennial ryegrass, and one in meadow fescue. Alkaloid profiles were consistent in the host plants for all isolates within a single isozyme phenotype and for most isolates within a taxonomic grouping. However, on the basis of morphological, cultural, and alkaloid profile criteria, only one of the three taxonomic groupings among the tall fescue endophytes fit the designation of *A. coenophlalum,* and only one of the two grouping of perennial ryegrass fit *A. lolii.* The *in vivo* alkaloid profiles within the taxonomic groupings are of particular interest because their variation will be the basis for the introduction of isolates that do not produce specific mammalian toxic alkaloids. For example, four out of eight genotypes (distinguished by isozyme phenotypes) of perennial ryegrass endophyte did not produce lolitrem B. Two of these nonproducers were classified as *A. lolii* on the basis of isozyme patterns. Two genotypes of tall fescue endophyte did not produce ergovaline, one of which was grouped with *A. coenophialum.* Host compatibility of isolates from the taxonomic groupings will be of considerable interest and should shed light on host specificity and the role of the host–endophyte interactions in alkaloid synthesis.

B. SELECTION OF GENETICALLY ENGINEERED ENDOPHYTES

The use of molecular genetics to manipulate isolates has specific advantages and disadvantages, over the selection of naturally occurring endophytes to enhance the function of symbiota. Host–endophyte incompatibility would be reduced because a host compatible isolate would be chosen as the candidate for manipulation. The introduction or elimination of a specific gene function(s) in an isolate from a well-characterized symbiota would facilitate testing for the desired trait, as well as the profile of other desirable properties previously found in the symbiota. Genetic manipulation is especially designed for well-characterized gene functions, e.g., synthesis of toxic alkaloids and the introduction of known foreign genes. The primary disadvantage is the cost of the research, the lack of biochemical information on selected natural and foreign beneficial genes, and the controversy associated with the use of genetically engineered organisms. The specifics of these techniques, and the ramifications of the molecular genetics approach for host improvement, will be discussed in Chapter 11.

C. METHODS OF ENDOPHYTE INTRODUCTION

Regardless of the source (natural vs. genetically modified) of the candidate isolate, it must be successfully introduced into the host as a persistent asymptomatic infection. The endophyte should have a high rate of seed dissemination, and the symbiota should express the new desired trait, without the loss of any characteristics that affect agronomic and animal production. It has not been possible to infect mature plants consistently, through stigma or wounds, using ascospores or conidia. However, persistent infections have been achieved by artificial introduction of mycelia and conidia into apical shoot meristems of seedlings and tillers and by inoculation of callus and somatic embryos. Additionally, endophytes are naturally transferred by seeds during maternal line breeding of grasses. The artificial and natural introduction of endophytes has been successfully used to study specific host–fungus interactions (pest resistance, alkaloid synthesis, drought tolerance, yield, host persistence, and phylogenetic relationships) and the establishment of infected grasses for turf and forage production. However, because the increase or transfer of endophytes into a population involves a range of host genotypes, it is possible that a specific endophyte biotype may not be compatible with all the host genomes.

1. Artificial Inoculation
a. Seedlings

The basic method of infecting seedling shoot apical meristems was described by Latch and Christensen in 1985.[53] The method is simple; large numbers of seedlings can be infected; infectivity rates are relatively high (if host and fungus are compatible), and consequently, production of infected plant populations is easily achieved. The disadvantage of the method is that only one endophyte biotype can be introduced into a single plant genotype at any one time. The procedure requires that seeds be surface sterilized with 2.62% sodium hypochlorite (15 to 30 min) and 50% sulfuric acid (15 to 20 min) and then washed three times with sterile water. Sterilization time depends on the grass species and the degree of microbial contamination. Seeds are germinated in the dark on water agar for 10 to 14 days at 20 to 22°C. At intervals, seeds showing bacterial and fungal contamination should be removed from the agar and discarded. The apical meristem, a swelling (approximately 1 mm wide) at the junction of mesocotyl and coleoptile of the seedling, is slit vertically with a sterile scalpel or punctured with an insect pin. A small amount of cultured, actively growing mycelia is placed into the wounded meristem, using a binocular stereomicroscope under sterile conditions in a laminar flow hood. Depending on the endophyte growth rate, seedlings are incubated for an additional 7 to 14 days in the dark. Seedlings are then placed under fluorescent light (12-h cycle) at 20 to 22°C for 5 to 7 days and then cut out of the agar and planted in potting soil in 5-cm plastic containers. The potted plants, covered with plastic domes to maintain high humidity, are incubated for 2 weeks under light at 20 to 22°C and then placed in the greenhouse. The plastic hoods are partially raised for 2 to 3 days and then removed, and the seedlings are grown for 6 to 8 weeks or until multiple tillers are established.

The presence of the endophyte is determined by staining leaf sheaths[16] or by tissue print immunoblot of stem sections.[1,38] Typically, there is a 25 to 50% loss of inoculated seedlings (at the preplant stage and/or after planting), due to damage to the meristem. Depending on the isolate and host, infection rates in compatible symbiota range from 10 to 65% of the surviving plants[53] (M. R. Siegel, 1992, unpublished data).

b. Callus and Somatic Embryo Cultures

Inoculation of callus and somatic embryo cultures has the advantage that the same plant genotype can be infected with more than one strain of endophyte at the same time. The disadvantages are complexity, time to produce an infected plant, a very low infection rate, and, most importantly, the potential for somaclonal variation. In addition, the method is not particularly suited for the establishment of an infected population of plants. A high rate of somaclonal variation for the callus culture inoculation method described by Johnson et al.[43] occurred because of the length of the callus phase (3 to 4 months). While the variation was considered a disadvantage at the time, the callus culture will produce genetically divergent plants with potentially different symbiota expression.

Infection of somatic embryos, as described by Kearney et al.,[45] involves a much shorter callus phase (approximately 1 month), minimizing somaclonal variation. However, the rate of successful introductions was low. In the somatic embryo method, leaf basal tissues (2 cm long) were surface sterilized in 70% alcohol (30 sec) and 0.525% sodium hypochlorite solution (15 min), and rinsed three times in sterile water, and the outer leaf tissue layers were peeled away; then the central, 1-cm-long explant section was placed on auxin-supplemented basal agar medium. Explant cultures were maintained at 25°C under fluorescent lights. After 28 days, calli were transferred to an auxin-free medium, to permit further development of the somatic embryos. Germinated plantlets from somatic embryos were transferred to a basal medium in Magenta boxes. After an additional period, the plantlets were planted in potting soil in 5-cm pots and placed in plastic dome-topped trays. Tray lids were progressively removed, and after two weeks the plants were transferred to the greenhouse. The optimum infection period, with *A. coenophialum* isolates, of calli from tall fescue occurred after week 3–4, on the auxin-supplemented medium (30 μ*M*; 2, 4, D), and after week 1, on the auxin-free medium. Inoculations of explants with isolates was accomplished by placing small amounts of mycelia from the outer region of an actively growing colony into a scalpel cut (2 to 3 mm) on the surface of the explant or growing callus. Plants were allowed to grow in the greenhouse for 8 weeks before leaf sheaths were tested by staining for the presence of endophyte. Infection rates for the optimum inoculation period were about 8%.

c. Meristem Culture

To completely overcome the potential for somatic variation occurring during infection of callus tissue and to increase the infection rate by endophyte isolates, O'Sullivan and Latch[64] devised a method of inoculating plantlets derived from single apical meristems. The basal 2 to 3 cm from actively growing mature tillers of tall fescue or perennial ryegrass plants were excised and surface sterilized with 95% ethanol (5 sec) and 20% sodium hypochlorite (5 min). The segments were washed three times with sterile water and dried in a laminar flow hood. Using sterile forceps and scalpel, the outer layers of sheath tiller piece was sliced away until the meristem was located. The meristem was excised and placed upright on mineral salt (MS) medium containing 0.05 ppm inositol, 0.002 ppm thiamine, 0.2 ppm iron, 3% sucrose, 0.3 ppm cytokinin, and 0.7% agar. Incubation was under light (12-h cycle) for 2 to 4 weeks at 22°C. Plantlets were removed, and the apical meristem was inoculated by inserting actively growing mycelia into a slit at the base of the plantlet. Plantlets were transferred to new medium and maintained at 15 to 22°C for approximately 3 weeks at a 12-h light cycle until roots had been formed and sufficient growth had occurred to allow transplanting into a potting mix. Plantlets were placed under lights with plastic lids, which were removed after an additional 3 weeks of growth. Plants were transferred to the greenhouse for a further 2 to 3 weeks of growth, and infection was determined by leaf-sheath staining.

This method resulted in infection rates that ranged from 0 to 56% for perennial ryegrass and 0 to 40% for tall fescue. Since the cultivars and lines of the two grass species were infected with an isolate that naturally infects that species, the rate of infection appeared to be host-genotype dependent, clearly indicating that some assemblages of supposedly compatible symbionts are more easily established than others. Infecting plantlets derived from tiller apical meristems allows for the possibility of infecting one host genotype with more than one endophyte biotype.

2. Natural Infections — Maternal Line Breeding

Turf grass breeders have used maternal-line-breeding techniques to increase an infection that was low to high levels and to transfer endophytes to breeding lines and cultivars of the same species. Most of the new endophyte-improved turf grass tall fescue and perennial ryegrass cultivars use naturally infected maternal parents. Here the objective is to have the maximum number of toxic alkaloid classes present in the infected cultivar, as mammalian toxicosis is not a factor for consideration, but insect resistance is important.[34,84] Other desirable traits, drought tolerance, and increased biomass may also be present in the improved cultivars, but they have not been adequately assessed.

D. CHARACTERIZATION OF NEW ASSOCIATIONS

Assessing the success of the newly created symbiota depends on an accurate characterization of host–endophyte genomic interactions. The method used to introduce the candidate isolate will determine the primary use of the resulting symbiota. Because a callus-somatic embryo culture and tiller meristem clone can easily be infected with more than one endophyte genotype, the new symbiota are ideally suited for studying host–fungus compatibility and the physiological and biochemical characterization of specific traits. For example, infection of clonal embryo culture and tiller meristems with specific biotypes of *A.*

coenophialum and/or *A. lolii* has been used to study the effect of the host genome on the production of ergovaline,[45] and host specificity,[64] respectively. The methods can be used to further characterize many of the currently known interactions (alkaloid synthesis, drought resistance, enhanced plant growth, pest tolerances, etc.), by manipulation of host and/or endophyte genomes. However, the methods are less suitable for establishing a cultivar where larger clonal populations are required for genetic diversity. The apparent method of choice for infecting populations of plants with a specific endophyte genome is by seedling inoculations. It is possible that tiller apical meristem inoculations could also be used to produce a population, but since multiple tiller meristems from clones would be required, the method appears to be suitable only if the original cultivar or breeding line has a very narrow genetic base. The method of cultivar establishment will be discussed in another section of this paper.

1. Sequence and Methods of Characterization

The sequence for establishing new asymptomatic seed-disseminated symbiota starts with testing for infection of the greenhouse-grown (6 to 8 weeks old, multitillered) inoculated seedlings. The plants should be retested after 4 to 6 months to prove that a stable symbiota has been established. It is important that several multiple tillers of each plant be tested to insure the uniformity of infection. Less-persistent infections will be indicated by the presence of endophyte-free plants that were previously infected and by infection of only a portion of the tillers. The plants, as populations or clones, should be divided to ensure active growth. Christensen (1993, personal communication) has suggested that if plants become pot bound, even persistent infections can be lost. Once persistent infections are demonstrated, the plants can be used in greenhouse experiments or placed in the field for further evaluation (e.g., persistence and seed dissemination).

However, before field testing is undertaken, the new symbiota should be tested under controlled conditions in the greenhouse, growth chamber, or laboratory, to assure the presence or loss of the trait under study. Assuming a positive response is measured under the controlled conditions, a sufficient number of plants can be generated for field studies, by subdividing whole plants into multitillers or individual tillers for field planting. The sequence of field testing depends on the traits being characterized, but initially involves small plots analysis followed by large plots analysis and may eventually lead to cultivar production.

Two important questions should be answered by field studies. Are the new symbiota persistent (stable) and is the added or missing trait of interest expressed in the individual and population of plants without the loss of previously known desirable characteristics? The tests to be described should ideally be undertaken for more than 1 year and at multiple locations, to assess the effect of biotic and abiotic environments on trait expression.

2. Selection of Plants and Methods of Testing

In order to answer the two previously proposed questions, proper comparison testing of host–endophyte responses are required. If populations are involved, tests should be devised that compare the new symbiota with naturally infected and noninfected isoline of the cultivar. If clonal plants are involved, the noninfected controls (not inoculated, but carried through the same procedure) should be developed in parallel with the symbiota. It is important that plants of the same size and age are compared in the greenhouse and/or field investigations.

Suitable comparison trait measurements include alkaloid profiles;[14,87] drought tolerance and regrowth, in the greenhouse[2] or field;[92,98] biomass accumulation in the field;[39] persistence of infection and seed dissemination;[88] yield, persistence, and competitiveness in the field;[9,39] insect tolerance in the laboratory, greenhouse, or field;[17,25,52] nematode tolerance in the greenhouse[36,50] or field;[94] fungitoxicity in agar culture;[14,85] disease resistance in the greenhouse[10,37] or field;[17] and animal assays (toxicity, daily intake, and weight gain, etc.).[73,83] It is also necessary that specific agronomic characteristics be tested in conjunction with seasonal changes. These include initial stand establishment, seed production, viability, palatability, and digestibility.

The experimental design depends on the characteristic measured and whether greenhouse trials or field trials are undertaken. In the greenhouse, space restriction can affect the number of clones tested. In addition, pot size, material (porosity), location on the bench, and location of the bench could affect results. Proper replication and randomization of pots is critical for statistical analysis.[2,41] Proper plot design and statistical analyses are also required in field measurements.

A modified systematic Nelder design for spacing experiments has been used by Hill et al.[39] to measure the effects of two *A. coenophialum*-infected tall fescue clones on competitiveness, the effect of density on growth and reproduction, and ergovaline synthesis. The advantage of this design is that intra- and intercompetition of infected and noninfected plants may be evaluated rapidly at several densities. The disadvantages are that greater precision is required in plant establishment, and it is difficult to avoid damaging plants in the high-density area near the center of the design.

Another very useful design is the honeycomb (focal plant or nearest neighbor) design[79] in which interactions between infected and noninfected plants may be evaluated. Kelrick et al.[46] found interactions between infected and noninfected plants to be neither symmetrical nor reciprocal. Focal plants growing in like neighborhoods performed better than those surrounded by plants of the other type. This observation would explain differences observed for individual infected or noninfected plants within a sward of all infected or noninfected plants.

3. Production of Improved Cultivars

The development of a population of a new symbiota from an established noninfected certified seed will require up to 100 individually infected seedlings, depending upon the breadth of the genetic base of the original cultivar or breeding line. If a cultivar does not exist, it may be necessary and desirable to infect seeds of the breeding lines needed to produce a cultivar. The infected plants can be increased in the greenhouse by separation of tillers or by planting separate tillers in the field. Randomization of the tillers, plant production, and seed production would be by standard breeding procedures. Plot isolation is required to ensure the production of polycross progeny and breeder seeds.

V. CONCLUSIONS: PROBLEMS AND PROGRESS

Considerable progress has been made in identifying and characterizing the benefits and detriments associated with endophyte-infected grasses, and the role of host and fungal genomes in trait expression and host specificity. Progress has resulted from the use of naturally infected grasses as well as the species and biotypes of *E. typhina-Acremonium* artificially introduced into host and nonhost grasses. Results, as previously discussed, would indicate that host–endophyte compatibility is more assured, but not guaranteed, by introducing indigenous endophytes into their host grasses. The use of indigenous isolates with desirable traits will still require direct experimentation using more than one host genome during artificial or natural methods of introduction.

While good progress has been made using naturally and artificially infected host clones, further characterization of physiological and biochemical mechanisms of the host–endophyte genome interaction is required for a rational approach to isolate introduction. This is especially needed for molecular genetic manipulation of the fungal genome. The identification of toxic alkaloids and their biosynthetic pathways is information that has facilitated the development of lolitrem and ergot minus symbiota. Similar information would certainly aid in the development of improved symbiota expressing other desirable traits.

The only example of a commercially produced symbiota is the artificial introduction of a naturally occurring *A. lolii* lolitrem-free isolate into perennial ryegrass cultivars released in New Zealand to control ryegrass staggers.[24,30,31] However, it now appears that the levels of ergovaline in some, but not all, of the cultivars is high, resulting in expression of toxicosis in grazing animals[26] and the withdrawal of specific cultivars (G. C. M. Latch, 1993, personal communication). In addition, all the lolitrem minus symbiota produce the lolitrem precursor, paxilline.[26] Paxilline is tremorgenic, but at higher concentrations than lolitrem, and may play a role in ryegrass staggers.[62,68]

Ergot minus isolates from tall fescue have been identified by Christensen et al.,[14] but have, as yet, not been released in cultivars. The success of lolitrem and ergot minus cultivars, as well as other new symbiota, will eventually be determined by the economics of the forage and animal production systems. Cost:benefit analysis should indicate whether any new symbiotum is better than the available naturally infected or noninfected cultivars. However, an additional problem may exist with the altered alkaloid symbiota. If one class of alkaloids is removed from a symbiotum, there is the potential for a reduced spectrum of insect toxicity, and the development of resistance is increased to one or both of the remaining alkaloids.[72,84] Even with the real and potential problems associated with toxic alkaloid minus cultivars, these new symbiota appear to be a major breakthrough in controlling animal toxicosis. The eventual production of other types of beneficial symbiota may also offer additional potential for long-term improvements of forage grasses.

ACKNOWLEDGMENTS

This work was supported by the U.S. Department of Agriculture NRICGP grant 92-3703-7612. This is Kentucky Agricultural Experiment Station Publication no. 9311-120.

REFERENCES

1. **An, Z. Q., J. S. Liu, M. R. Siegel, G. Bunge, and C. L. Schardl,** Diversity and origins of endophytic fungal symbionts of the North American grass *Festuca arizonica, Theor. Appl. Genet.,* 85:366–371, 1992.

2. **Arechavaleta, M., W. C. Bacon, C. S. Hoveland, and D. E. Radcliffe,** Effect of the tall fescue endophyte on plant response to environmental stress, *Agron. J.,* 81:83–90, 1989.

3. **Azevedo, M. D., R. E. Welty, A. M. Craig, and J. Bartlett,** Ergovaline distribution, total nitrogen and phosphorous content of two endophyte-infected tall fescue clones, In: D. E. Hume, G. C. M. Latch, and H. S. Easton, Eds., *Proc. 2nd Int. Symp.* Acremonium/*Grass Interactions,* AgResearch, Grasslands Research Centre, Palmerston North, New Zealand, 1993, 59.

4. **Bacon, C. W. and J. De Battista,** Endophytic fungi of grasses, In: D. K. Avora, B. Rai, K. G. Mukerji, and G. R. Knudsen, Eds., In: *Soil and Plants,* Marcel Dekker, New York, 1990, 231.

5. **Ball, O. J.-P., R. A. Prestidge, and J. M. Sprosen,** Effect of plant age and endophyte viability on peramine and lolitrem B concentration in perennial ryegrass seedlings, In: D. E. Hume, G. C. M. Latch, and H. S. Easton, Eds., *Proc. 2nd Int. Symp.* Acremonium/*Grass Interactions,* AgResearch, Grasslands Research Centre, Palmerston North, New Zealand, 1993, 63.

6. **Baxter, H. D., J. R. Owen, R. C. Buckner, R. W. Hemken, M. R. Siegel, L. P. Bush, and M. J. Montgomery,** Comparison of low alkaloid tall fescues and orchard grass for lactating Jersey cows, *J. Dairy Sci.,* 69:1329–1336, 1986.

7. **Belesky, D. P., J. A. Stuedemann, and S. R. Wilkinson,** Ergopeptine alkaloids in grazed tall fescue, *Agron. J.,* 80:209–212, 1986.

8. **Belesky, D. P., O. J. Devine, J. E. Pallas, Jr., and W. C. Stringer,** Photosynthetic activity of tall fescue as influenced by a fungal endophyte, *Photosynthetica,* 21:82–87, 1987.

9. **Bouton, J. H., R. N. Gates, D. P. Belesky, and M. Owsley,** Yield and persistence of tall fescue in the southeastern coastal plain after removal of its endophyte, *Agron. J.,* 85:52–55, 1993.

10. **Burpee, L. L. and J. H. Bouton,** Effect of eradication of the endophyte *Acremonium coenophialum* on epidemics of *Rhizoctonia* blight in tall fescue, *Plant Dis.,* 77:157–159, 1993.

11. **Bush, L., S. Gay, and W. Burhan,** Accumulation of pyrrolizidine alkaloids during growth of tall fescue, In: Proc. 17th Int. Grassland Congr., Palmerston North, New Zealand, in press, 1993.

12. **Bush, L. P., F. F. Fannin, M. R. Siegel, D. L. Dahlman, and H. R. Burton,** Chemistry, occurrence and biological effects of saturated pyrrolizidine alkaloids associated with endophyte-grass interactions, *Agric. Ecosyst. Environ.,* 44:81–102, 1993.

13. **Bush L. P. and P. B. Burrus, Jr.,** Tall fescue forage quality and agronomic performance as affected by the endophyte, *J. Prod. Agric.,* 1:55–60, 1988.

14. **Christensen, M. J., A. Leuchtmann, D. D. Rowan, and B. A. Tapper,** Taxonomy of *Acremonium* endophytes of tall fescue (*Festuca arundinacea*), meadow fescue (*F. pratensis*), and perennial ryegrass (*Lolium perenne*), *Mycolog. Res.,* 97:1083–1092, 1993.

15. **Chu-Chou, M., B. Guo, Z.-Q. An, J. W. Hendrix, R. S. Ferriss, M. R. Siegel, C. T. Dougherty, and P. B. Burrus,** Suppression of mycorrhizal fungi in fescue by the *Acremonium coenophialum* endophyte, *Soil Biol. Biochem.,* 24:633–637, 1992.

16. **Clark, E. M., J. F. White, and R. M. Patterson,** Improved histochemical techniques for the detection of *Acremonium coenophialum* in tall fescue and methods of in vitro culture of the fungus, *J. Microbiol. Methods,* 1:149–155, 1983.

17. **Clay, K.,** Endophytes as antagonists of plant pests, In: J. H. Andrews and S. S. Hirano, Eds., *Microbial Ecology of Leaves,* Springer-Verlag, New York, 1991, 331.

18. **Clay, K.,** Fungal endophytes of grasses, *Ann. Rev. Ecol. Syst.,* 21:275–295, 1990.

19. **Clay, K.,** Clavicepitaceous fungal endophytes of grasses: coevolution and the change from parasitism to mutualism, In: K. A. Pirozynski and D. Hawksworth, Eds., *Coevolution of Fungi with Plants and Animals,* Academic Press, London, 1988, 79.

20. **Clay, K.,** Fungal endophytes of grasses: a defensive mutualism between plants and fungi, *Ecology,* 69:10–16, 1988.

21. **Clay, K.,** Effects of fungal endophytes on the seed and seedling biology of *Lolium perenne* and *Festuca arundinacea, Oecologia,* 73:358–362, 1987.

22. **Clement, S. L., D. G. Lester, A. D. Wilson, and K. S. Pike,** Behavior and performance of *Diuraphis noxia* (Homoptera:Aphididae) on fungal endophyte-infected and uninfected perennial ryegrass, *J. Econ. Ent.,* 85:583–588, 1992.

23. **Collins, M., D. M. Tekrony, and J. C. Henning,** *Acremonium coenophialum* (Morgan-Jones and Gams) and cultivar effects on tall fescue seedling vigour and stand establishment, In: Proc. 17th Grasslands Congr., Palmerston North, New Zealand, in press, 1993.

24. **Cosgrove, G. P., C. B. Anderson, R. A. Mainland, C. J. Saunders, and D. E. Hume,** Fungal endophyte level and strain effects on pastures and cattle liveweight gain, In: H. E. Hume, G. C. M. Latch, and H. S. Easton, Eds., *Proc. 2nd Int. Symp.* Acremonium/*Grass Interactions*, AgResearch, Grasslands Research Centre, Palmerston North, New Zealand, 1993, 106.

25. **Dahlman, D. L., H. Eichenseer, and M. R. Siegel,** Chemical perspectives on endophyte-grass interactions and their impilations to insect herbivory, In: P. Barbosa, V. A. Krischik, and C. L. Jones, Eds., *Microbial Mediation of Plant-Herbivore Interactions*, Wiley-Interscience, New York, 1991, 227.

26. **Davies, E., G. A. Lane, G. C. M. Latch, B. A. Tapper, I. Garthwaite, N. R. Towers, L. R. Fletcher, and D. B. Pownall,** Alkaloid concentrations in field-grown synthetic perennial ryegrass endophyte associations, In: D. E. Hume, G. C. M. Latch, and H. S. Easton, Eds., *Proc. 2nd Int. Symp.* Acremonium/*Grass Interactions*, AgResearch, Grasslands Research Centre, Palmerston North, New Zealand, 1993, 72.

27. **Dougherty, C. T., L. M. Lauriault, N. W. Bradley, N. Gay, and P. L. Cornelius,** Induction of tall fescue toxicosis in heat-stressed cattle and its alleviation with thiamin, *J. Anim. Sci.,* 69:1008–1018, 1990.

28. **Eerens, J. P. J., J. G. H. White, and R. J. Lucas,** The influence of the *Acremonium* endophyte on the leaf extension rate of moisture stressed ryegrass plants, In: D. E. Hume, G. C. M. Latch, and H. S. Easton, Eds., *Proc. 2nd Int. Symp.* Acremonium/*Grass Interactions*, AgResearch, Grasslands Research Centre, Palmerston North, New Zealand, 1993, 200.

29. **Eichenseer, H., D. L. Dahlman, and L. P. Bush,** Influence of endophyte infection, plant age and harvest interval on *Rhopalosiphum padi* survival and its relation to quantity of N-formyl and N-acetyl loline in tall fescue, *Entomol. Exp. Appl.,* 60:29–38, 1991.

30. **Fletcher, L. R. and B. L. Sutherland,** Liveweight change in lambs grazing perennial ryegrasses with different endophytes, In: D. E. Hume, G. C. M. Latch, and H. S. Easton, Eds., *Proc. 2nd Int. Symp.* Acremonium/*Grass Interactions*, AgResearch, Grasslands Research Centre, Palmerston North, New Zealand, 1993, 125.

31. **Fletcher, L. R., I. Garthwaite, and N. R. Towers,** Ryegrass staggers in the absence of lolitrem B, In: D. E. Hume, G. C. M. Latch, and H. S. Easton, Eds., *Proc. 2nd Int. Symp.* Acremonium/*Grass Interactions*, AgResearch, Grasslands Research Centre, Palmerston North, New Zealand, 1993, 119.

32. **Fletcher, L. R., J. H. Hoglund, and B. L. Sutherland,** The impact of *Acremonium* endophytes in New Zealand, past, present and future, *Proc. N.Z. Grassland Assoc.,* 52:227–235, 1990.

33. **Fritz, J. O. and M. Collins,** Yield, digestibility, and chemical composition of endophyte free and infected tall fescue, *Agron. J.,* 83:537–541, 1991.

34. **Funk, C. R., R. H. White, and J. Breen,** Importance of *Acremonium* endophytes in turfgrass breeding and management, *Agric. Ecosyst. Environ.,* 44:215–232, 1993.

35. **Guo, B. Z., J. W. Hendrix, A.-Q. An, and R. S. Ferriss,** Role of *Acremonium* endophyte of fescue on inhibition of colonization and reproduction of mycorrhizal fungi, *Mycologia,* 84:882–885, 1992.

36. **Gwinn, K. D. and E. C. Bernard,** Interaction of endophyte-infected grasses with the nematodes *Meloidogyne marylandi* and *Praylenchus scribneri*, In: D. E. Hume, G. C. M. Latch, and H. S. Easton, Eds., *Proc. 2nd Int. Symp.* Acremonium/*Grass Interactions*, AgResearch, Grasslands Research Centre, Palmerston North, New Zealand, 1993, 156.

37. **Gwinn, K. D. and A. M. Gavin,** Relationship between endophyte infestation level of tall fescue seed lots and Rhizoctonia zeae seedling disease, *Plant Dis.,* 76:911–914, 1992.

38. **Gwinn, K. D., M. H. Shepard-Collins, and B. B. Reddick,** Tissue print-immunoblot: an accurate method for the detection of *Acremonium coenophialum* in tall fescue, *Phytopathology,* 81:747–748, 1991.

39. **Hill, N. S., D. P. Belesky, and W. C. Stringer,** Competitivness of tall fescue as influenced by *Acremonium coenophialum, Crop Sci.,* 31:185–195, 1991.

40. **Hill, N. S., W. A. Parrott, and D. D. Pope,** Ergopeptine alkaloid production by endophytes in a common tall fescue genotype, *Crop Sci.,* 31:1545–1547, 1991.

41. **Hill, N. S., W. C. Stringer, G. E. Rottinghaus, D. P. Belesky, W. A. Parrott, and D. D. Pope,** Growth, morphological, and chemical componet reponses of tall fescue to *Acremonium coenophliaum*, *Crop Sci.,* 30:156–161, 1990.

42. **Hoveland, C.,** Importance and economic significance of the *Acremonium* endophytes to performance of animals and grass plants, *Agric. Ecosyst. Environ.,* 44:2–12, 1993.

43. **Johnson, M. C., L. P. Bush, and M. R. Siegel,** Infection of tall fescue with *Acremonium coenophialum* by means of callus culture, *Plant Dis.,* 70:380–382, 1986.

44. **Johnson, M. C., D. L. Dahlman, M. R. Siegel, L. P. Bush, G. C. M. Latch, D. A. Potter, and D. R. Varney,** Insect feeding deterrents in endophyte-infected tall fescue, *Appl. Environ. Microbiol.,* 49:568–571, 1985.

45. **Kearney, J. F., W. A. Parrott, and N. S. Hill,** Infection of somatic embryos of tall fescue with *Acremonium coenophialum*, *Crop Sci.,* 31:979–984, 1991.

46. **Kelrick, M. I., N. A. Kasper, T. L. Bultman, and S. Taylor,** Direct interactions between infected and uninfected individuals of *Festuca arundinacea*: differential allocation to shoot and root biomass, In: S. S. Quisenberry and R. E. Joost, Eds., *Proc. Int. Symp.* Acremonium/*Grass Interactions*, Louisiana Agriculture Experiment Station, Baton Rouge, 1990, 21.

47. **Keogh, R. G. and R. J. Clements,** Grazing management: a basis for control of ryegrass staggers, In: D. E. Hume, G. C. M. Latch, and H. S. Easton, Eds., *Proc. 2nd Int. Symp.* Acremonium/*Grass Interactions*, AgResearch, Grasslands Research Centre, Palmerston North, New Zealand, 1993, 129.

48. **Keogh, R. G. and B. A. Tapper,** Acremonium lolii, lolitrem B, and peramine concentrations within vegetative tillers of perennial ryegrass, In: D. E. Hume, G. C. M. Latch, and H. S. Easton, Eds., *Proc. 2nd Int. Symp.* Acremonium/*Grass Interactions*, AgResearch, Grasslands Research Centre, Palmerston North, New Zealand, 1993, 81.

49. **Keogh, R. G. and T. M. Lawrence,** Influence of *Acremonium lolii* presence emergence and growth of ryegrass seedlings, *N.Z. J. Agric. Res.,* 30:507–510, 1987.

50. **Kimmons, C. A., K. D. Gwinn, and E. C. Bernard,** Nematode reproduction on endophyte-infected and endophyte-free tall fescue, *Plant Dis.,* 74:757–761, 1990.

51. **Kirfman, G. W., R. L. Brandenburg, and G. B. Garner,** Relationship between insect abundance and endophyte infestation level in tall fescue in Missouri, *J. Kansas Entomol. Soc.,* 59:552–554, 1986.

52. **Latch, G. C. M.,** Physiological interactions of endophytic fungi and their hosts: botic stress tolerance imparted to grasses by endophytes, *Agric. Ecosyst. Environ.,* 44:143–156, 1993.

53. **Latch, G. C. M. and M. J. Christensen,** Artifical infections of grasses with endophytes, *Ann. Appl. Biol.,* 107:17–24, 1985.

54. **Latch, G. C. M., W. F. Hunt and D. R. Musgrave,** Endophytic fungi affect growth of perennial ryegrass, *N.Z. J. Agric. Res.,* 28:165–168, 1985.

55. **Leuchtmann, A.,** Systematics, distribution, and host specificity of grass endophytes, *Nat. Toxins,* 1:150–162, 1992.

56. **Leuchtmann, A. and K. Clay,** Nonreciprocal conpatibility between *Epichloe typhina* and four host grasses, *Mycologia,* in press, 1993.

57. **Leuchtmann, A. and K. Clay,** Isozyme variation in the *Acremonium/Epichloe* fungal endophyte complex, *Phytopathology,* 80:1133–1139, 1990.

58. **Luu, K. T., A. G. Matches, and E. J. Peters,** Allelopathic effects of tall fescue on birdsfoot trefoil as influenced by N fertilization and seasonal changes, *Agron. J.,* 74:805–808, 1982.

59. **Lyons, P. C., J. J. Evans, and C. W. Bacon,** Effects of the fungal endophyte *Acremonium coenophialum* on nitrogen accumulation and metabolism in tall fescue, *Plant Physiol.,* 92:726–732, 1990.

60. **Maclean, B., C. Matthew, G. C. M. Latch, and D. J. Barker,** The effect of endophyte on drought resistance in tall fescue, In: D. E. Hume, G. C. M. Latch, and H. S. Easton, Eds., *Proc. 2nd Int. Symp.* Acremonium/*Grass Interactions*, AgResearch, Grasslands Research Centre, Palmerston North, New Zealand, 1993, 165.

61. **Marshall, D., L. R. Nelson, and B. Tunali,** The occurrence of *Acremonium* and other endophytic fungi in the indigenous wild cereals of Turkey, In: D. E. Hume, G. C. M. Latch, and H. S. Easton, Eds., *Proc. 2nd Int. Symp.* Acremonium/*Grass Interactions*, AgResearch, Grasslands Research Centre, Palmerston North, New Zealand, 1993, 8.

62. **Miles, O., R. T. Wilkens, A. D. Hawkes, S. C. Munday, and N. R. Towers,** Synthesis and tremorgenicity of paxitriols and lolitiol: possible biosynthetic precursors of lolitrem B, *J. Agric. Food Chem.,* 40:234–238, 1992.

63. **Muegge, M. A., S. S. Quisenberry, G. E. Bates, and R. E. Joost,** Influence of *Acremonium* infection and pesticide use on seasonal abundance of leafhoppers and froghoppers (Homoptera:Cicadellidae; Cercopidae) in tall fescue, *Environ. Ent.,* 20:1531–15336, 1991.

64. **O'Sullivan, B. D. and G. C. M. Latch,** Infection of plantlets, derived from ryegrass and fescue meristems, with *Acremonium* endophytes, In: D. E. Hume, G. C. M. Latch, and H. S. Easton, Eds., *Proc. 2nd Int. Symp.* Acremonium/*Grass Interactions*, AgResearch, Grasslands Research Centre, Palmerston North, New Zealand, 1993, 16.

65. **Oliver, J. W., R. G. Powell, L. K. Abney, R. D. Linnabary, and R. J. Petroski,** N-acetyl loline-induced vasoconstriction of the latral saphenous vein (cranial branch) of cattle, In: S. S. Quisenberry and R. E. Joost, Eds., *Proc. Int. Symp.* Acremonium/*Grass Interactions*, Louisiana Agriculture Experiment Station, Baton Rouge, 1990, 239.

66. **Pedersen, J. F., R. Rodriguez-Kabana, and R. A. Shelby,** Ryegrass cultivars and endophyte in tall fescue affect nematodes in grass and succeeding soybean, *Agron. J.,* 80:811–814, 1988.

67. **Pedersen, J. F., C. S. Hoveland, and R. L. Haaland,** *Performance of Tall Fescue Varieties in Alabama,* Alabama Agriculture Experiment Station, Auburn University, Circ. 262, 1982.

68. **Penn, J. I., I. Garthwaite, M. J. Christensen, C. M. Johnson, and N. R. Towers,** The importance of paxilline in screening for potentially tremogenic *Acremonium* isolates, In: D. E. Hume, G. C. M. Latch, and H. S. Easton, Eds., *Proc. 2nd Int. Symp.* Acremonium/*Grass Interactions*, AgResearch, Grasslands Research Centre, Palmerston North, New Zealand, 1993, 88.

69. **Peters, E. J.,** Toxicity of tall fescue to rape and birdsfoot trefoil seeds and seedling, *Crop Sci.,* 8:650–653, 1968.

70. **Petrini, O., T. N. Sieber, L. Toti, and O. Viret,** Ecology, metabolite production, and substrate utilization in endophytic fungi, *Nat. Toxins,* 1:185–196, 1992.

71. **Petroski, R. J., D. L. Dornbos, Jr., and R. G. Powell,** Germination and growth inhibition of annual ryegrass (*Loliurn multiflorum* L.) and alfalfa (*Medicago sativa* L.) by loline alkaloids and synthetic N-acyl loline derivatives, *J. Agric. Food Chem.,* 38:1716–1718, 1990.

72. **Pless, C. D., K. D. Gwinn, A. M. Cole, D. B. Chalkley, and V. C. Gibson,** Development of resistance by *Drosophila melanagaster* (Diptera:Drosophilidae) to toxic factors in powdered *Acremonium*-infected tall fescue, In: D. E. Hume, G. C. M. Latch, and H. S. Easton, Eds., *Proc. 2nd Int. Symp.* Acremonium/*Grass Interactions*, AgResearch, Grasslands Research Centre, Palmerston North, New Zealand, 1993, 170.

73. **Porter, J. K. and F. N. Thompson, Jr.,** Effects of fescue toxicosis on reproduction in livestock, *J. Anim. Sci.,* 70:1594–1603, 1992.

74. **Powell, R. G. and R. J. Petroski,** Alkaloid toxins in endophyte-infected grasses, *Nat. Toxins* 1:163–170, 1992.

75. **Prestidge, R. A. and O. J.-P. Ball,** The role of endophytes in alleviating plant biotic stress in New Zealand, In: D. E. Hume, G. C. M. Latch, and H. S. Easton, Eds., *Proc. 2nd Int. Symp. Plenary Papers* Acremonium/*Grass Interactions,* AgResearch, Grasslands Research Centre, Palmerston North, New Zealand, 1993, 88.

76. **Read, J. C. and B. J. Camp,** The effect of fungal endophyte *Acremonium coenophialum* in tall fescue on animal performance, toxicity, and stand maintenance, *Agron. J.,* 78:848–850, 1986.

77. **Richardson, M. D., C. W. Bacon, N. S. Hill, and D. M. Hinton,** Growth and water relations of *Acremonium coenophialum,* In: D. E. Hume, G. C. M. Latch, and H. S. Easton, Eds., *Proc. 2nd Int. Symp.* Acremonium/*Grass Interactions*, AgResearch, Grasslands Research Centre, Palmerston North, New Zealand, 1993, 181.

78. **Rottinghaus, G. E., G. B. Garner, C. N. Cornell, and J. L. Ellis,** HPLC method for quantitating ergovaline in endophyte-infected tall fescue: seasonal variation of ergovaline levels in stems with leaf sheaths, leaf blades, and seed heads, *J. Agric. Food Chem.,* 39:112–115, 1991.

79. **Saki, K. I.,** Competition in plants and its relation to selection, *Cold Spring Harbor Symp. Quant. Biol.,* 20:137–157, 1955.

80. **Schardl, C. L. and H.-F. Tsai,** Molecular biology and evolution of the grass endophytes, *Nat. Toxins,* 1:171–184, 1992.

81. **Schardl, C. L., J.-S. Liu, J. F. White, R. A. Finkel, Z. An, and M. R. Siegel,** Molecular phylogenetic relationships of nonpathogenic grass mycosymbionts and clavicipitaceous plant pathogens, *Plant Syst. Evol.,* 178:27–41, 1991.

82. **Schrnidt, D.,** Effects of *Acremonium uncinatum* and a Phialophora-like endophyte on vigour, insect and disease resistance of meadow fescue, In: D. E. Hume, G. C. M. Latch, and H. S. Easton, Eds., *Proc. 2nd Int. Symp.* Acremonium/*Grass Interactions*, AgResearch, Grasslands Research Centre, Palmerston North, New Zealand, 1993, 185.

83. **Schmidt, S. P. and T. G. Osborn,** Effects of endophyte-infected tall fescue on animal performance, *Agric. Ecosyst. Environ.,* 44:233–262, 1993.

84. **Siegel, M. R.,** Acremonium endophytes: our current state of knowledge and future directions for research, *Agric Ecosyst. Environ.,* 44:301–321, 1993.

85. **Siegel, M. R. and G. C. M. Latch,** Expression of antifungal activity in agar culture by isolates of grass endophytes, *Mycologia,* 83:525–537, 1991.

86. **Siegel, M. R. and C. L. Schardl,** Fungal endophytes of grasses: detrimental and beneficial associations, In: J. H. Andrew and S. S. Hirano, Eds., *Microbial Ecology of Leaves,* Springer-Verlag, Berlin, 1991, 198.

87. **Siegel, M. R., G. C. M. Latch, L. P. Bush, F. F. Fannin, D. D. Rowan, B. A. Tapper, C. W. Bacon, and M. C. Johnson,** Fungal endophyte-infected grasses: alkaloid accumulation and aphid response, *J. Chem. Ecol.,* 16:3301–3315, 1990.

88. **Siegel, M. R., M. C. Johnson, D. R. Varney, W. C. Nesmith, R. C. Buckner, L. P. Bush, P. B. Burrus, II, T. A. Jones, and J. A. Boling,** A fungal endophyte in tall fescue: incidence and dissemination, *Phytopathology,* 74:932–937, 1984.

89. **Sutherland, B. L. and J. H. Hoglund,** Effect of ryegrass containing the endophyte *Acremonium lolii* on the performance of associated white clover and subsequent crops, *Proc. N.Z. Grasslands Assoc.,* 50:265–2690, 1989.

90. **Testereci, H.,** Ergovaline, an ergopeptine alkaloid from toxic tall fescue, purification and intravenous infusion into the bovine to measure toxicity response, Ph.D. thesis, University of Missouri, Columbia (Diss. Abstr.), 1991.

91. **Welty, R. E., A. M. Craig, L. L. Blythe, M. D. Azevedo, J. Bartlett, D. K. Bilich, C. McNeal, M. E. Melbye, and G. Gingrich,** Endophyte and ergovaline content of seed and straw of tall fescue and perennial ryegrass, In: D. E. Hume, G. C. M. Latch, and H. S. Easton, Eds., *Proc. 2nd Int. Symp.* Acremonium/*Grass Interactions,* AgResearch, Grasslands Research Centre, Palmerston North, New Zealand, 1993, 94.

92. **West, C. P., E. Izekor, K. E. Turner, and A. A. Elmi,** Endophyte effects on growth and persistance of tall fescue along a water-supply gradient, *Agron. J.,* 85:1–7, 1993.

93. **West, C. P., D. M. Oosterhuis, and S. D. Wullschleger,** Osmotic adjustment in tissues of tall fescue in response to water deficit, *Environ. Exp. Bot.,* 30:149–156, 1990.

94. **West, C. P., E. Izekor, D. M. Oosterhuis, and R. T. Robbins,** The effect of *Acremonium coenophialum* on the growth and nematode infestation of tall fescue, *Plant Soil,* 112:3–6, 1988.

95. **White, J. F.,** Endophyte-host associations in grasses. XIX. A systematic study of some sympatric species of *Epichloe* in England, *Mycologia,* 85:444–455, 1993.

96. **White, J. F.,** Endophyte-host associations in forage grasses. XI. A proposal concerning origin and evolution, *Mycologia,* 80:442–446, 1988.

97. **White, J. F., Jr., G. Morgan-Jones, and A. C. Morrow,** Taxonomy, life cycle, reproduction and detection of *Acremonium endophytes, Agric. Ecosyst. Environ.,* 44:13–37, 1993.

98. **White, R. H., M. C. Engelke, S. J. Morton, J. M. Johnson-Circalese, and B. A. Ruemmele,** *Acremonium* endophyte effects on tall fescue drought tolerance, *Crop Sci.,* 32:1392–1396, 1992.

99. **Wilkinson, S. R.,** Influence of endophytic infection of KY 31 tall fescue on yield response to irrigation, cutting management and competition with Tifton 44 Bermuda grass, In: D. E. Hume, G. C. M. Latch, and H. S. Easton, Eds., *Proc. 2nd Int. Symp.* Acremonium/*Grass Interactions,* AgResearch, Grasslands Research Centre, Palmerston North, New Zealand, 1993, 189.

100. **Woodburn, O. J., J. R. Walsh, J. Z. Foot, and P. G. Heazlewood,** Seasonal ergovaline concentrations in perennial ryegrass cultivars of differing endophyte status, In: D. E. Hume, G. C. M. Latch, and H. S. Easton, *Proc. 2nd Int. Symp.* Acremonium/*Grass Interactions,* AgResearch, Grasslands Research Centre, Palmerston North, New Zealand, 1993, 100.

Chapter 11

Molecular and Genetic Methodologies and Transformation of Grass Endophytes

Christopher L. Schardl

CONTENTS

I. INTRODUCTION

The *Acremonium* endophytes — those fungi with a close phylogenetic relationship to the ascomycetous genus *Epichloë* (Clavicipitaceae) — have a remarkable life cycle that, though perhaps not unique, is nevertheless unusual among plant-biotrophic fungi. Many species of *Epichloë* and *Acremonium* endophytes (see Chapter 1) are capable of clonal dissemination via endophytic infection of host seeds, during which they neither produce external spores nor cause demonstrable disease on their hosts.[5] The extreme circumstance typifies the symbionts that I will refer to as endophytes *sensu stricto*, which have never been known to produce infectious spores in nature and, apparently, rely entirely on seed transmission. Because of this life cycle, they are confined to host matrilines in much the same way as are cellular organelles (e.g., mitochondria and plastids). However, they differ from organelles in that the endophytes grow intercellularly, rather than intracellularly, and are not required for basic life processes of their host grasses. The dependence of a host on an endophyte is ecological, being based on the profound enhancement of stress tolerance, that the endophyte can confer on its host.

There is considerable evidence that some species of grasses require their endophytes, for maximal fitness under various environmental conditions characterized by naturally occurring stresses.[14] The combination of highly efficient seed transmission and strong selective pressure in favor of the symbiotic systems causes them to be maintained through many host generations. Thus, the endophytes with their associated genomes are heritable and often very stable. A grass–fungus symbiotic entity can be conceptualized as a single unit, otherwise known as a symbiotum (pl. symbiota), with a definable genotype.[64] Thus, symbiotum genotypes can be manipulated in a remarkable fashion. The endophytes can be removed;[17] they can be grown in culture, mutated, or transformed;[43,72] and they can be introduced either back into their normal host grasses[36] or into other grass species if the combinations are compatible.[33,63] The ability to take apart and assemble symbiota provides an underpinning and a rationale for endophyte biotechnology.

The present state of research on grass endophytes has begun to open the door to possible biotechnological applications. One important advance is the advent of transformation systems whereby novel genes may be introduced into these organisms.[43,72] Although this is a crucial achievement, other technological advances are also needed for an informed program of genetic manipulation of the endophytes. For example, much more information is required on the biosynthetic pathways and regulation of the pathways, for synthesis of the various alkaloid types (see Chapter 10). Knowledge of the biochemical basis for enhanced drought tolerance by the host when infected with the endophyte remains elusive (see Chapter 7). Even fundamental research on the structure and expression of genes in the Clavicipitaceae has, so far, received little attention. Aspects of the genome structures and genome sizes of the endophytes are beginning to be explored (Schardl, Tsai, and Staben, unpublished data). Systems for both sexual and parasexual genetic manipulation are only now beginning to be developed for this group of organisms.[57] All of these efforts should be tied in with basic biological studies. There is some work on the compatibility relationships between endophyte genotypes and host species (see Chapter 2), and the possibility of gene-for-gene interactions has not yet been broached. The few published studies[33,40] suggest that host compatibility is a matter of considerable complexity. Likewise, the biochemical, morphological, and genetic characteristics of the endophytes are highly variable,[13,56,63,83] but the connections between these characteristics and host specificity have not yet been explored in detail.

All of these studies must account for the multiorganismal nature of the systems. Plant characteristics can undoubtedly be affected by fungal genotype, and the expression of fungal traits can be affected by plant genotype. A remarkable example of this is the expression of ergovaline, a fungal product, in tall fescue–endophyte associations. Even though it is synthesized from common constituents by the fungal symbiont,[4] the host genotype can significantly affect ergovaline levels.[27] Another example is the loline alkaloid class (saturated aminopyrrolizidines). These compounds are produced at very high levels in some symbiota, but have not been observed either in uninfected plants or in fungal culture (see Chapter 10).

This review will touch on areas of endophyte research that may lead to the most promising biotechnological applications. Among them is, of course, the prospect of using DNA-mediated transformations and related technologies to modify the endophytes. However, numerous other approaches can be taken, either individually or in combination, and these will also be discussed.

II. NATURALLY OCCURRING ENDOPHYTES

Naturally occurring endophytes may come from three types of source material: commonly used cultivars of turf and forage grasses, wild populations of the same grass species, or different grass species. The most commonly used are endophytes already infecting cultivars. If it is desired to maintain these infections, then care should be taken to ensure viability. Treatment, age, and storage conditions of the seed are crucial factors.[51] Endophytes that produce a broad array of alkaloids in abundance[63] are highly desirable for turf, since all antiherbivore activities, including antimammalian activity, may be of benefit. However, toxicosis to livestock is a major concern if the grass is used for forage.[35,46]

A second approach is to identify endophytes from wild populations and move them into cultivars of the same host species. Interestingly, endophytes of tall fescue and perennial ryegrass, commonly used in North America and Australia, are producers of high levels of both anti-insect and antimammalian alkaloids,[63] yet a survey of endophytes in native regions of Europe and North Africa demonstrated considerably more variation in alkaloid profiles.[13] It may be speculated that those symbiota that most strongly deter grazing have performed best in the pastoral environments, so that certain genotypes were selectively spread by colonists throughout North America, Australia, and New Zealand. Evidence supporting this speculation is the common occurrence, in wild grass populations in Europe and North Africa, of endophyte genotypes that produce less of the known antimammalian alkaloids.[13] In the future the use of endophytes (either natural or genetically engineered) with little or no antimammalian activity might render the hosts more subject to overgrazing. However, knowing the endophyte characteristics in forage grass cultivars should permit more careful and appropriate pastoral management.

Useful endophytes with the desired activities and secondary product profiles may also be identified in other grass species. There has been little published information indicating the feasibility of this approach. A recent study,[33] and unpublished results from M. R. Siegel (personal communication), so far suggest that endophytes that are moved into new host species are often less apt to be maintained than in their normal hosts. Therefore, this option presently appears less desirable than the use of endophytes from the same host species.

The possibility of using multiple endophyte genotypes in combination in forage grass cultivars (like crop multilines) has so far received little consideration. Because the endophytes are important for defense, such an approach might help minimize the effects of catastrophic biotic assaults (disease epidemics). In fact, antagonistic parasites have been viewed as a major, and perhaps the most important, selective force for genotypic variation in plants and animals, but such variation is normally provided by sexual recombination.[37] If this is true, the biology of strictly seedborne endophytes presents an apparent paradox because their asexual nature should reduce their potential for genetic variation. If parasite populations are able to respond to, and overcome, resistance factors conferred by the mutualistic endophytes (e.g., alkaloids), then benefits of the symbionts to the hosts may be eroded.

Recent studies of genetic variation in endophytes have presented a possible resolution of this paradox. There is actually much genotypic variation even among endophytes of individual host species. This variation was observed at the level of DNA sequences[1] and allozyme profiles of the endophytes, and at the level of alkaloid profiles of the symbiota.[13] For example, among the tall fescue symbiota, some produce ergot alkaloids, lolitrems, and peramine; others produce ergot alkaloids, peramine, and saturated pyrrolizidines (lolines); and others produce subsets of these (e.g., lolines and peramine). Wild populations or agricultural multilines that have such variation may provide a mixed set of antiherbivore and antiparasite activities, thus reducing the potential for, or severity of, epidemics.

The observed diversity of endophytes leads to the question of their origin. Clearly, *Acremonium* endophytes are derived from sexual states of *Epichloë* species.[57] Phylogenetic evidence indicates that, in some cases, the evolution of endophyte genotypes involves infections of *Epichloë* endophytes across host species lines.[56] That this appears to be a fairly common process[1,41,56] is surprising because relatively few attempts to move *Epichloë* between hosts have succeeded in generating stable associations (M. R. Siegel, personal communication, but see Leuchtman and Clay[40]). Recently, a more detailed phylogeny of endophytes of tall fescue[71] and of perennial ryegrass[55] has indicated that *Epichloë* host transfers have often involved genetic interaction between the infecting *Epichloë* strains and endophytes already resident in the plants or seed. The consequence is genetic hybridization of the endophyte with the *Epichloë* strain that has invaded the symbiotum. Such a process, though not related to the sexual cycle of *Epichloë*, nevertheless should have the effect of providing new and radically different genotypes of endophytes (essentially new imperfect species) that can then be inherited by maternal-line transmission to succeeding generations of the grass.

III. ENDOPHYTE BREEDING

Although the sexual states of the *Epichloë/Acremonium* endophytes are not strict mutualists (they can cause choke disease), many isolates of *Epichloë* may behave predominantly as seed-disseminated endophytes.[54,80] The sexual stage of *Epichloë* occurs only during pathogenesis. The leaf sheaths around immature inflorescences become surrounded by mycelial structures (stromata), which then give rise to sterigmata and produce spermatia. The spermatia can be transferred between stromata, a process that normally is facilitated by a symbiotic insect, *Phorbia phrenione*.[10] If the spermatia and the recipient fungus are compatible, then perithecia (fruiting structures) develop, within which are the developing meiotic spores (ascospores).

The compatibility of two parents in a sexual mating depends upon two factors. First, because *Epichloë* exhibits bipolar heterothallism,[81,82] the parents must be of opposite mating type. Inoculations of a stroma of one mating type (e.g., *mat-1*) with spermatia of the other (e.g., *mat-2*) soon results in a discernible change in the morphology of the recipient stroma.[57] Within days a thicker, white mycelial growth begins at the point of inoculation, and raised white bumps emerge. These reactions are observed whether or not the cross will ultimately lead to fertile perithecia, but are never seen in test matings between individuals of the same mating type. Even if a cross involves opposite mating types, further development of the perithecia may stop because the parents do not meet another criterion; to complete the cycle, they must be of the same biological species, also referred to as a mating population.[57,81]

The true test of sexual compatibility is the ultimate production of ascospores. These are forcibly ejected from mature stromata. Therefore, collection of the spores can be done by taping a mature stroma to the lid of a water agar plate, inverting the plate, and periodically checking the area above the stroma for the filamentous ascospores. Sometimes these are approximately 300 µm in length, but other times they can be much shorter due to septation and fragmentation of the spores in the asci prior to ejection.[81] The ascospores have sufficient reserves to undergo germination on water agar. Microscopic inspection of the process[6,7] has revealed that the spores first give rise to many phialides with conidia (vegetative spores) and that the conidia subsequently germinate to also produce phialides and conidia. Only after two or three rounds of this iterative germination process are vegetative hyphae produced. The significance of the iterative germination process, and indeed the specific role of the ascospores, has not been demonstrated in nature. It is presumed that the process is associated with infections either of vegetative plant tissues or, via the stigmata, of the ovules and developing seeds of the host grasses.[7]

Identification of the sexual cycle and the mating types of an *Epichloë* endophyte opens the way for genetic analysis. Determinants of host range and specificity, and genes that regulate alkaloid levels, are areas that clearly lend themselves to genetic analysis. Ultimately, it is important to reestablish symbiosis with host plants. Even though the natural infection cycle has never been demonstrated, there are artificial means to introduce *Epichloë* or endophytes into seedlings[36] or into embryogenic callus cultures, of the host grass.[27] Once this is accomplished, alkaloid profiles can be determined. Production of some alkaloids depends on the association, whereas other alkaloids may not be reliably produced in culture, and their production *in planta* is much more dependable.[4,63]

It may sometimes prove difficult to reintroduce *Epichloë* endophytes and other endophytes into hosts. Often, only a minority of inoculated plants will become infected (particularly in inoculations with asexual endophytes).[36] Therefore, a sufficient number of inoculations must be undertaken to allow a statistical evaluation of compatibility.[38] If the *Acremonium* or *Epichloë* endophyte is genetically modified or altered in any way (e.g., mutation, transformation, or sexual recombination), the introduction of the original strain or parental types into the natural host(s) must also be done as a positive control.

IV. TRANSFORMATION OF ENDOPHYTES WITH EXOGENOUS DNA

Transformation systems have now been described for two asexual endophyte species. An endophyte of perennial ryegrass, designated LpTG-2 (taxonomically distinct from *Acremonium lolii* Latch, Christensen & Samuels) isolate 187BB (= Lp1), was transformed to hygromycin resistance,[43] using a chimeric *hph* gene[23,32] in the plasmid pAN7-1.[45] This plasmid contains the *Emericella (Aspergillus) nidulans* (Eidam) Vuillemin *TrpC* 5' region to promote transcription of the *Escherichia coli hph* gene. The effect of the introduction of the chimeric gene into fungal genomes is resistance to the antibiotic hygromycin B.

In the same study, a similarly modified *uidA* gene (but with the *Em. nidulans gpd* promoter) was introduced into this endophyte. The *uidA* gene encodes the widely used histological marker, *beta*-glucuronidase.[50] Use of *uidA* provided two advantages. First, the frequency of cotransformation of two plasmids containing two different and easily identifiable markers could be determined. The frequency was found to be rather high (up to 80%). The second advantage was that expression of the gene could be demonstrated *in planta*. This is important for future work if genes are to be constructed and designed especially for expression by the endophyte in its natural niche.

As discussed later, some objectives of transformation are the introduction of novel characteristics for anti-insect activity, or the expression of antisense RNA for the purpose of down-regulating expression of genes for alkaloid biosynthesis. These will require knowledge of *in planta* expression of the transcription promoters of gene transcription.

The means of introducing DNA into isolate 187BB was a modification of the standard protocols for transforming fungal protoplasts.[77,84] First, cell walls were removed by treatment with the enzyme preparation Novozyme 234 in a high-osmosis solution. The resulting cell protoplasts (spheroplasts) were purified by centrifugation and then treated first with $CaCl_2$ and plasmid DNA (containing the selectable, modified *hph* gene) and then with polyethyleneglycol (PEG 8000) in sorbitol osmoticum. Finally, they were plated in top agar onto nutrient agar plates containing hygromycin B (200 µg ml^{-1}). This proved to be a highly efficient procedure, yielding up to 500 transformants per microgram of plasmid DNA per 2.5×10^7 protoplasts.

An endophyte of tall fescue, *Acremonium coenophialum* Morgan-Jones & Gams, was transformed by Tsai et al. using a very different procedure. Electroporation was used to introduce the DNA into protoplasts.[72] Although the transformation frequency by electroporation may be lower, this technique has the advantage of less manipulation once the protoplasts have been prepared. Also, in some fungi, electroporation might enhance homologous integration or gene conversion,[26,48] processes that allow specific genes to be modified or inactivated in the fungal genome.[8,69] Whether homologous integration will occur to an appreciable extent in an electroporated endophyte is presently being investigated.

Optimization of electroporation parameters involved varying each independently while maintaining other parameters at standard settings.[72] Ultimately, each of the optimum parameters was close enough to the standard settings that it is plausible that a global optimum was identified for *A. coenophialum* transformation with this particular apparatus (the Gene Pulser™ of BioRad, Richmond, CA) and the electroporation buffer. However, because of the number of parameters involved and the difficulty in obtaining large numbers of *A. coenophialum* protoplasts, it remains infeasible to prove that this was the best possible combination of conditions. The optima obtained were 78 µg ml⁻¹ DNA, 2 × 10⁹ protoplasts per milliliter in buffer consisting of 1.2 *M* D-sorbitol, 10 m*M* tris-HCl, pH 7.5, 10 m*M* CaCl$_2$ (STC: alternative buffers were not explored), with settings of 25 µF and 5000 v cm⁻¹. Serial resistance settings of 100 to 200Ω were best, but in order to avoid damage to the instrument, lower resistances were never used. Under the optimized conditions, the killing rates were 81 to 93% (7 to 19% survival) of the originally viable protoplasts (determined under nonselective conditions).

The efficiencies of electroporative transformation showed considerable dependence upon the transformation vector.[72] A series of vectors were constructed using *hph* as the selectable marker, and fungal promoters to drive its transcription. The overall strategy was to use regions of the *E. typhina beta*-tubulin gene (*tub2*), extending from various points 5′ of the *tub2* start codon and ending after codon 3, in a translational fusion with the *hph* sequence from codon 4 to the termination codon (Figures 1 and 2). Such a construct gave active product because this hygromycin B phosphotransferase does not require its native *N*-terminus for activity, even though its *C*-terminus is critical.[9] A 235-basepair segment of the *tub2* upstream promoter in pKAES101 (Figure 1) was sufficient for transcription of the chimeric *tub2-hph* gene in the transformed fungus. Higher transformation levels were achieved with approximately 1100 basepairs in pKAES105 (Figure 1). Inclusion of the 3′ region of the *trpC* region of *E. nidulans* also helped increase transformation, perhaps because it allowed efficient 3′-terminal processing of the transcript, ultimately enhancing the stability of the mRNA. The highest frequencies of transformation to hygromycin resistance were obtained with plasmid pKAES105 and ranged from 10 to 20 µg⁻¹ DNA. Such frequencies make feasible the generation of 1000 or more independent transformants in a single experiment.

Having accomplished endophyte transformation, several avenues of genetic engineering are now open for investigation. Among these are disruption of genes for biosynthesis of ergot alkaloids and tremorgens, antisense RNA production to interfere with expression of such genes, and introduction of new genes into endophyte genomes to enhance their beneficial qualities.

V. ALTERING SECONDARY METABOLISM

A. IDENTIFICATION OF GENES FOR ALKALOID BIOSYNTHESIS

The biosynthetic pathways of even the simpler alkaloids are likely to involve more than one step from common precursors. Ergoline alkaloids have particularly complex pathways. Several steps in the clavine and ergoline alkaloid synthesis pathway may be appropriate for eventual disruption, but the first determinant and rate-limiting step is a logical choice. This step is dimethylallyl pyrophosphate: L-tryptophan dimethylallyltransferase, also called dimethylallyltryptophan synthase (DMAT synthase).[34,47] The enzyme has been isolated from *Claviceps purpurea* Fr. (Tul.) and *C. fusiformis* Loveless,[21,61] and because the endophytes are related to these fungi, they probably also have this enzyme (although activity has not yet been reported from endophytes).

Other alkaloids that have been implicated as potent antimammalian factors are of the indolediterpenoid class, which includes the tremorgens paxilline and lolitrem B.[52] There is less information regarding the biosynthesis of these alkaloids, but other fungi may serve as appropriate model systems.[60] One species noted for producing high levels of paxilline is *Penicillium paxilli* Bainier.[15] In the future, such systems should be a valuable source of experimental information that allows the synthesis pathway to be elucidated and the corresponding enzymes to be isolated.

Identification of the gene encoding DMAT synthase and other enzymes for alkaloid biosynthesis may be accomplished by one of several approaches. One is to obtain peptide sequences from fragments of the enzyme (C. D. Poulter, personal communication), which can then be applied to a strategy involving the polymerase chain reaction (PCR) called the mixed oligonucleotide primer amplification of cDNA (MOPAC).[22] In this procedure a mixture of oligonucleotides corresponding to all possible coding sequences of one peptide segment, plus another mixture corresponding to the complement of the coding sequence for another peptide, can be used for PCR amplification of the cDNA sequence between them (cDNA is initially generated from a mRNA template, by the oligonucleotide-primed activity of reverse transcriptase). The MOPAC protocol both selects for those primers with the best match and amplifies the authentic sequence from the cDNA, thus providing an appropriate probe for screening genomic or cDNA libraries.

A second alternative is to use antibodies for screening cDNA-expression libraries.[3] There are a number of vectors, most based on bacteriophage lambda, for cloning of cDNA and expression of the polypeptides

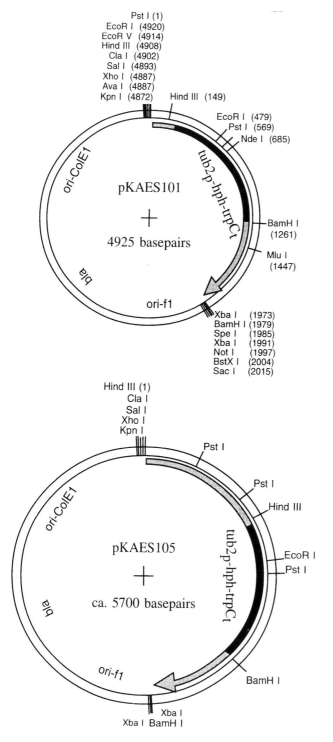

Figure 1 Maps of pKAES101 and pKAES105. The arrows on the maps indicate the chimeric gene constructs consisting of the bacterial *hph* gene (in black), flanked by controlling elements. The orientation of the arrow is 3′ to 5′, the direction of transcription. The 3′ flanking sequence is derived from the *Epichloë typhina tub2* locus and includes the transcription promoter. The 5′ flanking portion includes the *trpC* terminator from *Emericella nidulans*. Other genes for maintenance and selection in the bacterium *Escherichia coli* are the ColE1 origin of replication and the *bla* gene for beta-lactamase (conferring resistance to ampicillin), respectively. The f1 phage origin of replication allows bacteriophage-mediated rescue of single-stranded DNA from the plasmids, for the purpose of sequence analysis.[75] The total length of pKAES101 is 4925 basepairs (bp), and the length of pKAES105 is approximately 5700 bp.

encoded by the corresponding mRNA. These proteins are released from the cells of the *E. coli* host, upon lysis by the lambda phage. They can then be fixed to membranes and screened using antibodies raised against the protein of interest. As with the previous approach, purification of substantial amounts of the protein is first required. In this case adequate amounts are needed to raise antisera.

Clones of biosynthesis genes may also be detected by virtue of their expression and function. If a mutant lacks one of the steps and is thus incapable of synthesizing the metabolite, then it may be

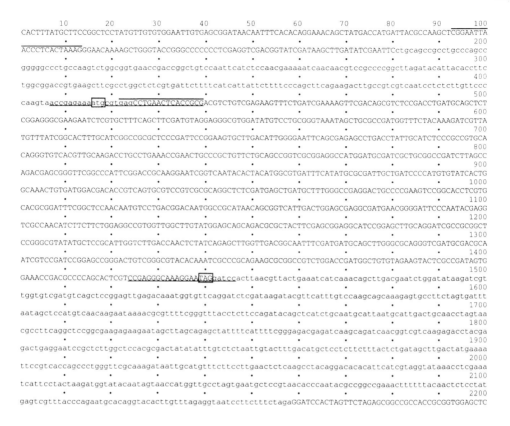

Figure 2 Sequence of the modified *hph* gene in the fungal transformation vector, pKAES101. The sequence is inferred from published sequences of its components.[11,23,42] Alternating upper-case and lower-case letters delineate different regions. The cloning vector, pBluescriptKS+, constitutes 2974 bp of the sequence, a small portion of which is shown here (positions zero to 186 and positions 2159 to 2200). The *E. typhina tub2* 5′ region extends from position 181 to 424, past the third codon. The *hph* reading frame extends from position 423 to 1441 and includes codons 4 to 341 and the stop codon. The *Em. nidulans trpC* 5′ region, including the transcription terminator, extends from position 1441 to 2164. Boxed sequences correspond to the start codon (ATG) and stop codon (TAG) of the chimeric *hph* gene predicted to be expressed in transformed fungal cells. Overbars indicate sequences homologous to oligonucleotide primers, and underlined sequences are complementary to primers, used in simultaneous amplification and splicing (SOE) of the chimeric gene (see Figure 4). The total length of the plasmid is 4925 bp.

complemented by clones from a clone library. The main difficulty in this strategy is obtaining the appropriate mutants. The expression of secondary metabolites by clavicipitaceous fungi in culture is notoriously unreliable.[4,65] However, screening for the mutants *in planta* is tedious because plant inoculation methods remain inefficient. Furthermore, there is a good chance that the majority of the presumptive mutants would be unaltered in biosynthesis genes and may simply have undergone an epigenetic change, silencing the expression of these alkaloids. Finally, if the genes are redundant, rather than single copy, it may be virtually impossible to generate the desired class of mutant. This is a concern in light of recent evidence that most tall fescue endophytes tend to have multiple gene copies, due to their hybrid nature.[71]

A second expression system may be modeled on that used by Weltring et al.[79] to clone the gene for pisatin demethylase from *Nectria haematococca* Berk & Broome. A specific enzyme of choice may be assayed from a number of transformants of a fungus that normally does not express that enzyme. A complete genomic DNA library would be prepared with defined insert sizes: for example, several thousand cosmids, each containing approximately 30 kilobasepairs of genomic DNA of *C. purpurea* or *A. coenophialum*. The library may then be moved into another species, such as *Neurospora crassa* Sheer

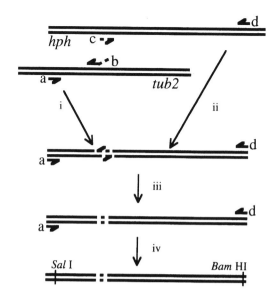

Figure 4 Construction of a chimeric *tub2-hph* gene, using a variation of SOE. Four oligonucleotide primers, designated a, b, c, and d (half arrows), are used. Primers a and b are used for PCR amplification of the 5′ segment, which, in this example, contains the *tub2* promoter of *Epichloë typhina*. Primers c and d are to amplify the 3′ segment, the *hph* gene in this case. The region to be joined is that of primers b and c. To facilitate their joining, the 5′ region of primer b (dashed line) is complementary to the whole of primer c (and not to the target site in *tub2*). Extensions on the 5′ ends of primers a and d can also be used to introduce restriction endonuclease cleavage sites for cloning the amplified, chimeric gene. The entire amplification and splicing is performed in one reaction. Primers b and c are kept in limiting concentration (1 pmol each per 100 μl reaction), and primers a and d are in excess (25 pmol each). For several cycles of primer annealing followed by extension by the action of a thermostable DNA polymerase, two segments of DNA are amplified (i and ii). As primers b and c are used up, the amplified products anneal to one another, and their 3′ ends are extended by the polymerase (iii). This results in the chimeric product, which, in subsequent cycles, is further amplified by extension of primers a and d (iv). If appropriate restriction endonuclease recognition sites (*Sal*I and *Bam*HI in this example) were included in primer a and d sequences, the product may be digested with these enzymes, to generate sticky ends that facilitate ligation into cloning vectors.[3]

derived from a bacterium or a distantly related eukaryote, or when the gene of interest is normally expressed at low levels. In these cases chimeric genes are constructed with controlling elements derived from the same, or a related, fungus. Generally, several hundred basepairs of the regions flanking both the 5′ end and 3′ end of the coding sequence are used. The fungal promoter is most critical, but sometimes inclusion of a fungal 3′ region enhances expression.[72]

Prior to the PCR technique, very cumbersome steps were required to construct such chimeric genes. A PCR-based method, called splicing by overhanging extensions (SOE), was recently described for construction of chimeric genes.[29] A simplification of this method, summarized in Figure 4, was used to construct the *tub2* 5′-*hph* fusion for expression of hygromycin phosphotransferase in *A. coenophialum*.[72] In the simplified version, primers bounding the desired final product (primers a and d) are in excess, and the others (primers b and c) are in limiting concentration. As primers b and c get used up in the PCR reaction, the products themselves act as primers and are extended. Finally, the desired product is amplified (because of the excess of primers a and d) and becomes the predominant product, which can then be ligated into cloning vectors by various means.[30] If appropriate restriction-endonuclease recognition sites have been included in the primers, then the products can be cut with the enzymes, to facilitate cloning. It is best to use different cleavage sites at each end (i.e., incorporated into primers a and d), to enhance the production of recombinant clones and to predetermine the orientation of the clones.

As mentioned earlier, a translational fusion of the *tub2* 5′ region with the selectable marker *hph* was constructed by this method.[72] The complete sequence of a chimeric transformation marker (*tub2* 5′-*hph*-

trpC 3′) is shown in Figure 2. In constructing chimeric genes, if translational fusions are to be used, then it must be determined whether the gene product will be functional when modified in such a fashion.[9] (However, if the constructs are meant for expression of antisense RNA, then production of a functional protein is not required or desirable.)

B. ENHANCING PROTECTIVE BENEFITS

The strategies and considerations involved in the construction of selectable markers for transformation are applicable to the design of other novel genes for introduction into endophytes and other fungi. Genes encoding peptides or proteins that enhance resistance to insects or other parasitic organisms may be of great benefit, particularly if the removal of alkaloid production capabilities results in some loss of resistance of the symbiotum against insects and other pests.

In recent years several genes have been constructed, or modified, and introduced into plants to help confer activity against insects, viruses, or fungal pathogens. Conceivably, some of these may be useful as additional genetic capabilities introduced into the endophytes. The advantage to using an endophyte is that it can be readily introduced into many different breeding lines of its host plant species, generating, in less than a year, new genetic resources for cultivar development. In essence, transformed endophytes can be used as surrogates for transgenic grass plants.[43] For example, transgenic endophytes may encode insecticidal deltatoxins[28] or the antinutritional proteinase inhibitors[31] that have already been demonstrated to provide enhanced insect resistance to transgenic plants.[53,73]

For a fungal gene to cause a demonstrable effect on fitness of the symbiotum, it must direct adequate production of the protein or metabolite, in the appropriate tissues, and the product should act against significant pests of the grass. Preliminary experiments should first be undertaken to determine the toxicity of the gene products, such as particular deltatoxins from *Bacillus thuringiensis*, to insects problematic to tall fescue or other forage grasses that harbor endophytes. An example would be fall or southern armyworm (*Spodoptera* spp.). It is not widely appreciated that the deltatoxins constitute a group of related proteins from *B. thuringiensis* strains. Gene sequences for 13 distinct toxins are known and can be grouped into four classes, as distinguished by their activities against insect species in the families *Lepidoptera* and *Diptera*.[28] Furthermore, even if the enhanced insect resistance provided by one such gene breaks down,[67] other deltatoxin genes may prove to be effective.[16] Thus, *B. thuringiensis* deltatoxins may constitute an appropriate model to test this strategy.

There are also fungal genes known to encode antifungal proteins.[66,68,70] One could introduce such a gene into an endophyte in an effort to enhance the antifungal activity and, thus, disease resistance conferred by the endophyte. Of course, the endophyte itself must either be inherently tolerant to the protein, or a mechanism of tolerance or resistance must be engineered into it. However, such genes for tolerance may not be hard to come by, since they are usually present, as a means of self-protection, in the same fungi that produce the antifungal proteins. Some of the endophytes, as well as other grass symbionts, produce antifungal activities.[62] The nature of these has yet to be elucidated, but once they are known, several possibilities can be considered, such as increasing expression of the genes for antifungal factors, or moving them into other endophytes or into plants.

C. SECONDARY METABOLITES

Most known plant defensive molecules are not primary gene products, but are the end products of complex biosynthetic pathways. Because multiple genes are involved, there has been little work on the manipulation of these pathways. Some require a single step from common plant precursors[58] and have been tested for activity by transformation of new plant species,[59] but most other secondary metabolites remain, for now, beyond our immediate capabilities to manipulate. Eventually it will be desirable to do so, particularly if the products can move systemically in the plant and travel a significant distance from the site of synthesis (namely, the endophyte).

The prospect of introducing into plants some of the endophyte genes for biosynthesis of a protective metabolite is also worth considering. The recent development of a transformation system for tall fescue[78] is a step toward introducing new and important genetic capacities to this plant. Eventually, once genes have been identified for synthesis of loline, peramine, and other compounds responsible for enhanced fitness (including drought tolerance), tall fescue and related grasses may have these capabilities transferred directly into their genetic backgrounds. Before this can be accomplished, these pathways must be elucidated. The molecular genetic manipulations that can now be undertaken with endophytes and their hosts will almost certainly help the effort.

VII. CONCLUSIONS

Endophyte biotechnology depends on recognition of the diversity of endophytes, their host specificities and interactions, and the ability to culture, mutate, and genetically engineer them. In general, the endophytes are used as agents of biological protection from both biotic and abiotic stresses encountered by the host. Their use in agriculture has been extensive, but, for millennia, has also been unwitting. Now that we have the capability to both identify *Epichloë*-type endophytes and characterize their specific genotypes, we can take a more cognizant approach to their use. Four types of modifications were considered: (1) the beneficial characteristics of endophytes may be augmented by introduction of new genes; (2) using gene replacement or gene disruption techniques, or expression of antisense RNA, some of the characteristics of endophytes may need to be eliminated, even though these may constitute protective, ecologically advantageous traits (particularly antimammal activity); (3) novel combinations of host and fungus genotypes may be produced (this may even involve moving endophytes between host species) as another way to tailor the symbiota to specific needs or environments; (4) endophyte genotypes may be combined in multilines of agricultural grass species, in much the same way that different host genotypes may be combined to enhance the overall fitness of a population.

ACKNOWLEDGMENTS

I thank A. D. Byrd and W. Hollin for their help and expertise. This work was funded by a National Science Foundation New Zealand Program grant number INT-912083, U.S. Department of Agriculture NRICGP grant 92-00694, the University of Kentucky Graduate School and the Kentucky Agriculture Experiment Station (publication 93-11-162).

REFERENCES

1. **An, Z.-Q., J.-S. Liu, M. R. Siegel, G. Bunge, and C. L. Schardl,** Diversity and origins of endophytic fungal symbionts of the North American grass *Festuca arizonica, Theor. Appl. Genet.,* 85:366–371, 1992.
2. **Asch, D. K. and J. A. Kinsey,** Relationship of vector insert size to homologous integration during transformation of *Neurospora crassa* with the cloned *am* (GDH) gene, *Mol. Gen. Genet.,* 221:37–43, 1990.
3. **Ausubel, F. M., R. Brent, R. E. Kingston, D. D. Moore, J. G. Seidman, J. A. Smith, and K. Struhl,** Eds., *Current Protocols in Molecular Biology,* John Wiley & Sons, New York, 1993.
4. **Bacon, C. W.,** Procedure for isolating the endophyte from tall fescue and screening isolates for ergot alkaloids, *Appl. Environ. Microbiol.,* 54:2615–2618, 1988.
5. **Bacon, C. W. and J. De Battista,** Endophytic fungi of grasses, In: D. K. Arora, B. Rai, K. G. Mukerji, and G. R. Knudsen, Eds., *Handbook of Applied Mycology. Soil and Plants,* Vol. 1, Marcel Dekker, New York, 1990, 231.
6. **Bacon, C. W. and D. M. Hinton,** Ascosporic iterative germination in *Epichloe typhina, Trans. Br. Mycol. Soc.,* 90:563–569, 1988.
7. **Bacon, C. W. and D. M. Hinton,** Microcyclic conidiation cycles in *Epichloe typhina, Mycologia,* 83:743–751, 1991.
8. **Barnes, D. A. and J. Thorner,** Use of the LYS2 gene for gene disruption, gene replacement, and promoter analysis in *Saccharomyces cerevisiae,* In: J. W. Bennett and L. L. Lasure, Eds., *Gene Manipulations in Fungi,* Academic Press, New York, 1985, 197.
9. **Bilang, R., S. Iida, A. Peterhans, I. Potrykus, and J. Paszkowski,** The 3′-terminal region of the hygromycin-B resistance gene is important for its activity in *Escherichia coli* and *Nicotiana tabacum, Gene,* 100:247–250, 1991.
10. **Bultman, T. L. and J. F. White,** "Pollination" of a fungus by a fly, *Oecologia,* 75:317–319, 1988.
11. **Byrd, A. D., C. L. Schardl, P. J. Songlin, K. L. Mogen, and M. R. Siegel,** The beta-tubulin gene of *Epichloe typhina* from perennial ryegrass (*Lolium perenne*), *Curr. Genet.,* 18:347–354, 1990.
12. **Cambareri, E. B., B. C. Jensen, E. Schabtach, and E. U. Selker,** Repeat-induced G-C to A-T mutations in *Neurospora, Science,* 244:1571–1575, 1989.
13. **Christensen, M. J., A. Leuchtmann, D. D. Rowan, and B. A. Tapper,** Taxonomy of *Acremonium* endophytes of tall fescue (*Festuca arundinacea*), meadow fescue (*F. pratensis*), and perennial ryegrass (*Lolium perenne*), *Mycolog. Res.,* 97:1083–1092, 1993.

14. **Clay, K.,** Fungal endophytes of grasses, *Ann. Rev. Ecol. Syst.,* 21:275–295, 1990.

15. **Cole, R. J., J. W. Kirksey, and J. M. Wells,** A new tremorgenic metabolite from *Penicillium paxilli, Can. J. Microbiol.,* 20:1159–1162, 1974.

16. **Crickmore, N., C. Nicholls, D. J. Earp, T. C. Hodgman, and D. J. Ellar,** The construction of *Bacillus thuringiensis* strains expressing novel entomocidal delta endotoxin combinations, *Biochem. J.,* 270:133–136, 1990.

17. **De Battista, J. P., J. H. Bouton, C. W. Bacon, and M. R. Siegel,** Rhizome and herbage production of endophyte-removed tall fescue clones and populations, *Agron. J.,* 82:651–654, 1990.

18. **Delauney, A. J., Z. Tabaeizadeh, and D. P. S. Verma,** A stable bifunctional antisense transcript inhibiting gene expression in transgenic plants, *Proc. Natl. Acad. Sci. U.S.A.,* 85:4300–4304, 1988.

19. **Frederick, G. D., D. K. Asch, and J. A. Kinsey,** Use of transformation to make targeted sequence alterations at the *am* (GDH) locus of *Neurospora, Mol. Gen. Genet.,* 217:294–300, 1989.

20. **Gebler, J. C. and C. D. Poulter,** Purification and characterization of dimethylallyltryptophan synthase from *Claviceps purpurea, Arch. Biochem. Biophys.,* 296:308–313, 1992.

21. **Gebler, J. C., A. B. Woodside, and C. D. Poulter,** Dimethylallyltryptophan synthase. An enzyme catalyzed electrophilic aromatic substitution, *J. Am. Chem. Soc.,* 114:7354–7360, 1992.

22. **Griffin, L. D., G. R. MacGregor, D. M. Muzny, J. Harter, R. G. Cook, and E. R. B. McCabe,** Synthesis and characterization of a bovine hexokinase 1 cDNA probe by mixed oligonucleotide primed amplification of cDNA using high complexity primer mixtures, *Biochem. Med. Metab. Biol.,* 41:125–131, 1989.

23. **Gritz, L. and J. Davies,** Plasmid-encoded hygromycin B resistance: the sequence of hygromycin B phosphotransferase gene and its expression in *Escherichia coli* and *Saccharomyces cerevisiae, Gene,* 25:179–188, 1983.

24. **Hamilton, A. J., G. W. Lycett, and D. Grierson,** Antisense gene that inhibits synthesis of the hormone ethylene in transgenic plants, *Nature,* 346:284–287, 1990.

25. **Hampel, A., R. Tritz, M. Hicks, and P. Cruz,** 'Hairpin' catalytic RNA model: evidence for helices and sequence requirement for substrate RNA, *Nucl. Acids Res.,* 18:299–304, 1990.

26. **Higgins, D. R. and J. N. Strathern,** Electroporation-stimulated recombination in yeast, *Yeast,* 7:823–831, 1991.

27. **Hill, N. S., W. A. Parrott, and D. D. Pope,** Ergopeptine alkaloid production by endophytes in a common tall fescue genotype, *Crop Sci.,* 31:1545–1547, 1991.

28. **Hofte, H. and H. R. Whiteley,** Insecticidal crystal proteins of *Bacillus thuringiensis, Microbiol. Rev.,* 53:242–255, 1989.

29. **Horton, R. M., Z. Cai, S. N. Ho, and L. R. Pease,** Gene splicing by overlap extension: tailor-made genes using the polymerase chain reaction, *Biotechniques,* 8:528–535, 1990.

30. **Innis, M. A., D. H. Gelfand, J. J. Sninsky, and T. J. White, Eds.,** *PCR Protocols: A Guide to Methods and Applications,* Academic Press, San Diego, 1990.

31. **Johnson, R., J. Narvaez, G. An, and C. Ryan,** Expression of proteinase inhibitors I and II in transgenic tobacco plants: effects on natural defense against *Manduca sexta* larvae, *Proc. Natl. Acad. Sci. U.S.A.,* 86:9871–9875, 1989.

32. **Kaster, K. R., S. G. Burgett, R. N. Rao, and T. D. Ingolia,** Analysis of a bacterial hygromycin B resistance gene by transcriptional and translational fusions and by DNA sequencing, *Nucl. Acids Res.,* 11:6895–6911, 1983.

33. **Koga, H., M. J. Christensen, and R. J. Bennett,** Cellular interactions of some grass/*Acremonium* endophyte associations, *Mycolog. Res.,* 97:1237–1244, 1993.

34. **Krupinski, V. M., J. E. Robbers, and H. G. Floss,** Physiological study of ergot: induction of alkaloid synthesis by tryptophan at enzymatic level, *J. Bacteriol.,* 125:158–165, 1976.

35. **Lacey, J.,** Natural occurrence of mycotoxins in growing and conserved forage crops, In: J. E. Smith and R. S. Henderson, Eds., *Mycotoxins and Animal Foods,* CRC Press, Boca Raton, FL, 1991, 363.

36. **Latch, G. C. M. and M. J. Christensen,** Artificial infections of grasses with endophytes, *Ann. Appl. Biol.,* 107:17–24, 1985.

37. **Law, R. and D. H. Lewis,** Biotic environments and the maintenance of sex — some evidence form mutualistic symbioses, *Biol. J. Linnean Soc.,* 20:249–276, 1983.

38. **Leuchtmann, A. and K. Clay,** Experimental evidence for genetic variation in compatibility between the fungus *Atkinsonella hypoxylon* and its three host grasses, *Evolution,* 43:825–834, 1989.

39. **Leuchtmann, A. and K. Clay,** Isozyme variation in the *Acremonium/Epichloe* fungal endophyte complex, *Phytopathology,* 80:1133–1139, 1990.

40. **Leuchtmann, A. and K. Clay,** Nonreciprocal compatibility between *Epichloe* and four host grasses, *Mycologia,* 85:157–163, 1993.

41. **Liu, J.-S.,** Evolution of the fungal tribe Balansieae (Clavicipitaceae) and related endophytes of grasses, Ph.D. thesis, University of Kentucky, Lexington, 1993.

42. **Mullaney, E. J., J. E. Hamer, K. A. Roberti, M. M. Yelton, and W. E. Timberlake,** Primary structure of the *trpC* gene from *Aspergillus nidulans, Mol. Gen. Genet.,* 199:37–45, 1985.

43. **Murray, F. R., G. C. M. Latch, and D. B. Scott,** Surrogate transformation of perennial ryegrass, *Lolium perenne,* using genetically modified *Acremonium* endophyte, *Mol. Gen. Genet.,* 233:1–9, 1992.

44. **Orr-Weaver, T. L. and J. W. Szostak,** Fungal recombination, *Microbiol. Rev.,* 49:33–58, 1985.

45. **Punt, P. J., R. P. Oliver, M. A. Dingemanse, P. H. Pouwels, and C. A. M. J. J. van den Hondel,** Transformation of Aspergillus based on the hygromycin B resistance marker from *Escherichia coli, Gene,* 56:117–124, 1987.

46. **Raisbeck, M. F., G. E. Rottinghaus, and J. D. Kendall,** Effects of naturally occurring mycotoxins on ruminants, In: J. E. Smith and R. S. Henderson, Eds., *Mycotoxins and Animal Foods,* CRC Press, Boca Raton, FL, 1991, 647.

47. **Rehacek, Z.,** Physiological controls and regulation of ergot alkaloid formation, *Folia Microbiol.,* 36:323–342, 1991.

48. **Richey, M. G., E. T. Marek, C. L. Schardl, and D. A. Smith,** Transformation of filamentous fungi with plasmid DNA by electroporation, *Phytopathology,* 79:844–847, 1989.

49. **Rine, J. and M. Carlson,** *Saccharomyces cerevisiae* as a paradigm for modern molecular genetics of fungi, In: J. W. Bennett and L. L. Lasure, Eds., *Gene Manipulations in Fungi,* Academic Press, New York, 1985.

50. **Roberts, I. N., R. P. Oliver, P. J. Punt, and C. A. M. J. J. van den Hondel,** Expression of the *Escherichia coli* b-glucuronidase gene in industrial and phytopathogenic filamentous fungi, *Curr. Genet.,* 15:177–180, 1989.

51. **Rolston, M. P., M. D. Hare, K. K. Moore, and M. J. Christensen,** Viability of *Lolium* endophyte fungus in seed stored at different moisture contents and temperatures, *N.Z. J. Agric. Res.,* 14:297–300, 1986.

52. **Rowan, D. D.,** Lolitrems, paxilline and peramine: mycotoxins of the ryegrass/endophyte interaction, *Agric. Ecosyst. Environ.,* 44:103–122, 1993.

53. **Ryan, C. A.,** Protease inhibitors in plants — genes for improving defenses against insects and pathogens, *Ann. Rev. Phytopathol.,* 28:425–449, 1990.

54. **Saha, D. C., J. M. Johnson-Cicalese, P. M. Halisky, M. I. van Heemstra, and C. R. Funk,** Occurrence and significance of endophytic fungi in fine fescues, *Plant Dis.,* 71:1021–1024, 1987.

55. **Schardl, C. L., A. Leuchtmann, H.-F. Tsai, M. Collett, D. M. Watt, and D. B. Scott,** Origin of a fungal symbiont of perennial ryegrass by interspecific hybridization of a mutualist with the ryegrass choke pathogen, *Epichloe typhina,* submitted, 1993.

56. **Schardl, C. L., J.-S. Liu, J. F. White, R. A. Finkel, Z. An, and M. R. Siegel,** Molecular phylogenetic relationships of nonpathogenic grass mycosymbionts and clavicipitaceous plant pathogens, *Plant Syst. Evol.,* 178:27–41, 1991.

57. **Schardl, C. L. and H.-F. Tsai,** Molecular biology and evolution of the grass endophytes, *Nat. Toxins,* 1:171–184, 1992.

58. **Schroder, G., J. W. S. Brown, and J. Schroder,** Molecular analysis of resveratrol synthase cDNA, genomic clones and relationship with chalcone synthase, *Eur. J. Biochem.,* 172:161–169, 1988.

59. **Schroder, J. and G. Schroder,** Stilbene and chalcone synthases: related enzymes with key functions in plant-specific pathways, *Z. Naturforsch.,* 45c:1–8, 1990.

60. **Scott, B. and C. Schardl,** Fungal symbionts of grasses: evolutionary insights and agricultural potential, *Trends Microbiol.,* 1:196–200, 1993.

61. **Shibuya, M., H.-M. Chou, M. Fountoulakis, S. Hassam, S.-U. Kim, K. Kobayashi, H. Otsuka, E. Rogalska, J. M. Cassady, and H. G. Floss,** Stereochemistry of the isoprenylation of tryptophan catalyzed by 4-(g,g-dimethylallyl)tryptophan synthase from *Claviceps,* the 1st pathway-specific enzyme in ergot alkaloid biosynthesis, *J. Am. Chem. Soc.,* 112:297–304, 1990.

62. **Siegel, M. R. and G. C. M. Latch,** Expression of antifungal activity in agar culture by isolates of grass endophytes, *Mycologia,* 83:525–537, 1991.

63. **Siegel, M. R., G. C. M. Latch, L. P. Bush, F. F. Fannin, D. D. Rowan, B. A. Tapper, C. W. Bacon, and M. C. Johnson,** Fungal endophyte-infected grasses: alkaloid accumulation and aphid response, *J. Chem. Ecol.,* 16:3301–3315, 1990.

64. **Siegel, M. R. and C. L. Schardl,** Fungal endophytes of grasses: detrimental and beneficial associations, In: J. H. Andrew and S. S. Hirano, Eds., *Microbial Ecology of Leaves*, Springer-Verlag, Berlin, 1991, 198.

65. **Spalla, C.,** Genetic problems of production of ergot alkaloids in saprophytic and parasitic conditions, In: Z. Vanek, Z. Hostalek, and J. Cudlin, Eds., *Genetics of Industrial Microorganisms*, Elsevier, New York, 1973, 393.

66. **Stark, M. J. R. and A. Boyd,** The killer toxin of *Kluyveromyces lactis*: characterization of the toxin subunits and identification of the genes which encode them, *EMBO J.,* 5:1995–2002, 1986.

67. **Stone, T. B., S. R. Sims, and P. G. Marrone,** Selection of tobacco budworm for resistance to a genetically engineered *Pseudomonas fluorescens* containing the d-endotoxin of *Bacillus thuringiensis* subsp. *Kurstaki, J. Invertebr. Pathol.,* 53:228–232, 1989.

68. **Tao, J., I. Ginsberg, N. Banerjee, W. Held, Y. Koltin, and J. A. Bruenn,** *Ustilago maydis* KP6 killer toxin: structure, expression in *Sacchromyces cerevisiae*, and relationship to other cellular toxins, *Mol. Cell. Biol.,* 10:1373–1381, 1990.

69. **Timberlake, W. E.,** Cloning and analysis of fungal genes, In: J. W. Bennett and L. L. Lasure, Eds., *More Gene Manipulations in Fungi*, Academic Press, San Diego, 1991, 51.

70. **Tipper, D. J. and K. A. Bostian,** Double-stranded ribonucleic acid killer system in yeasts, *Microbiol. Rev.,* 48:125–156, 1984.

71. **Tsai, H.-F., J.-S. Liu, C. Staben, M. J. Christensen, G. C. M. Latch, M. R. Siegel, and C. L. Schardl,** Evolutionary diversification of fungal endophytes of tall fescue grass by hybridization with *Epichloe* species, submitted, 1993.

72. **Tsai, H.-F., M. R. Siegel, and C. L. Schardl,** Transformation of *Acremonium coenophialum*, a protective fungal symbiont of the grass *Festuca arundinacea, Curr. Genet.,* 22:399–406, 1992.

73. **Vaeck, M., A. Reynaerts, H. Hofte, S. Jansens, B. M. De, C. Dean, M. Zabeau, M. M. Van, and J. Leemans,** Transgenic plants protected from insect attack, *Nature,* 328:33–37, 1987.

74. **van der Krol, A. R., P. E. Lenting, J. Veenstra, I. M. van der Meer, R. E. Koes, A. G. M. Gerats, J. N. M. Mol, and A. R. Stuitje,** An anti-sense chalcone synthase gene in transgenic plants inhibits flower pigmentation, *Nature,* 333:866–869, 1988.

75. **Vieira, J. and J. Messing,** Production of single-stranded plasmid DNA, *Methods Enzymol.,* 153:3–11, 1987.

76. **Vining, L. C.,** Physiological aspects of alkaloid production by *Claviceps* species, In: Z. Vanek, Z. Hostalek, and J. Cudlin, Eds., *Genetics of Industrial Microorganisms, Vol. II, Actinomycetes and Fungi,* Elsevier, New York, 1973, 405.

77. **Vollmer, S. J. and C. Yanofsky,** Efficient cloning of genes in *Neurospora crassa, Proc. Natl. Acad. Sci. U.S.A.,* 83:4869–4873, 1986.

78. **Wang, Z. Y., T. Takamizo, V. A. Iglesias, M. Osusky, J. Nagel, I. Potrykus, and G. Spangenberg,** Transgenic plants of tall fescue (*Festuca arundinacea* Schreb.) obtained by direct gene transfer to protoplasts, *Biotechnology,* 10:691–699, 1992.

79. **Weltring, K.-M., B. G. Turgeon, O. C. Yoder, and H. D. VanEtten,** Isolation of a phytoalexin-detoxification gene from the plant pathogenic fungus *Nectria haematocca* by detecting its expression in *Aspergillus nidulans, Gene,* 68:335–344, 1988.

80. **White, J. F.,** Endophyte-host associations in forage grasses. XI. A proposal concerning origin and evolution, *Mycologia,* 80:442–446, 1988.

81. **White, J. F., Jr.,** Endophyte-host associations in grasses. XIX. A systematic study of some sympatric species of *Epichloe* in England, *Mycologia,* 85:444–455, 1993.

82. **White, J. F. and T. L. Bultman,** Endophyte-host associations in forage grasses. VIII. Heterothallism in *Epichloe typhina, Am. J. Bot.,* 74:1716–1721, 1987.

83. **White, J. F., Jr., G. Morgan-Jones, and A. C. Morrow,** Taxonomy, life cycle, reproduction and detection of *Acremonium* endophytes, *Agric. Ecosyst. Environ.,* 44:13–38, 1992.

84. **Yelton, M. M., J. E. Hamer, and W. E. Timberlake,** Transformation of *Aspergillus nidulans* by using a *trpC* plasmid, *Proc. Natl. Acad. Sci. U.S.A.,* 81:1470–1474, 1984.

Section VI

Utilization of Endophyte-Infected Grasses Based on Agronomic Characteristics

Utilization of Endophyte-Infected Perennial Ryegrasses for Increased Insect Resistance

Daryl D. Rowan and Garrick C. M. Latch

CONTENTS

I. INTRODUCTION

Insect resistance associated with endophyte infection of a grass was first reported by Prestidge et al.[86] in 1982 after regrowth of grass in plots of endophyte-free perennial ryegrass (*Lolium perenne* L.) was observed to be greatly inferior to that in endophyte-infected plots. Inspection of the endophyte-free plants revealed they were heavily infested with Argentine stem weevil (*Listronotis bonariensis*) and that the majority of tillers had died as a result of feeding by stem-weevil larvae. Little damage from the Argentine stem weevil was observed in the plots containing endophyte-infected perennial ryegrass. This initial observation of the protection afforded to a grass by its endophyte aroused much interest and provided the explanation for the maternally inherited resistance of turf ryegrasses to sod web worm (*Crambus* sp.) subsequently reported by Funk et al.[40]

Many other species of insects have now been found to be adversely affected by the presence of endophytes in ryegrasses and other grasses. Latch[55] lists some 24 diverse insect species adversely affected by the presence of endophytes in perennial ryegrass and tall fescue (*Festuca arundinacea* Schreb.). New records of insects affected by endophytes in ryegrass, since this publication, are listed in Table 1. A number of recent reviews have considered the insect resistance of endophyte-infected grasses, from agronomic or ecological perspectives.[21-23,35,78,104] The use of fungal endophytes to render plants resistant to insect herbivores is both novel and timely in a world that is becoming increasingly aware of pesticide pollution. Not only can the use of pesticides be reduced on endophyte-infected turf grass species, but

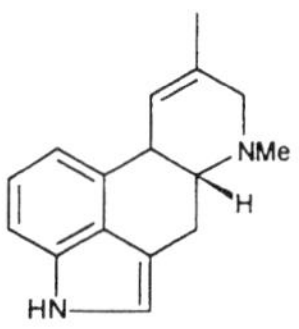

Agroclavine

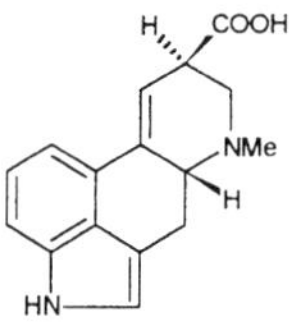

Lysergic acid

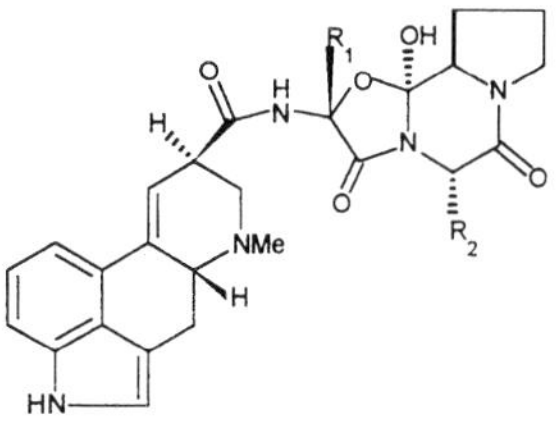

Ergopeptine alkaloids

Ergovaline R_1 = Me, R_2 = isopropyl
Ergotamine R_1 = Me, R_2 = benzyl

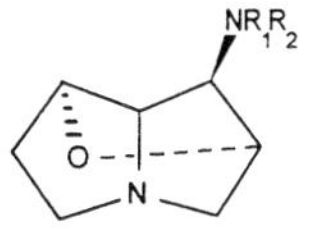

Loline alkaloids

N-Acetyl norloline R_1= H, R_2= CH_3CO
N-Formyl loline R_1= Me, R_2= CHO
N-Acetyl loline R_1= Me, R_2= CH_3CO

Figure 2 Structural formulae of the three types of ergot alkaloids and of the major ergopeptine and loline alkaloids, identified in endophyte-infected ryegrass and tall fescue.

reduced feeding and growth of the armyworm. Lysergic acid derivatives and members of the ergoline series can be found when endophytes are grown in culture[61,79] and, presumably, also in endophyte-infected ryegrass.

C. PERAMINE

The third alkaloid identified as important in the insect resistance of perennial ryegrass has been named peramine (Figure 3). Peramine may be considered as a cyclic dipeptide formally derived from proline and arginine[93] and is the only example of this structural type identified in endophyte-infected grasses. Peramine occurs widely in endophyte-infected grasses[99] and seems to play an important role in the resistance of ryegrasses to the Argentine stem weevil, to the sod web worm, and to the aphid *Schizaphis graminum*.[99] A number of other graminivorous insects are, however, insensitive to peramine. Peramine seems to function solely as a feeding deterrent against stem weevil,[91] with low chronic toxicity to both the weevils and to mice (Rowan, unpublished). In view of the widespread occurrence of peramine in endophyte-infected grasses,[99] it is tempting to speculate that with sensitive insects, such as the stem weevil, this metabolite may serve as a marker to enable the insect to avoid other, more potent endophyte toxins. A number of analogues of peramine have been obtained by chemical synthesis, and some of these are active as feeding deterrents against adult stem weevil. A minimal, but definite, heterocyclic ring system was found to be essential for feeding-deterrent activity[91] (Figure 3), but to obtain the full biological response, an appropriately distributed positive charge on the side chain also seems to be required.[90] Thus, the synthetic analogue homoperamine (Figure 3) was as active as peramine itself in feeding-deterrence bioassays.

III. RESISTANCE TO THE ARGENTINE STEM WEEVIL

Endophyte-induced resistance of grasses to insect pests has been most intensively studied for the interaction of perennial ryegrass with the Argentine stem weevil. This weevil is a major pest of perennial ryegrass pastures in New Zealand and also occurs in Australia and South America. Adult weevils produce characteristic window-like feeding scars on leaves, but generally cause little permanent damage to established plants. Larvae, however, are more destructive because they tunnel into the middle of the grass tiller, eating the innermost rolled leaf, growing point, and then the base of the outer leaves. As they

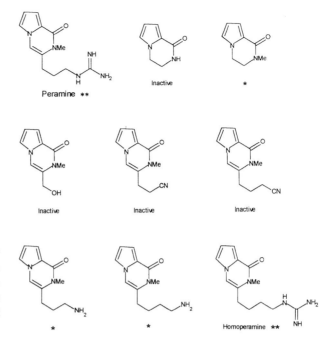

Figure 3 Feeding deterrent activity of peramine, and some of its analogues, against adult Argentine stem weevil (* = active at 10 μg/g; ** = active at 1 μg/g; no asterisk = not active >33 μg/g).

mature, larvae transfer between tillers, killing from three to five more tillers as they develop through their four instars. The effects of stem weevil predation are particularly apparent in drier areas where tiller death due to high weevil numbers may prevent pasture regrowth after autumn rain.

A. ADULT-STAGE RESISTANCE

Mortimer and di Menna[69] recorded three- to fourfold greater dry-matter production during regrowth of high-endophyte ryegrass plots than in the low-endophyte plots affected by predation of the Argentine stem weevil. In the field, the population dynamics of Argentine stem weevil are closely linked with the availability of endophyte-free ryegrass tillers.[11] Endophyte-infected ryegrasses have been shown to suffer less adult feeding and larval damage and to have fewer eggs laid on them than do endophyte-free ryegrasses.[10,13,46] Prolonged exposure of the adult weevil to endophyte-infected grass resulted in increased flight-muscle development and in the inhibition or regression of reproductive development.[11] These negative effects also occurred to weevils under crowded conditions, but were strongest in the presence of *A. lolii*.

The prospect of exploiting the weevil resistance conferred by the endophyte has provided much of the impetus for the extensive research conducted into this system. The endophyte-mediated resistance of perennial ryegrass to the Argentine stem weevil is affected primarily through the mobile adult stage, which exhibits a marked preference for feeding and ovipositing on endophyte-free ryegrass.[10,46] Systematic fractionation of extracts from endophyte-infected ryegrass has identified the fungal alkaloid peramine as a highly active deterrent to feeding by adult weevils.[92] Peramine was found to deter adult weevils from feeding, at concentrations as low as 0.1 μg/g of basal diet.[91] However, when the feeding stimulus provided by endophyte-free ryegrass was incorporated into the diet, 1- and 10-ppm peramine was required to significantly reduce feeding in choice and no-choice tests, respectively.[77] Concentrations of peramine in the plant may reach 100 μg/g dry matter, but values of 10 to 40 μg/g would be typical.[8,89,90]

While peramine has been suggested as the major feeding deterrent to the adult stem weevil in endophyte-infected ryegrass,[84,85,92] a number of other endophyte metabolites also show activity against adult weevils in choice bioassays. The ergopeptine alkaloids ergovaline and ergotamine are active at concentrations comparable to peramine[33,77] and at concentrations similar to those at which they occur in ryegrass.[95] The indole diterpene paxilline is also a deterrent to feeding at 1 μg/g, but concentrations in the grass are not generally known (but see Garthwaite et al.).[45] Resistance of endophyte-infected ryegrasses to stem weevils is related to reduced egg laying on infected plants. Only high concentrations of peramine directly reduce oviposition by gravid females,[77] and it therefore seems likely that the reduction in egg laying recorded on endophyte-infected grass is a result of reduced adult feeding and residence on infected plants.

B. LARVAE-STAGE RESISTANCE

Endophyte-infected ryegrass is also resistant to larvae of the Argentine stem weevil,[10,75] with effects again mediated through the deterrency of peramine at concentrations above 2 µg/g.[91] Peramine showed no evidence of toxicity to larvae, at concentrations up to 25 ppm, whereas both lolitrem B[77,84] at a concentration of 5 µg/g and paxilline at 10 µg/g[44] reduced growth and delayed development. Ryegrasses infected with endophytes that produce peramine, ergot alkaloids, and indole diterpenes, other than the lolitrems, are also resistant to the stem weevil,[36] suggesting that not all metabolites are required for insect resistance.

C. RESISTANCE IN SEEDS AND SEEDLINGS

Endophyte infection of seeds also protects seedlings from predation by Argentine stem weevils, and transient protection occurs even if the endophyte in the seeds is not viable. Stewart[101] observed that seedlings of perennial ryegrass, grown from seeds containing nonviable endophytes, were protected from weevils, but that this protection was lost as the plants matured. He postulated that a chemical deterrent present in the seed and translocated into the developing seedling was subsequently diluted or metabolized as the plant matured. This protection may be attributed to the presence of peramine, which, following germination, moves from the seed into the developing plantlet. Comparative studies of seeds containing viable and nonviable endophytes[7] show that in each case both lolitrem B and peramine move from the seed into the developing seedling. A higher proportion of the peramine present was translocated, and more rapidly, than the lolitrem B, in accord with the hydrophilic character of the peramine molecule. When the endophyte in the seed was not viable, the quantity of peramine and lolitrem in the seedlings declined to zero some 50 days after planting. With a viable endophyte, lolitrem concentrations peaked at day 25 before declining to near zero, while peramine declined steadily to a minimum at 50 days after planting. *De novo* synthesis of peramine and lolitrems by the endophyte began some 30 and 50 days, respectively, after planting.[7] Weevil resistance prior to this was attributed to metabolites translocated from the seed into the developing seedling.

Annual ryegrasses infected with an *Acremonium*-like endophyte also show resistance, as seedlings, to Argentine stem weevil, but this resistance disappears as the plants grow and peramine levels drop to below 1 µg/g dry weight.[32] Lolitrems may also be present in these plants, but the levels appear too low to affect insect resistance.[82]

IV. OTHER INSECT SPECIES AFFECTED BY ENDOPHYTE-INFECTED RYEGRASS

A. FALL ARMYWORMS

A number of other insect species are also affected by the presence of endophytes in perennial ryegrass[55] (Table 1). In most cases the chemical basis of these effects is not understood. Larvae of the fall armyworm (*S. frugiperda*), a generalist on species of Gramineae, show reduced growth, lower survival, and delayed development, when confined to endophyte-infected ryegrasses.[25,48] Early larval instars prefer to feed on endophyte-free plants, but by the fourth instar, this preference is no longer apparent.[49] Greater susceptibility of early instars of other insects to endophyte-infected ryegrasses has also been recorded by Quigley et al.[87] and Potter et al.[80] Both antibiotic and antifeedant effects to fall armyworm have been demonstrated for various ergot alkaloids, but at relatively high concentrations.[24] Indole diterpene mycotoxins, including paxilline, reduce the growth of fall armyworm larvae, at microgram concentrations,[31] while the loline alkaloids, present in high concentrations in various *Lolium* and *Festuca* species, were also toxic,[88] but are not present in endophyte-infected perennial ryegrass.

B. APHIDS

Three aphid species, *Rhopalosiphum maidis*,[18,51] *Schizaphis graminum*,[17,99] and *Diuraphis noxia*,[26,27,54,100] are affected by *Acremonium* endophytes in perennial ryegrass. Both toxicity and deterrence are involved. The lolines and peramine are mediators of the effects of endophyte infection on *S. graminum*.[17,99] However, only antibiosis factors appear to be involved in the response of *Diuraphis noxia*, the Russian wheat aphid, to *Acremonium*-infected perennial ryegrass.[26,54] No apparent preference for endophyte-free perennial ryegrass was seen in laboratory tests, but populations rapidly declined when confined to the infected grass.[26]

Aphid responses to endophyte infection of ryegrass may also depend on the strain of insect involved. Latch et al.[57] found that the aphid *Rhopalosiphum padi* was not deterred from feeding on endophyte-infected cultivars of New Zealand perennial ryegrasses, whereas Christensen (1992, personal communication) has observed that this aphid was deterred from feeding on the leaves of the same cultivars in Japan. Presumably, the *R. padi* observed in Japan is a different biotype that is sensitive to some component of the endophyte-infected grass. Three species of aphid, *Metopholophium dirhodum, Sitobion fragariae,* and *Macrosiphon avenae,* have been reported to be unaffected by endophytes.[51,57]

C. OTHER INSECTS AND VERTEBRATES

Fewer larvae, adults, and eggs of the sod webworm (*Crambus* spp.)[40] and the bluegrass billbug (*Sphenophorus parvulus*)[3] were found on *A. lolii*-infected perennial ryegrass. Survival of adults of the black beetle (*Heteronychus arator*) was also reduced by the presence of the endophyte in perennial ryegrass.[6] The ergot alkaloids, at relatively low concentrations, were found to deter feeding by this species, while peramine, lolitrem B, and paxilline had no effect.[83] Populations of the sap-sucking hairy chinch bug (*Blissus leucopterous hirtus*) were lower on endophyte-infected perennial ryegrass[64] and fine fescues.[97] Survival and population growth of the flour beetle (*Tribolium castaneum*) was reduced by the presence of endophytes in ground seeds of perennial ryegrass.[21]

Limited evidence suggests that endophyte infection may provide protection against some root-feeding insects. The loline alkaloids of endophyte-infected tall fescue were deterrent[71] and toxic[98] to the Japanese beetle (*P. japonica*); however, neither Oliver et al.[70] nor Potter et al.[80] obtained clear evidence of an effect of endophytes, when investigating the feeding ecology of the grubs. Several fescue–endophyte combinations reduced survival of third-instar larvae of the grass grub (*Costelytra zealandica*).[76] Survival and fecundity of the nongraminivorous species *Folsomia candida*[15] and *Drosophila melanogaster*[28] were also affected by root tissue of tall fescue. However, there is, as yet, only one report[83] of an insect being affected by feeding on the roots of endophyte-infected ryegrass.

Predation of seeds by birds and small mammals may also be reduced by the presence of endophytes. Feeding trials with five species of passerines showed these birds preferred endophyte-free seeds, and if forced to eat infected seeds, they lost weight and had difficulty walking and maintaining balance.[63] Fungal alkaloids in the endophyte-infected seeds seem the most likely cause of these effects. There are indications that small mammals may also be affected by endophyte-infected grasses. Trapping of mice and moles in fields of endophyte-infected and endophyte-free tall fescue indicated that these animals were much more abundant in endophyte-free fields.[72] Chemical extracts from endophyte-infected tall fescue have been shown to affect rats and rabbits.[29,30]

V. EXPRESSIONS OF ENDOPHYTE-MEDIATED INTERACTIONS

A. SPECIES VARIATION

Endophyte infection of perennial ryegrass does not provide protection against all insect species that may damage ryegrass.[83] Adults of the frit fly (*Oscinella frit*) are not deterred from laying eggs on endophyte-infected ryegrass,[59] presumably because the adult, unlike the stem weevil, does not feed upon the grass and uses other cues to determine suitable oviposition sites. A second dipteran, the wheat-sheath miner (*Ceredontha australis*), which mines the leaf sheath and stem of grasses, is also unaffected by endophytes.[12] In New Zealand the distribution of four species of leafhoppers on ryegrass was unrelated to endophyte infection,[81] whereas Muegge et al.[67] found leafhoppers were affected by the presence of endophytes in tall fescue.

There are other cases where an insect may prefer endophyte-infected grasses. Argentine stem weevil showed more feeding on ryegrass infected with a *Gliocladium*-like endophyte than on endophyte-free grass.[47] It is likely that many other species of insects will be found to be unaffected, or only subtly affected, by the presence of endophytes, and it is not known how this might affect damage to the plant. If insect numbers are high, little protection may be afforded by endophytes, even against a species normally regarded as sensitive. Larvae of the southern armyworm (*Spodoptera eridania*) are sensitive to endophyte-infected ryegrass,[2] but when larval populations are high, severe pasture damage occurs.

B. ACUTE TOXICITY

Both toxic and deterrent factors are implicated in the responses of most insects to endophytes. Two species, however, the southern armyworm and the house cricket, *Acheta domesticus,* appear highly sensitive to toxins in endophyte-infected perennial ryegrass, but are not deterred from feeding by the presence of the endophyte.[1,2] For each species, death occurred after feeding on basal leaf-sheath material where the endophyte mycelia is concentrated. In the case of house crickets, gross morphological changes to the gut were observed,[1] but no such changes were observed with *S. eridania*. Intriguingly, endophyte-infected tall fescue was not toxic to *S. eridania*.[51] There do not appear to be any other reports of rapid and complete mortality of insects feeding on endophyte-infected ryegrasses or their metabolites, although endophytes have been implicated as the cause of acute mortality in house crickets fed tall fescue forage.[4] Generally, insect species that succumb to toxic factors, without apparently being deterred from feeding, would not be expected to utilize the grass as a natural host. Deterrence, however, is not involved in the response of the polyphagous southern armyworm and may not be a factor in the response of the graminivorous billbugs (*Sphenophorus* spp.). Johnson-Cicalese and Funk[52] observed little difference in adult feeding on endophyte-infected and uninfected ryegrass, but reported a significantly greater mortality of billbug adults on the infected grass.

C. OVERCOMING RESISTANCE

Why then have some graminivorous insects not succeeded in overcoming the endophyte-induced resistance? Research, principally with the Argentine stem weevil, has suggested several factors that may be necessary for the retention of insect resistance in grass–endophyte systems. First, the insect is not an obligate grass feeder and can find and use other endophyte-free Graminae, such as cereals, as secondary hosts. Secondly, the insect is highly mobile, with a long-lived adult stage, aiding its ability to find or oviposit on endophyte-free individuals (for stem weevil[13] and for sod webworm[40]). Finally, multiple toxins and feeding deterrents with divergent chemical structures and different biosynthetic origins are involved in affecting resistance. Thus, resistance or tolerance, on the part of an insect, to one group of metabolites need not provide resistance to all groups. This may help explain how the resistance of endophyte-infected ryegrass to the Argentine stem weevil is maintained. Stem weevils, insensitive to the feeding-deterrent peramine, have been obtained under laboratory conditions in just two generations (A. J. Popay, unpublished data), but such insensitive individuals may choose inappropriate oviposition sites on endophyte-infected plants in the field and not persist. In addition, synergistic interactions between the alkaloids present in endophyte-infected grasses can produce higher mortality than would otherwise be caused by each compound alone.[24,71,106]

VI. FACTORS AFFECTING ENDOPHYTE-INDUCED INSECT RESISTANCE

A. BIOTIC INFLUENCES

A number of other factors have been found to affect the expression of endophyte-induced insect resistance. The life stage of the insect may affect susceptibility to endophytes, with juveniles being generally more sensitive. Fall armyworm,[49] the common armyworm (*Mythimna convecta*), and the common cutworm (*Agrotis infusa*)[39,62] are deterred by the presence of endophytes, as early-, but not late-, stage larvae. Nymphs of the Russian wheat aphid are more susceptible to endophytes than are the adults.[54] More complex interactions involving other organisms may also occur. The level of resistance to Argentine stem weevil afforded by *A. lolii*-infected ryegrass is reduced by the presence of the vesicular-arbuscular mycorrhizal fungus, *Glomus fasciculatum,* whereas mycorrhizal infection alone had no effect on the weevil.[9]

B. ABIOTIC INFLUENCES

The nutrient status of the host affects the production of alkaloids by endophyte-infected grasses[14,55] and, presumably, influences the degree of insect resistance conferred by the endophyte. Temperature has a similar effect. Latch et al. (unpublished) observed that maximum production of the insect-feeding-deterrent peramine was obtained when endophyte-infected perennial ryegrass plants were grown at 20°C, the least production was at 26°C, and an intermediate level of production was at 14°C. The preference of *S. graminum* for endophyte-free perennial ryegrass was greater at 14 and 21°C than at 7 and 28°C,[17] which would correlate with increased alkaloid production in the infected plants. In spring, when endophyte and alkaloid concentrations are low, oviposition by Argentine stem weevil, on endophyte-infected ryegrass, can be similar to oviposition on endophyte-free ryegrass.[75] The effect of water stress on ryegrass plants was investigated by Barker et al.[8] They reported higher levels of ergovaline in ryegrasses that had been subjected to several cycles of drought than in plants that had not been stressed. There is scope to examine the effect of other environmental factors on the production of insect-toxic compounds by endophyte-infected ryegrasses.

VII. SELECTION OF ENDOPHYTE STRAINS FOR INSECT RESISTANCE

As yet, very few of the worlds grasses have been examined for the presence of endophytic fungi, and all grasses may not necessarily be hosts to endophytes. Attempts have been made to find endophytes in the important turf species *Poa pratensis*, but to date, these have been unsuccessful[102] (Latch, unpublished). Endophytes have been exchanged between the genera *Lolium* and *Festuca* quite satisfactorily.[56,99] While incompatibility exists between some grasses and species and isolates of endophytes, it may be possible to introduce an endophyte from another *Poa* species into *P. pratensis*.

A. NONALKALOID-PRODUCING ISOLATES

Studies on the effects of endophyte infection on insect resistance have generally not considered the variety of endophyte isolates that occur in perennial ryegrass plants.[19,20,58] These ryegrass isolates differ greatly in their morphology and growth characteristics in culture, in their isozyme analyses, and in the types and concentrations of insect-active metabolites and antibiotics that might be produced. Some isolates, for example, produce neither peramine, lolitrems, nor ergovaline, in their grass host, and hence, these plants should be susceptible to stem weevil damage. The continuing existence of such "nonalkaloid"-producing isolates in the field suggests they must confer some, as yet unknown, advantage to their host.

B. NATURAL AND GENETICALLY ALTERED ALKALOID-PRODUCING ISOLATES

While it may be difficult to control the environmental factors affecting endophyte-infected ryegrasses and, hence, their production of insect toxins, it is possible to select natural isolates of ryegrass endophytes, that consistently produce altered profiles of metabolites. Ideally, such isolates would be capable of producing a desired range of compounds deterrent or toxic to insects, at appropriate concentrations when infected into new host grasses. Although some insect-deterrent compounds, such as peramine and the loline alkaloids, seem essentially nontoxic to mammalian herbivores, others, such as the ergot alkaloids and the indole diterpenes, are harmful at concentrations normally encountered in endophyte-infected grasses. While toxicity to animals may be of little importance in a turf situation, it is obviously crucial if the grasses are used for grazing livestock. In addition, for economic and ecological reasons, seed production of turf grasses is often combined with grazing of the pasture or of the stubble after harvest, with consequent risk of animal toxicoses. Thus, a balance must be kept between the numbers and quantities of compounds necessary to maintain insect resistance, and the need to minimize animal toxicity.

Such an approach has been adopted with perennial ryegrass and a search has been made for isolates of *A. lolii* that produce peramine, but little or no lolitrem or ergovaline, in their natural host grasses. A number of isolates that do not produce lolitrem have been found and one such isolate has been inoculated into two New Zealand ryegrass cultivars. These new cultivars show negligible ryegrass staggers toxicity to stock and retain their resistance to the Argentine stem weevil.[36,37] One of these cultivars shows evidence of a host-specific interaction between the endophyte and its environment, which leads to levels of ergovaline that are probably too high for commercial use. The same endophyte isolate inoculated into the second cultivar has, so far, shown normal levels of ergovaline, under the same environmental conditions. A search is now being made for further isolates that produce less ergovaline. This suggests that host-specific interactions must now be considered in all future evaluations of selected endophyte isolates.

It may not be possible to find natural endophyte isolates that will produce the compounds required to control particular insect pests and that will not affect the health of farm animals. An option is to genetically manipulate the endophyte, either to remove the genes responsible for production of the undesirable toxins or to produce new compounds required to protect the grass from insects. In this case the endophyte is being used as a surrogate host in order to introduce new genes into the grass. Murray et al.[68] have genetically transformed *A. lolii* and infected ryegrasses with these transformants. Using such methods it should be possible to introduce into the endophyte genome specific forms of, for example, the Bt gene, protease or amylase inhibitors, or any other future vehicles of insect resistance.

VIII. CONCLUSION

The use of endophytes has opened up a new, environmentally friendly means of controlling insect pests of grasses. The initial results reported in this chapter indicate that significant improvements to pasture performance can be made by the use of appropriate strains of endophytes. This work has also emphasized the complexity of the endophyte–grass interaction and, in particular, the potential for host-specific plant–endophyte–environment interactions to lead to unexpectedly high concentrations of fungal metabolites. While animal toxicity must, for practical reasons, be assessed initially by chemical assays of presumed toxins, insect resistance can be determined by conventional selection methods in the field or glasshouse, without knowledge of the chemical or biochemical mechanisms involved. This offers the chance to quickly screen the wide range of endophyte–host combinations now available against a complete range of insect pests. Use of this approach should give new sources of insect resistances, which can be used for the development of improved grass cultivars.

REFERENCES

1. **Ahmad, S., S. Govindarajan, C. R. Funk, and J. M. Johnson-Cicalese,** Fatality of house crickets on perennial ryegrass infected with a fungal endophyte, *Entomol. Exp. Appl.,* 39:183–190, 1985.
2. **Ahmad, S., S. Govindarajan, J. M. Johnson-Cicalese, and C. R. Funk,** Association of a fungal endophyte in perennial ryegrass with antibiosis to larvae of the southern armyworm, *Spodoptera eridania, Entomol. Exp. Appl.,* 43:287–294, 1987.
3. **Ahmad, S., J. M. Johnson-Cicalese, W. K. Dickson, and C. R. Funk,** Endophyte-enhanced resistance in perennial ryegrass to the bluegrass billbug, *Sphenophorus parvulus, Entomol. Exp. Appl.,* 41:3–10, 1986.
4. **Asay, K. H., T. R. Minnick, G. B. Garner, and B. W. Harmon,** Use of crickets in a bioassay of forage quality in tall fescue, *Crop Sci.,* 15:585–588, 1975.
5. **Bacon, C. W., P. C. Lyons, J. K. Porter, and J. D. Robbins,** Ergot toxicity from endophyte-infected grasses: a review, *Agron. J.,* 78:106–116, 1986.
6. **Ball, O. J.-P. and R. A. Prestidge,** The effect of the endophytic fungus *Acremonium lolii* on adult black beetle (*Heteronychus arator*) feeding, *Proc. 45th N. Z. Plant Prot. Conf.,* 45:201–204, 1992.
7. **Ball, O. J.-P., R. A. Prestidge, and J. M. Sprosen,** Effect of plant age and endophyte viability on peramine and lolitrem B concentration in perennial ryegrass seedlings, In: D. E. Hume, G. C. M. Latch, and H. S. Easton, H. S, Eds., *Proc. 2nd Int. Symp.* Acremonium/*Grass Interactions*, AgResearch, Grasslands Research Centre, Palmerston North, New Zealand, 1993, 63.
8. **Barker, D. J., E. Davies, G. A. Lane, G. C. M. Latch, H. M. Nott, and B. A. Tapper,** Effect of water deficit on alkaloid concentrations in perennial ryegrass endophyte associations, In: D. E. Hume, G. C. M. Latch, and H. S. Easton, Eds., *Proc. 2nd Int. Symp.* Acremonium/*Grass Interactions*, AgResearch, Grasslands Research Centre, Palmerston North, New Zealand, 1993, 67.
9. **Barker, G. M.,** Mycorrhizal infection influences *Acremonium*-induced resistance to Argentine stem weevil in ryegrass, *Proc. N.Z. Weed Pest Control Conf.,* 40:199–203, 1987.
10. **Barker, G. M., R. P. Pottinger, and P. J. Addison,** Effect of *Lolium* endophyte fungus infections on survival of larval Argentine stem weevil, *N.Z. J. Agric. Res.,* 27:279–281, 1984a.
11. **Barker, G. M., R. P. Pottinger, and P. J. Addison,** Population dynamics of the Argentine stem weevil (*Listronotus bonariensis*) in pastures of Waikato, New Zealand, *Agric. Ecosyst. Environ.,* 26:79–115, 1989.
12. **Barker, G. M., R. P. Pottinger, P. J. Addison, and E. H. A. Oliver,** Pest status of *Cerodontha* spp. and other shoot flies in Waikato pasture, *Proc. N.Z. Weed Pest Control Conf.,* 37:96–100, 1984b.
13. **Barker, G. M., R. P. Pottinger, P. J. Addison, and R. A. Prestidge,** Effect of *Lolium* endophyte fungus infections on behavior of adult Argentine stem weevil, *N.Z. J. Agric. Res.,* 27:271–277, 1984c.
14. **Belesky, D. P., J. A. Stuedemann, R. D. Plattner, and S. R. Wilkinson,** Ergopeptine alkaloids in grazed tall fescue, *Agron. J.,* 80:209–312, 1988.
15. **Bernard, E. C., A. M. Cole, J. B. Oliver, and K. D. Gwinn,** Survival and fecundity of *Folsomia candida* (Collembola) fed tall fescue tissues or ergot peptide-amended yeast, In: S. S. Quisenberry and R. E. Joost, Eds., *Proc. 1st Int. Symp.* Acremonium/*Grass Interactions*, Louisiana Agriculture Experiment Station, Baton Rouge, 1990, 125.

16. **Betina, V.,** Indole derived tremorgenic toxins, In: V. Betina, Ed., *Mycotoxins Production, Isolation, Separation and Purification. Developments in Food Science*, Vol. 8, Elsevier, New York, 1984, 415.

17. **Breen, J. P.,** Temperature and seasonal effects on expression of *Acremonium lolii* enhanced resistance to greenbug, *Schizaphis graminum* (Rondani) (Homoptera:Aphididae), In: S. S. Quisenberry and R. E. Joost, Eds., *Proc. 1st Int. Symp.* Acremonium/*Grass Interactions*, Louisiana Agriculture Experiment Station, Baton Rouge, 1990, 12.

18. **Buckley, R. J., P. M. Halisky, and J. P. Breen,** Variation in feeding deterrence of the corn leaf aphid related to *Acremonium* endophytes in grasses, *Phytopathology,* 81:120–121, 1991.

19. **Christensen, M. J., G. C. M. Latch, and B. A. Tapper,** Variation within isolates of *Acremonium* endophytes from perennial rye-grasses, *Mycol. Res.,* 95:918–923, 1991.

20. **Christensen, M. J., A. Leuchtmann, D. D. Rowan, and B. A. Tapper,** Taxonomy of *Acremonium* endophytes of tall fescue (*Festuca arundinacea*), meadow fescue (*F. pratensis*) and perennial rye-grass (*Lolium perenne*), *Mycol. Res.,* 97:1083–1092, 1993.

21. **Clay, K.,** Fungal endophytes of grasses: a defensive mutualism between plants and fungi, *Ecology,* 69:10–16, 1988a.

22. **Clay, K.,** Clavicipitaceous endophytes of grasses: coevolution and the change from parasitism to mutualism, In: D. L. Hawksworth and K. Pirozynski, Eds., *Co-evolution of Fungi with Plants and Animals,* Academic Press, London, 1988b, 79.

23. **Clay, K.,** Clavicipitaceous endophytes of grasses: their potential as biocontrol agents, *Mycol. Res.,* 92:1–12, 1989.

24. **Clay, K. and G. P. Cheplick,** Effect of ergot alkaloids from fungal endophyte-infected grasses on fall armyworm (*Spodoptera frugiperda*), *J. Chem. Ecol.,* 15:169–182, 1989.

25. **Clay, K., T. N. Hardy, and A. M. Hammond, Jr.,** Fungal endophytes of grasses and their effects on an insect herbivore, *Oecologia,* 66:1–5, 1985.

26. **Clement, S. L., D. G. Lester, A. D. Wilson, and K. S. Pike,** Behavior and performance of *Diuraphis noxia* (Homoptera:Aphididae) on fungal endophyte-infected and uninfected perennial ryegrass, *J. Econ. Entomol.,* 85:583–588, 1992.

27. **Clement, S. L., K. S. Pike, W. J. Kaiser, and A. D. Wilson,** Resistance of endophyte-infected plants of tall fescue and perennial ryegrass to the Russian wheat aphid (Homoptera:Aphididae), *J. Kansas Entomol. Soc.,* 63:646–648, 1990.

28. **Cole, A. M., C. D. Pless, and K. D. Gwinn,** Survival of *Drosophila melanogaster* (Diptera:Drosophilidae) on diets containing roots or leaves of *Acremonium*-infected or non-infected tall fescue, In: S. S. Quisenberry and R. E. Joost, Eds., *Proc. 1st Int. Symp.* Acremonium/*Grass Interactions*, Louisiana Agriculture Experiment Station, Baton Rouge, 1990, 128.

29. **Daniels, L. B., T. S. Nelson, and J. N. Beasley,** Effects of extracts of toxic fescue given orally to rats, *Can. J. Comp. Med.,* 45:173–176, 1981.

30. **Daniels, L. B., A. Ahmed, T. S. Nelson, and J. N. Beasley,** Physiological responses in pregnant white rabbits given a chemical extract of toxic tall fescue, *Nutr. Rep. Int.,* 29:505–510, 1984.

31. **Dowd, P. F., R. J. Cole, and R. F. Vesonder,** Toxicity of selected tremorgenic mycotoxins and related compounds to *Spodoptera frugiperda* and *Heliothis zea, J. Antibiot.,* 41:1868–1872, 1988.

32. **Dymock, J. J., G. C. M. Latch, and B. A. Tapper,** Novel combinations of endophytes in ryegrasses and fescues and their effects on Argentine stem weevil (*Listronotus bonariensis*) feeding, In: P. P. Stahle, Ed., *Proc. 5th Australasian Conf. Grassland Invert. Ecol.,* D & D Printing, Victoria, Australia, 1989a, 28.

33. **Dymock, J. J., D. D. Rowan, and I. R. McGee,** Effects of endophyte-produced mycotoxins on Argentine stem weevil and the cutworm, *Graphania mutans*, In: P. P. Stahle, Ed., *Proc. 5th Australasian Conf. Grassland Invert. Ecol.,* D & D Printing, Victoria, Australia, 1989b, 35.

34. **Eichenseer, H., D. L. Dahlman, and L. P. Bush,** Influence of endophyte infection, plant age and harvest interval on *Rhopalosiphum padi* survival and its relation to quantity of N-formyl and N-acetyl loline in tall fescue, *Entomol. Exp. Appl.,* 60:29–38, 1991.

35. **Fletcher, L. R., J. H. Hoglund, and B. L. Sutherland,** The impact of *Acremonium* endophytes in New Zealand, past, present and future, *Proc. N.Z. Grasslands Assoc.,* 52:227–235, 1990.

36. **Fletcher, L. R., A. J. Popay, and B. A. Tapper,** Evaluation of several lolitrem-free endophyte/ perennial ryegrass combinations, *Proc. N.Z. Grasslands Assoc.,* 53:215–219, 1991.

37. **Fletcher, L. R. and B. L. Sutherland,** Liveweight change in lambs grazing perennial ryegrasses with different endophytes, In: D. E. Hume, G. C. M. Latch, and H. S. Easton, Eds., *Proc. 2nd Int. Symp.* Acremonium/*Grass Interactions*, AgResearch, Grasslands Research Centre, Palmerston North, New Zealand, 1993, 125.

38. **Frost, W. E.,** Role of the perennial ryegrass endophyte *Acremonium lolii* in population development of cereal rust mite, In: *Proc. 6th Australasian Conf. Grassland Invert. Ecol.*, 1993, 178.

39. **Frost, W. E., P. E. Quigley, and P. J. Cunningham,** Research into the effect of *Acremonium lolii* on the feeding behaviour and nutrition of principal Australian pests of perennial ryegrass, In: S. S. Quisenberry and R. E. Joost, Eds., *Proc. 1st Int. Symp.* Acremonium/*Grass Interactions*, Louisiana Agriculture Experiment Station, Baton Rouge, 1990, 144.

40. **Funk, C. R., P. M. Halisky, M. C. Johnson, M. R. Siegel, A. V. Stewart, S. Ahmad, R. H. Hurley, and I. C. Harvey,** An endophytic fungus and resistance to sod webworms: association in *Lolium perenne* L., *Biotechnology*, 1:189–191, 1983.

41. **Funk, C. R., R. H. White, and J. P. Breen,** Importance of *Acremonium* endophytes in turfgrass breeding and management, *Agric. Ecosyst. Environ.*, 44:215–232, 1993.

42. **Gallagher, R. T. and A. D. Hawkes,** The potent tremorgenic neurotoxins lolitrem B and aflatrem: a comparison of the tremor response in mice, *Experientia*, 42:823–825, 1986.

43. **Gallagher, R. T., A. D. Hawkes, P. S. Steyn, and R. Vleggaar,** Tremorgenic neurotoxins from perennial ryegrass causing ryegrass staggers disorder of livestock: structure elucidation of lolitrem B, *J. Chem. Soc. Chem. Commun.*, 1984:614–616, 1984.

44. **Gallagher, R. T. and R. A. Prestidge,** Structure-activity studies on indole diterpenes, including lolitrems and related indoles and tremorgens, In: S. S. Quisenberry and R. E. Joost, Eds., *Proc. 1st Int. Symp.* Acremonium/*Grass Interactions*, Louisiana Agriculture Experiment Station, Baton Rouge, 1990, 80.

45. **Garthwaite, I., C. O. Miles, and N. R. Towers,** Immunological detection of the indole diterpenoid tremorgenic mycotoxins, In: D. E. Hume, G. C. M. Latch, and H. S. Easton, Eds., *Proc. 2nd Int. Symp.* Acremonium/*Grass Interactions*, AgResearch, Grasslands Research Centre, Palmerston North, New Zealand, 1993, 77.

46. **Gaynor, D. L. and W. F. Hunt,** The relationship between nitrogen supply, endophytic fungus, and Argentine stem weevil resistance in ryegrasses, *Proc. N.Z. Grasslands Assoc.*, 44:257–263, 1983.

47. **Gaynor, D. L., D. D. Rowan, G. C. M. Latch, and S. Pilkington,** Preliminary results on the biochemical relationship between adult Argentine stem weevil and two endophytes in ryegrass, *Proc. Weed Pest Control Conf.*, 36:220–224, 1983.

48. **Hardy, T. N., K. Clay, and A. M. Hammond, Jr.,** Fall armyworm (Lepidoptera:Noctuidae): a laboratory bioassay and larval preference study for the fungal endophyte of perennial ryegrass, *J. Econ. Entomol.*, 78:571–575, 1985.

49. **Hardy, T. N., K. Clay, and A. M. Hammond, Jr.,** Leaf age and related factors affecting endophyte-mediated resistance to fall armyworm (Lepidoptera:Noctuidae) in tall fescue, *Environ. Entomol.*, 15:1083–1089, 1986.

50. **Huizing, H. J., W. van der Molen, W. Kloek, and A. P. M. den Nijs,** Detection of lolines in endophyte-containing meadow fescue in the Netherlands and the effect of elevated temperature on induction of lolines in endophyte-infected perennial ryegrass, *Grass Forage Sci.*, 46:441–445, 1991.

51. **Johnson, M. C., D. L. Dahlman, M. R. Siegel, L. P. Bush, G. C. M. Latch, D. A. Potter, and D. R. Varney,** Insect feeding deterrents in endophyte-infected tall fescue, *Appl. Environ. Microbiol.*, 49:568–571, 1985.

52. **Johnson-Cicalese, J. M. and C. R. Funk,** Host range of, and effect of endophyte on, four species of billbug (*Sphenophorus* spp.) found on New Jersey turfs, *Agron. Abst.*, 152, 1988.

53. **Kanda, K., H. Koga, Y. Hirai, K. Hasegawa, T. Uematu, and T. Tukiboshi,** Resistance of *Acremonium* endophyte-infected perennial ryegrass and tall fescue to bluegrass webworm, *Parapediasia teterrella*, Abstr. Trans. Proc. Phythopathol. Soc. Japan, Morioka, Iwate, May 1992.

54. **Kindler, S. D., J. P. Breen, and T. L. Springer,** Reproduction and damage by Russian wheat aphid (Homoptera:Aphididae) as influenced by fungal endophytes and cool-season turfgrasses, *J. Econ. Entomol.*, 84:685–692, 1991.

55. **Latch, G. C. M.,** Physiological interactions of endophytic fungi and their hosts. Biotic stress tolerance imparted to grasses by endophytes, *Agric. Ecosyst. Environ.*, 44:143–156, 1993.

56. **Latch, G. C. M. and M. J. Christensen,** Artificial infection of grasses with endophytes, *Ann. Appl. Biol.,* 107:17–24, 1985.

57. **Latch, G. C. M., M. J. Christensen, and D. L. Gaynor,** Aphid detection of endophyte infection in tall fescue, *N.Z. J. Agric. Res.,* 28:129–132, 1985.

58. **Latch, G. C. M. and B. A. Tapper,** *Lolium* endophytes — problems and progress, *Proc. Jpn. Assoc. Mycotoxicol.,* (Suppl. 1):220–223, 1988.

59. **Lewis, G. C. and R. O. Clements,** A survey of ryegrass endophyte (*Acremonium loliae*) in the U. K. and its apparent ineffectuality on a seedling pest, *J. Agric. Sci.,* 107:633–638, 1986.

60. **Lewis, G. C., J. F. White, and J. Bonnefont,** Evaluation of grasses infected with fungal endophytes against locusts. *Tests of agrochemicals and cultivars, Ann. Appl. Biol.,* 14(Suppl.):142–143, 1993.

61. **Lyons, P. C., R. D. Plattner, and C. W. Bacon,** Occurrence of peptide and clavine ergot alkaloids in tall fescue grass, *Science,* 232:487–489, 1986.

62. **McDonald, G.,** Development and survival of *Mythimna convecta* (Walker) and *Persectania ewingii* (Westwood) (Lepidoptera:Noctuidae) on cereal and grass hosts, *J. Aust. Entomol. Soc.,* 30:295–302, 1991.

63. **Madej, C. W. and K. Clay,** Avian seed preference and weight loss experiments: the effect of fungal endophyte-infected tall fescue seeds, *Oecologia,* 88:296–302, 1991.

64. **Mathias, J. K., R. H. Ratcliffe, and J. L. Hellman,** Association of an endophytic fungus in perennial ryegrass and resistance to the hairy chinch bug (Hemiptera:Lygaeidae), *J. Econ. Entomol.,* 83:1640–1646, 1990.

65. **Miles, C. O., S. C. Munday, A. L. Wilkins, R. M. Ede, A. D. Hawkes, P. P. Embling, and N. R. Towers,** Large scale isolation of lolitrem B, structure determination of some minor lolitrems, and tremorgenic activities of lolitrem B and paxilline in sheep, In: D. E. Hume, G. C. M. Latch, and H. S. Easton, Eds., *Proc. 2nd Int. Symp.* Acremonium/*Grass Interactions,* AgResearch, Grasslands Research Centre, Palmerston North, New Zealand, 1993, 85.

66. **Miles, C. O., A. L. Wilkins, R. T. Gallagher, A. D. Hawkes, S. C. Munday, and N. R. Towers,** Synthesis and tremorgenicity of paxitriols and lolitriol: possible biosynthetic precursors of lolitrem B, *J. Agric. Food Chem.,* 40:234–238, 1992.

67. **Muegge, M. A., S. S. Quisenberry, G. E. Bates, and R. E. Joost,** Effect of endophyte mediated resistance and management strategies on population dynamics of leafhoppers in tall fescue, In: S. S. Quisenberry and R. E. Joost, Eds., *Proc. 1st Int. Symp.* Acremonium/*Grass Interactions,* Louisiana Agriculture Experiment Station, Baton Rouge, 1990, 163.

68. **Murray, F. R., G. C. M. Latch, and D. B. Scott,** Surrogate transformation of perennial ryegrass, *Lolium perenne,* using genetically modified *Acremonium* endophyte, *Mol. Gen. Genet.,* 233:1–9, 1992.

69. **Mortimer, P. H. and M. E. di Menna,** Ryegrass staggers: further substantiation of a *Lolium* endophyte aetiology and the discovery of weevil resistance of ryegrass pastures infected with *Lolium* endophyte, *Proc. N.Z. Grasslands Assoc.,* 44:240–243, 1983.

70. **Oliver, J. B., C. D. Pless, and K. D. Gwinn,** Effect of endophyte, *Acremonium coenophialum* in 'Kentucky 31' tall fescue, *Festuca arundinaceae,* on survival of *Popilla japonica,* In: S. S. Quisenberry and R. E. Joost, Eds., *Proc. 1st Int. Symp.* Acremonium/*Grass Interactions,* Louisiana Agriculture Experiment Station, Baton Rouge, 1990, 173.

71. **Patterson, C. G., D. A. Potter, and F. F. Fannin,** Feeding deterrency of alkaloids from endophyte-infected grasses to Japanese beetle grubs, *Entomol. Exp. Appl.,* 61:285–289, 1991.

72. **Pelton, M. R., H. A. Fribourg, J. W. Laundre, and T. D. Reynolds,** Preliminary assessment of small wild mammal populations in tall fescue habitats, *Tenn. Farm Home Sci.,* 160:68–71, 1991.

73. **Penn, J., I. Garthwaite, M. J. Christensen, C. M. Johnson, and N. R. Towers,** The importance of paxilline in screening for potentially tremorgenic *Acremonium* isolates, In: D. E. Hume, G. C. M. Latch, and H. S. Easton, Eds., *Proc. 2nd Int. Symp.* Acremonium/*Grass Interactions,* AgResearch, Grasslands Research Centre, Palmerston North, New Zealand, 1993, 88.

74. **Petroski, R. J., S. G. Yates, D. Weisleder, and R. G. Powell,** Isolation, semisynthesis, and NMR spectral studies of loline alkaloids, *J. Nat. Prod.,* 52:810–817, 1989.

75. **Popay, A. J. and R. A. Mainland,** Seasonal damage by Argentine stem weevil to perennial ryegrass pastures with different levels of *Acremonium lolii, Proc. N.Z. Weed Pest Control Conf.,* 44:171–175, 1991.

76. **Popay, A. J., R. A. Mainland, and C. J. Saunders,** The effect of endophytes in fescue grass on growth and survival of third instar grass grub larvae, In: D. E. Hume, G. C. M. Latch, and H. S. Easton, Eds., *Proc. 2nd Int. Symp.* Acremonium/*Grass Interactions*, AgResearch, Grasslands Research Centre, Palmerston North, New Zealand, 1993, 174.

77. **Popay, A. J., R. A. Prestidge, D. D. Rowan, and J. J. Dymock,** The role of *Acremonium lolii* mycotoxins in insect resistance of perennial ryegrass (*Lolium perenne*), In: S. S. Quisenberry and R. E. Joost, Eds., *Proc. 1st Int. Symp.* Acremonium/*Grass Interactions*, Louisiana Agriculture Experiment Station, Baton Rouge, 1990, 44.

78. **Popay, A. J. and D. D. Rowan,** Endophytic fungi as mediators of plant-insect interactions, In: E. Bernays, Ed., *Insect-Plant Interactions*, Vol. 5, CRC Press, Boca Raton, FL, in press, 1993.

79. **Porter, J. K., C. W. Bacon, J. D. Robbins, and D. Betowski,** Ergot alkaloid identification in Clavicipitaceae systemic fungi of pasture grasses, *J. Agric. Food Chem.,* 29:653–657, 1981.

80. **Potter, D. A., C. G. Patterson, and C. T. Redmond,** Influence of turfgrass species and tall fescue endophyte on feeding ecology of Japanese beetle and southern masked chafer grubs (Coleoptera:Scarabaeidae), *J. Econ. Entomol.,* 85:900–909, 1992.

81. **Prestidge, R. A.,** Preliminary observations on the grassland leafhopper fauna of the central North Island volcanic plateau, *N. Z. Entomol.,* 12:54–57, 1989.

82. **Prestidge, R. A.,** Susceptibility of Italian ryegrasses (*Lolium multiflorum* Lam.) to Argentine stem weevil (*Listronotus bonariensis* (Kuschel)) feeding and oviposition, *N.Z. J. Agric. Res.,* 34:119–125, 1991.

83. **Prestidge, R. A. and O. J.-P. Ball,** The role of endophytes in alleviating plant biotic stress in New Zealand, In: D. E. Hume, G. C. M. Latch, and H. S. Easton, Eds., *Proc. 2nd Int. Symp.* Acremonium/ *Grass Interactions: Plenary Papers*, AgResearch, Grasslands Research Centre, Palmerston North, New Zealand, 1993, 121.

84. **Prestidge, R. A. and R. T. Gallagher,** Lolitrem B — a stem weevil toxin isolated from *Acremonium*-infected ryegrass, *Proc. N.Z. Weed Pest Control Conf.,* 38:38–40, 1985.

85. **Prestidge, R. A. and R. T. Gallagher,** Endophyte fungus confers resistance to ryegrass: Argentine stem weevil studies, *Ecol. Entomol.,* 13:429–435, 1988.

86. **Prestidge, R. A., R. P. Pottinger, and G. M. Barker,** An association of *Lolium* endophyte with ryegrass resistance to Argentine stem weevil, *Proc. N.Z. Weed Pest Control Conf.,* 35:119–122, 1982.

87. **Quigley, P., X. Li, G. McDonald, and A. Noske,** Effects of *Acremonium lolii* on mixed pastures and associated insect pests in south-eastern Australia, In: D. E. Hume, G. C. M. Latch, and H. S. Easton, Eds., *Proc. 2nd Int. Symp.* Acremonium/*Grass Interactions*, AgResearch, Grasslands Research Centre, Palmerston North, New Zealand, 1993, 177.

88. **Riedell, W. E., R. E. Kieckhefer, R. J. Petroski, and R. G. Powell,** Naturally-occurring and synthetic loline alkaloid derivatives: insect feeding behaviour modification and toxicity, *J. Entomol. Sci.,* 26:122–129, 1991.

89. **Rottinghaus, G. E., D. D. Rowan, and B. A. Tapper,** Alkaloids in pasture grasses, In: H. F. Linsken and J. F. Jackson, Eds., *Modern Methods of Plant Analysis*, Vol. 15, *Alkaloids*, Springer-Verlag, Heidelberg, in press, 1994.

90. **Rowan, D. D.,** Lolitrems, paxilline and peramine: mycotoxins of the ryegrass/endophyte interaction, *Agric. Ecosyst. Environ.,* 44:103–122, 1993.

91. **Rowan, D. D., J. J. Dymock, and M. A. Brimble,** Effect of fungal metabolite peramine and analogs on feeding and development of Argentine stem weevil (*Listronotus bonariensis*), *J. Chem. Ecol.,* 16:1683–1695, 1990a.

92. **Rowan, D. D. and D. L. Gaynor,** Isolation of feeding deterrents against Argentine stem weevil from ryegrass infected with the endophyte *Acremonium loliae*, *J. Chem. Ecol.,* 12:647–658, 1986.

93. **Rowan, D. D., M. B. Hunt, and D. L. Gaynor,** Peramine, a novel insect feeding deterrent from ryegrass infected with the endophyte *Acremonium loliae*, *J. Chem. Soc. Chem. Commun.,* 1986:935–936, 1986.

94. **Rowan, D. D. and G. J. Shaw,** Detection of ergopeptine alkaloids in endophyte-infected perennial ryegrass by tandem mass spectrometry, *N.Z. Vet. J.,* 35:197–198, 1987.

95. **Rowan, D. D., B. A. Tapper, N. L. Sergejew, and G. C. M. Latch,** Ergopeptine alkaloids in endophyte-infected ryegrasses and fescues in New Zealand, In: S. S. Quisenberry and R. E. Joost, Eds., *Proc. 1st Int. Symp.* Acremonium/*Grass Interactions*, Louisiana Agriculture Experiment Station, Baton Rouge, 1990b, 97.

96. **Rutschmann, J. and P. A. Stadler,** Chemical background, In: B. Berde and H. O. Schild, Eds., *Ergot Alkaloids and Related Compounds*, Springer-Verlag, Berlin, 1978, 29.

97. **Saha, D. C., J. M. Johnson-Cicalese, P. M. Halisky, M. I. Van Heemstra, and C. R. Funk,** Occurrence and significance of endophytic fungi in the fine fescues, *Plant Dis.,* 71:1021–1024, 1987.

98. **Siegel, M. R., D. L. Dahlman, and L. P. Bush,** The role of endophytic fungi in grasses: new approaches to biological control of pests, In: A. R. Leslie and R. L. Metcalf, Eds., *Integrated Pest Management for Turfgrass and Ornamentals,* U.S. Environmental Protection Agency, Washington, D.C., 1989, 169.

99. **Siegel, M. R., G. C. M. Latch, L. P. Bush, F. F. Fannin, D. D. Rowan, B. A. Tapper, C. W. Bacon, and M. C. Johnson,** Fungal endophyte-infected grasses: alkaloid accumulation and aphid response, *J. Chem. Ecol.,* 16:3301–3315, 1990.

100. **Springer, T. L. and S. D. Kindler,** Endophyte-enhanced resistance to the Russian wheat aphid and the incidence of endophytes in fescue species, In: S. S. Quisenberry and R. E. Joost, Eds., *Proc. 1st Int. Symp.* Acremonium/*Grass Interactions*, Louisiana Agriculture Experiment Station, Baton Rouge, 1990, 194.

101. **Stewart, A. V.,** Perennial ryegrass seedling resistance to Argentine stem weevil, *N.Z. J. Agric. Res.,* 28:403–407, 1985.

102. **Sun, D. and J. P. Breen,** Inhibition of *Acremonium* endophyte in Kentucky Bluegrass, In: D. E. Hume, G. C. M. Latch, and H. S. Easton, Eds., *Proc. 2nd Int. Symp.* Acremonium/*Grass Interactions*, AgResearch, Grasslands Research Centre, Palmerston North, New Zealand, 1993, 19.

103. **TePaske, M. R., J. B. Gloer, D. T. Wicklow, and P. F. Dowd,** Aflavazole: a new antiinsectan carbazole metabolite from the sclerotia of *Aspergillus flavus, J. Org. Chem.,* 55:5299–5301, 1990.

104. **van Heeswijck, R. and G. McDonald,** *Acremonium* endophytes in perennial ryegrass and other pasture grasses in Australia and New Zealand, *Aust. J. Agric. Res.,* 43:1683–1709, 1992.

105. **Weedon, C. M. and P. G. Mantle,** Paxilline biosynthesis by *Acremonium loliae;* a step towards defining the origin of lolitrem neurotoxins, *Phytochemistry,* 26:969–971, 1987.

106. **Yates, S. G., J. C. Fenster, and R. J. Bartelt,** Assay of tall fescue seed extracts, fractions, and alkaloids using the large milkweed bug, *J. Agric. Food Chem.,* 37:354–357, 1989.

Chapter 13

Acremonium Endophytes in Germplasms of Major Grasses and Their Utilization for Insect Resistance

Stephen L. Clement, Walter J. Kaiser, and Herb Eichenseer

CONTENTS

I. INTRODUCTION

Fungi in the tribe Balansieae (Clavicipitaceae) infect several hundred species of grasses and a few species of sedges and rushes. The Balansieae can be divided into two groups: teleomorphic forms that produce a sexual stage (teleomorph) on stromata and may induce "choke diseases" of hosts; and related anamorphic derivatives (endophytes) which lack a sexual stage and are not considered to induce diseases of hosts.[4,24,91] Although the taxonomic status of many fungi in the tribe Balansieae is uncertain,[67] those producing teleomorphs are, presently, most often placed in the genera *Epichloë* (Fr.) Tul., *Balansia* Speg., *Atkinsonella* Diehl, and *Myriogenospora* Atk.[4] The genus *Acremonium* Link. sect. *Albo-lanosa* Morgan-Jones & W. Gams was erected to accommodate the related anamorphic states.[66] Other seedborne endophytes that are responsible for asymptomatic infections of grasses, but that are unrelated to *Acremonium* anamorphs, belong to the genera *Gliocladium* Corda, *Phialophora* Medlar, and *Pseudocercosporella* Deighton.[60,107] Several recent reviews[4,5,21,22,24,25,51,67,91,106,107,109] and Chapters 1 and 2 in this volume provide more details on the taxonomy and classification of fungi in the Balansieae, as well as information concerning their host-plant relations and their role in causing toxicity to grazing animals.

Host grasses infected with clavicipitaceous fungi are also associated with increased insect resistance. Endophyte-conferred resistance may be the result of the fungus or the grass–fungus interaction, producing specific metabolites, such as alkaloids.[4,5,21,22,24,25,57,91,104]

Although some grasses and sedges infected with teleomorphic[27,28,89,98] and unidentified (sterile) endophytes[18] adversely affect the behavior and/or performance (growth and survival) of insects, compared to uninfected plants, most of the existing information on the anti-insect properties of endophyte-infected plants comes from work on tall fescue, *Festuca arundinacea* Schreb., and perennial ryegrass, *Lolium perenne* L., harboring *Acremonium coenophialum* Morgan-Jones & W. Gams and *A. lolii* Latch, Christensen & Samuels, respectively. Indeed, *Acremonium* species appear to be more applicable, and offer greater potential for exploitation in the development of pest-resistant grasses than teleomorphic endophytes for the following reasons:[23]

1. A large taxonomic diversity of insects and other pests are negatively affected by infected grasses.
2. Anamorphic forms do not cause disease symptoms on their hosts. (However, some isolates of *A. typhinum* Morgan-Jones & W. Gams cause choke in fine fescue species.[108])

3. They are self perpetuated through maternal transmission in seeds.
4. With the development of artificial inoculation techniques,[58] synthetic combinations of grasses and anamorphic endophytes suitable for pest resistance can possibly be developed.[93]

In addition, the recent demonstration by Murray et al.[70] that an *Acremonium* endophyte can be used as a surrogate host to introduce foreign genes into grasses bodes well for the future establishment of new grass–endophyte associations for pest resistance. The reader is referred to Siegal and Schardl[94] and Schardl (Chapter 11) for general information on the application of various molecular techniques to genetic engineering of anamorphic endophytes for specific purposes.

This chapter reviews progress in determining the incidence of *Acremonium* and other anamorphic endophytes in cool-season grasses and in cereal grasses, with the emphasis on surveys of germplasm seedbanks. These repositories may contain new sources of endophytes that could be important in improving grasses for increased pest resistance. Additionally, this chapter reviews our current state of knowledge of grass–endophyte–insect relationships and provides an up-to-date listing of reported cases of *Acremonium*-conferred resistance to insects. Before ending this chapter with remarks about potential uses of anamorphic endophytes and their associated metabolites, for insect control, we draw attention to the need to look upon these fungi as components of biodiversity, which, like other types of genetic building material, must be conserved for research and development.

II. INCIDENCE OF ENDOPHYTES IN GRASS GERMPLASM COLLECTIONS

A. COOL-SEASON GRASSES

Although few surveys of germplasm collections of cool-season grasses have been made, they have provided information on the incidence of anamorphic endophytes in these grasses. In one systematic survey, Halisky et al.[47] found endophytes in seeds of *Festuca* L. and *Lolium* L. from several sources. Likewise, Saha et al.[88] detected anamorphic endophytes in seeds of several cultivars and selections of *Festuca* species. Latch et al.[61] detected endophytes in some of the *Lolium* and *Festuca* seed collections held at the Welsh Plant Breeding Station, Aberystwyth, Wales. In addition, Christensen et al.[19] and Christensen and Latch[20] isolated endophytes from several cultivars and ecotypes of perennial ryegrass and tall fescue, respectively. During a survey of 962 germplasm entries in eight species of *Poa* L., researchers at Rutgers University found endophyte (*A. typhinum*) in only one entry (*P. ampla* Merr.).[99]

The Western Regional Plant Introduction Station (WRPIS) at Pullman, WA, stores the largest working seed collections of cool-season grasses: in the U.S. National Plant Germplasm System. A recent survey of 14% of the *Lolium* accessions in the WRPIS collection revealed the presence of *Acremonium* fungi in seeds of 28 of 85 accessions from five species: *L. perenne*, *L. multiflorum* Lam., *L. persicum* Boiss & Hohen., *L. rigidum* Gaudin, and *L. temulentum* L.[111] Information on specific endophyte-infected and uninfected plant introduction (PI) lines of *Lolium* is available from the Germplasm Resources Information Network (GRIN) database. Springer and Kindler[96] reported finding endophytes (presumably *Acremonium* fungi) in seedlings of accessions of 17 of 41 *Festuca* species held by the WRPIS; however, they provided no information on the endophyte status of individual PI lines. A recent screening of WRPIS fescue collections by West et al.[105] led to the discovery of *Acremonium* endophytes in 49 of 529 accessions of tall fescue and in 26 of 198 accessions of meadow fescue (*F. pratensis* Huds.).

The endophytes found in most of the aforementioned surveys of tall fescue and perennial ryegrass resembled *A. coenophialum* and *A. lolii*, respectively. That isolates of *Acremonium* fungi exhibit wide variation in cultural, morphological, and physiological characteristics was borne out by the findings of Christensen et al.[19] and Christensen and Latch.[20] A *Gliocladium*-like endophyte also has been isolated from *L. perenne* (PI 205278) stored in the WRPIS seedbank.[16,29]

B. CEREAL GRASSES

Latch[56] (personal communication) did not detect endophytes in seeds of rice (*Oryza* spp.) and small grains (*Triticum* L., *Hordeum* L.) stored at the International Rice Research Institute in the Philippines, and at the U.S. National Small Grains Collection in Aberdeen, ID, respectively. However, Wilson et al.[110] found *Acremonium* endophytes in 21% of the seed samples of 77 wild barley (*Hordeum*) lines from the National Small Grains Collection. Endophytes were found in three perennial species (Table 1), which are in the tertiary genepool.[101] Isolates from *Acremonium*-infected *H. bogdanii* Wilensky and *H. brevisubulatum* ssp. *violaceum* (Boissier & Hohenacker) Tzvelev exhibit wide variation in cultural and morphological

Table 1 *Acremonium*-infected accessions from *Hordeum* germplasm in the U.S. national small grains collection

Species	Accession (Plant Introduction No.)	Country of Origin	Infected Seeds (%)[a]
Hordeum bogdanii	269406	Afghanistan	47
	314696	Former USSR	62
	440413	Former USSR	88
	440414	Former USSR	80
	499499	China	18
	499500	China	52
	499501	China	77
	499643	China	47
	499644	China	99
	499645	China	94
	499646	China	97
H. brevisubulatum	401386	Iran	68
ssp. *violaceum*	440420	Former USSR	98
H. comosum	264404	Argentina	86
	264405	Argentina	92
	269648	Argentina	74

[a] Percentages are based on examination of 100 seeds.

characteristics,[16,29] indicating that wild barley may be a source of great endophyte diversity. Seed samples of *H. brachyantherum* Nevski, *H. bulbosum* L., *H. chilense* Roemer & Schultes, *H. jubatum* L., *H. marinum* Hudson, *H. murinum* L., and *H. stenostachys* Godron were endophyte free.[110]

More recently, Marshall et al.[63] detected *Acremonium*-like fungi in two diploid cereal grass species, *Aegilops mutica* Boiss. [= *Amblyopyrum muticum* (Boiss.) Eig] and *Ae. uniaristata* Vis. Seed of this and other wild cereals was collected in Turkey in 1992. This is the first discovery of *Acremonium* endophytes in an annual grass species (*Ae. mutica*) that can hybridize with cultivated wheat (*T. aestivum* L.).[63]

C. SUMMARY

To date, most surveys undertaken have screened small portions of the total available accessions of any one collection. Usually, a small percentage of examined accessions have been found to be infected with endophytes. Although few extensive surveys have been conducted, they, nonetheless, reveal that several strains and even separate species of *Acremonium* fungi can infect an accession and a given host species, respectively. Systematic surveys of seeds stored in seedbanks around the world[13] may lead to the discovery of new sources of anamorphic endophytes. Moreover, explorations for endophyte-infected grasses will likely lead to the discovery of new and diverse types of *Acremonium* fungi.[63,71,103]

III. *ACREMONIUM*-INFECTED GRASSES AND INSECT RELATIONSHIPS

A. ENDOPHYTE-CONFERRED RESISTANCE TO INSECTS

The first reports linking anamorphic fungal endophytes in grasses to enhanced insect resistance appeared in the early 1980s. These studies involved *A. lolii* in perennial ryegrass resistance to Argentine stem weevil [*Listronotus bonariensis* (Kuschel)][83] and sod webworms (*Crambus* spp.).[42] By 1987 at least nine more species of insects were found to be adversely affected by *Acremonium*-infected grasses or by metabolites associated with infected grasses.[91] Recently, Dahlman et al.[34] and Latch[57] reported that 18 to 23 species of insects are negatively affected by tall fescue and/or perennial ryegrass infected with *Acremonium* endophytes, or by metabolites associated with endophyte infection in these two grasses. Today, this "list" has grown to at least 40 species of insects (Table 2) in six orders (Coleoptera, Lepidoptera, Diptera, Hemiptera, Homoptera, and Orthoptera). All of the insects listed in Table 2 are pests on grasses, except *Tribolium* sp., *T. castaneum* Herbst, *Oncopeltus fasciatus* (Dallas), and *Drosophila melanogaster* Meigen. The house cricket [*Acheta domesticus* (L.)] is not usually thought of as a grass pest, although it feeds on grasses.

Table 2 **Insects reported to be aversely affected by *Acremonium*-infected grasses and metabolites associated with infected grasses**

Insect	Grass[a]	Effects on Insect	Ref.
Orthoptera			
Teleogryllus commodus (Black field cricket)	L.p.	Feeding deterrence	85
T. oceanicus (Queensland field cricket)	L.p.	Feeding deterrence	85
Acheta domesticus (House cricket)	L.p.	Reduced survival	2
Hemiptera			
Blissus leucopterus hirtus (Hairy chinch bug)	L.p., F.l., F.r.c.	Feeding deterrence, reduced survival (L.p.); lower densities (F.l., F.r.c.)	64,88
Draculacephala spp. (Sharpshooter leafhopper)	F.a.	Lower densities	55,68
Endria inimica (Painted leafhopper)	F.a.	Lower densities	55
Agallia constricta (Leafhopper)	F.a.	Lower densities	55
Exitianus exitiosus (Graylawn leafhopper)	F.a.	Lower densities	68
Graminella nigrifrons (Back-faced leafhopper)	F.a.	Lower densities	68
Prosapia bicincta (Two-lined spittlebug)	F.a.	Lower densities	68
Oncopeltus fasciatus (Large milkweed bug)	F.a.	Feeding deterrence, high mortality	52,112
Homoptera			
Balanoccus poae (Pasture mealy bug)	L.p.	Lower densities	74
Rhopalosiphum padi (Bird cherry-oat aphid)	F.a.	See Table 3	20,38,39,45, 52,59,93
R. maidis (Corn leaf aphid)	L.p.	See Table 3	17
Schizaphis graminum (Greenbug aphid)	L.p., F.a.	See Table 3	14,38,52, 86,93
Diuraphis noxia (Russian wheat aphid)	L.p., F.a., H.b.v.	See Table 3; reduced survival (H.b.v.)	30–32,54, 96,111
Sipha flava (Yellow sugarcane aphid)	L.p.	Feeding deterrence	43
Aploneura lentisci (Root aphid)	F.p.	No development	90
Rhopalomyzus poae (Leaf aphid)	F.p.	Reduced development	90
Coleoptera			
Listronotus bonariensis (Argentine stem weevil)	L.p., F.a.	Feeding and oviposition deterrence (L.p., F.a.); reduced survival/ovarian development, lower densities (L.p.)	7–12,37, 79,81–83, 87,97
Heteronychus arator (Black beetle)	L.p.	Feeding deterrence, lower densities	6,79,84
Sphenophorus parvulus (Bluegrass billbug)	L.p., F.a.	Lower densities (L.p.); reduced survival (L.p., F.a.)	1,53

Table 2 (Cont.) **Insects reported to be aversely affected by *Acremonium*-infected grasses and metabolites associated with infected grasses**

Insect	Grass[a]	Effects on Insect	Ref.
S. venatus	L.p., F.a.	Reduced survival	53,69
S. inaequalis	L.p., F.a.	Reduced survival	53
S. minimus	L.p., F.a.	Reduced survival	53,69
Chaetocnema pulicaria (Corn flea beetle)	F.a.	Lower densities	55
Tribolium sp. (Flour beetle)	F.a.	Lower densities	34
T. castanuem (Red flour beetle)	L.p., F.a.	Reduced survival and population growth rates	18
Popilla japonica (Japanese beetle)	F.a.	Feeding deterrence, lower densities	34,69,72, 73,78
Cyclocephala lurida (Southern masked chafer)	F.a.	Lower weight gain	78
Costelytra zealandica (Grass grub)	F.a.	Reduced survival	76
Lepidoptera			
Crambus spp. (Sod webworms)	L.p., F.a.	Lack of feeding, lower densities (L.p.); reduced survival (F.a.)	34,41,42,69
Parapediasia teterrella (Bluegrass webworm)	L.p.	Feeding deterrence, reduced survival, lower weight gain	65
Ostrinia nubilalis (European corn borer)	F.a.	Altered feeding behavior, lower weight gain	86
Spodoptera frugiperda (Fall armyworm)	L.p., F.a., F.l., F.r.c., P.a.	Feeding deterrence, reduced survival, lower weight gain, delayed development (L.p., F.a.); lower weight gain, high mortality (F.l., F.r.c.); high mortality (P.a.)	15,26,27,48, 49,86,99
S. eridania (Southern armyworm)	L.p.	Reduced survival	3,36
Mythimna convecta (Common armyworm)	L.p.	Feeding deterrence, delayed development	40,85
Agrotis infusa (Cutworm)	L.p.	Feeding deterrence	85
Graphania mutans (Cutworm)	L.p.	Feeding deterrence, reduced survival	37
Diptera			
Drosophila melanogaster (Fruit fly)	F.a.	Reduced survival	33,46,75

[a] L.p., *Lolium perenne* (perennial ryegrass); F.a., *Festuca arundinacea* (tall fescue); P.a., *Poa ampla*; F.l., *F. longifolia* (hard fescue); F.r.c., *F. rubra* ssp. *commutata* (Chewings fescue); H.b.v., *Hordeum brevisubulatum* ssp. *violaceum*; F.p., *F. pratensis* (meadow fescue).

It should also be noted that distinct life stages of some insects listed in Table 2 may be affected differently by exposure to endophyte-infected grasses. A good example is fall armyworm (*Spodoptera frugiperda* (J. E. Smith); neonate larvae prefer feeding on endophyte-free tall fescue, whereas fourth-instar larvae exhibit no preference for endophyte-infected or endophyte-free tall fescue.[48] More recent examples are provided by Quigley et al.,[85] who observed that common armyworm [*Mythimna convecta* (Walker)] and field crickets [*Teleogryllus commodus* (Walker) and *T. oceanicus* (Le Guillou)] are most sensitive to *A. lolii*-infected perennial ryegrass during early stages of development.

As indicated earlier, most prior studies on grass–endophyte–insect interactions have involved tall fescue and perennial ryegrass, infected with *A. coenophialum* and *A. lolii*, respectively. Table 2 clearly demonstrates the large extent to which these particular grass–*Acremonium* associations have been used to study the effect of endophyte infection on insect behavior and performance. The source of the plant material used in many of the studies referenced in Table 2 were cultivars of tall fescue (e.g., "Arid," "Chieftain," "Forager," "Georgia-5," "Johnston," "Kentucky 31," "Mustang") and perennial ryegrass (e.g., "Citation ll," "Ellet," "Grasslands Nui," "Grasslands Ruanui," "Repell"), with and without endophytes. Additionally, insect responses have been measured on agronomically adapted selections and single-plant progenies of these two grasses.

There are a few reports of *Acremonium*-enhanced resistance to insects involving host grasses other than tall fescue and perennial ryegrass (Table 2). In one study fall armyworm larvae feeding on *A. typhinum*-infected tillers of big bluegrass (*P. ampla*, cultivar "Service") experienced high mortality compared to larvae feeding on endophyte-free tillers.[99] Additionally, Saha et al.[88] reported that chinch bug [*Blissus leucopterus hirtus* (Montandon)] infestations were lower in field plots with high percentages (>84%) of *Acremonium*-infected plants of hard fescue (*F. longifolia* Thuill, cultivar "SR3000") and Chewings fescue (*F. rubra* L. ssp. *commutata* Gaud., cultivar "Longfellow"), compared to plots with cultivars of low endophyte (<10% infected plants). Moreover, *Acremonium*-infected hard and Chewings fescue are resistant to fall armyworm.[15] Schmidt[90] showed that *A. uncinatum* W. Gams, Petrini, & Schmidt infection of meadow fescue (*F. pratensis*) imparts resistance to the aphids *Aploneura lentisci* Passerini and *Rhopalomyzus poae* Gillette. Recent research by one of us (S.L.C.) has shown that *Acremonium* endophyte in *Hordeum brevisubulatum* ssp. *violaceum* (PI 440420) imparts increased plant resistance to Russian wheat aphid [*Diuraphis noxia* (Mordvilko)] (Figure 1).

B. MECHANISMS AND FACTORS INFLUENCING RESISTANCE

Some insects are repelled from endophyte-infected hosts or choose not to feed on infected plants if they come in contact with them (often referred to as nonpreference behavior or feeding deterrency in the literature). For other species of insects (see Table 2), performance (measured as population density, growth rate, developmental time, or weight of adults and immature stages) is negatively affected by the bioactive metabolites associated with endophyte infection. Preference behavior and poor performance in insects are equivalent to antixenotic and antibiotic resistance, respectively, in plants.[95] For extensive reviews in which *Acremonium* toxins (ergot, loline, peramine, lolitrem B) might be responsible for observed antixenotic and antibiotic activity against specific insects, the reader is referred to Dahlman et al.[34] and Chapters 8 and 12 of this volume.

Peramine is most likely responsible for antixenosis resistance in perennial ryegrass to Argentine stem weevil. Interestingly, peramine may also increase the susceptibility of polyphagous insect pests [e.g., *Spodopters eridania* (Cramer)] to some commonly used insecticides (e.g., carbaryl) by synergizing the effect of the insecticides.[36] Loline is probably involved in antixenotic and/or antibiotic resistance in tall fescue to bird cherry-oat aphid (*Rhopalosiphum padi* L.) (see Dahlman et al.[34] and references therein). Also, some loline alkaloids exhibit insecticidal activity toward insects.[34,52,86,112] Recently, *N*-formylloline and ergot alkaloids (ergovaline, ergosine, ergotamine) were detected in *Acremonium*-infected *H. brevisubulatum* ssp. *violaceum* (PI 440420) (TePaske, Powell, and Clement, unpublished data), but it is not known if any of these alkaloids are responsible for enhanced resistance to Russian wheat aphids (Figure 1).

The expression of resistance, and the type of plant resistance (antixenosis or antibiosis), to insects depends on the host plant species, the *Acremonium* species and strain, and the insect species involved. For instance, *A. coenophialum*-infected tall fescue may impart increased resistance to Japanese beetle grubs (*Popillia japonica* Newman).[72,73] However, *A. lolii*-conferred resistance in perennial ryegrass has not been detected.[41,42] Breen[15] detected no antibiosis resistance in *Acremonium*-infected blue fescue., *F. glauca* Lam., to fall armyworm, but she did find strong antibiosis in hard fescue and Chewings fescue infected with *Acremonium* spp. Johnson-Cicalese and White[53] found that four species of billbugs had no preference for endophyte-free tall fescue and perennial ryegrass over *Acremonium*-infected conspecifics of both grasses, but mortality of all species was greater on endophyte-infected plants of both grasses. Table 3 summarizes the differential sensitivities of seven aphid species to different *Acremonium*–grass associations and to different strains of *Acremonium* endophytes in tall fescue. Host plant genotype, the concentration of the *Acremonium* fungus or its metabolites in plants, infection by other endophytes, ambient temperature, and the level of nitrogen fertilization are other factors that can influence the effect endophyte-infected plants have upon insects.[8,15,57,84]

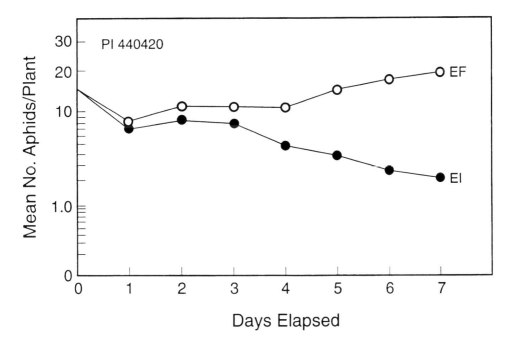

Figure 1 Population trends of *Diuraphis noxia* on endophyte-free (EF) and endophyte-infected (EI) clones of *Hordeum brevisubulatum* ssp. *violaceum* (PI 440420). To obtain genetically identical plants, with and without *Acremonium* endophyte, propiconazole fungicide was used to remove endophytes from one set of clones. There were 15 EF and 15 EI clones: Fifteen aphids were placed on each clone, after which the number of live aphids on each clone was recorded daily for 7 d. A repeated-measures ANOVA revealed significant differences in aphid mortality among EI and EF clones, and over time on these two plant types (F = 26.81; df = 1, 28; P > F = 0.0001). Also, there was a significant time × endophyte interaction (F = 36.91; df = 6, 168; P > F = 0.0001), indicating that the effect of time on *D. noxia* mortality was different on EI and EF clones.

C. INSECTS UNAFFECTED BY *ACREMONIUM* ENDOPHYTES

A link between *Acremonium* infection of grasses and increased plant resistance or toxicity to insects is not always indicated in particular associations. For example, three aphid species [*Metopolophium dirhodum* (Walker), *Sitobion fragariae* (Walker), and *S. avenae* (F.)] have been recorded as being insensitive to endophyte-infected tall fescue and perennial ryegrass, compared to uninfected plants of both grasses (Table 3). Greenbug aphid [*Schizaphis graminum* (Rondani)] did not exhibit a preference for *A. typhinum*-free tillers of *P. ampla* over infected tillers from conspecific plants.[99] In other studies *S. eridania*,[52] *Pseudaletia unipuncta* (Haworth),[34] *Melanoplus differentialis* (Thomas),[34] and some leafhopper species[55,68] were not adversely affected by endophyte-infected tall fescue. Interestingly, Breen[15] found that larvae of *S. eridania* preferred and had improved development on endophyte-infected tall fescue. *Oscinella* sp.,[62] *Wiseana cervinata* Walker,[100] and several Hemipteran species[84] are seemingly unaffected by endophyte-infected perennial ryegrass. In yet another study researchers observed no consistent response when larvae of *Manduca sexta* L. and *Heliothis virescens* F. were fed artificial diets containing ground plant material or plant extracts from endophyte-infected tall fescue, as compared with diet preparations from endophyte-free tall fescue.[52] Although *P. japonica* and *Cyclocephala lurida* Bland are listed in Table 2 as being adversely affected by endophyte-infected tall fescue, researchers have obtained mixed results when investigating the effects of *A. coenophialum*-infected tall fescue on grubs of these two scarab beetles.[72,73]

D. EXPERIMENTAL METHODS

Information on how endophyte infection influences the behavior and performance of insects has been generated predominately through the use of conventional research approaches, such as (1) laboratory or glasshouse experiments where insects do not have a choice as to the type of plant

Table 3 Differential effects of *Acremonium*-infected tall fescue and perennial ryegrass on aphid pests

| Aphid | Grass[a] | | Effects[b] |
	F.a.	L.p.	
Rhopalosiphum padi (Bird cherry-oat aphid)	+[c]	–	Deterred by F.a., but not L.p.; reduced survival and reproduction on F.a.
R. maidis (Corn leaf aphid)	–	+	Deterred by L.p., but not F.a.
Schizaphis graminum (Greenbug)	+	+	Deterred by F.a. and L.p.; reduced survival on F.a.
Diuraphis noxia (Russian wheat aphid)	+[d]	+	Feeding preference unaffected by L.p. and F.a.; reduced survival on F.a., L.p.
Metopolophium dirhodum (Rose-grass aphid)	–	–	Feeding preference unaffected by F.a. and L.p.
Sitobion fragariae (Strawberry aphid)	–	–	Feeding preference unaffected by F.a. and L.p.
S. avenae (English grain aphid)	–	–	Feeding preference unaffected by F.a. and L.p.

Note: + = adverse effect on aphid; – = no adverse effect.

[a] F.a., *Festuca arundinaceae* (tall fescue); L.p., *Lolium perenne* (perennial ryegrass); [b] References listed in Table 2 were sources of information on effects of *Acremonium* infection on *R. padi*, *R. maidis*, *S. graminum*, and *D. noxia*. Information on *M. dirhodum*, *S. fragariae*, and *M. avenae* summarized from Johnson et al.[52] and Latch et al.;[59] [c] Tall fescue infected by endophyte strains with smaller conidia than *A. coenophialum* failed to deter *R. padi*;[20] [d] Tall fescue with *Acremonium* endophyte that did not resemble *A. coenophialum* was susceptible to *D. noxia*.[54]

they feed upon,[2,3,12,15,18,27,28,30-32,38-40,45,48,49,52-54,64,65,76,78,81,89,93,96,98,99] (2) laboratory or glasshouse experiments where insects are free to discriminate between endophyte-infected and uninfected plant material,[7,11,14,15,17,20,30-32,38-40,48,49,52-54,59,64,85,90,96,98,99,111] and (3) field observations in which insect diversity and abundance are correlated with the presence or absence of endophytes in plants or with the varying endophyte infection levels in pastures.[1,6,7,10,32,41,42,55,62,68,69,72,74,78,83,88] In some cases laboratory choice and no-choice tests conducted with artificial diets impregnated with dried or ground tissues from endophyte-infected plants or with endophyte-associated metabolites and their derivatives have proven useful for determining the relationship between endophyte infection and insect toxicity or enhanced plant resistance.[33,36,37,46,52,73,75,81,84,86,87,104] In addition, insects have been exposed to endophyte-associated metabolites or their derivatives, leaf disks soaked in solutions of individual endophyte-associated metabolites, seeds (ground or whole) of endophyte-infected grasses, or chemical extracts of infected seeds and plants.[18,26,52,81,82,86,112]

E. SUMMARY

Preliminary reports, abstracts, and references to unpublished data in review articles were sources for a significant amount of the information summarized in the preceding sections on insects. It is also important to note that most prior studies on grass–endophyte–insect interactions were conducted without considering the possible confounding effects of host-plant genetic resistance and endophyte infection status. Moreover, few studies conducted have used set experimental designs in the establishment of replicated field plots. Notwithstanding these limitations, most reports show that fungal endophytes alter plant suitability to insects. The insects best known to be adversely affected by *Acremonium* infection in one or more grasses are Argentine stem weevil, billbugs, sod webworms, armyworms (*Spodoptera* spp.), and some aphid species, including the greenbug, bird cherry-oat aphid, and the Russian wheat aphid (Tables 2 and 3).

Clay et al.[27] recognized the importance of using genetically identical plants, both infected with and free of *Acremonium* endophytes, to test the hypothesis that endophyte infection alone is responsible for enhanced resistance to an insect species. Researchers are beginning to make use of identical grass genotypes, with and without endophytes (see Figure 1).[14,15,30,53] That researchers are addressing the need for controlled experiments to study grass–endophyte–insect relationships comes from recently conducted "reproducible" field studies in which set designs (i.e., randomized and completely randomized block designs) were used to establish research plots.[32,68,72]

Finally, knowledge that both antixenosis and antibiosis are involved in *Acremonium*-based resistance to insects (Tables 2 and 3) and that different endophyte-associated alkaloids are likely responsible for the expression of these two resistance modalities means that plant scientists may have the option of developing cultivars with combinations of different endophyte-based resistance factors. The effect of this would be to prolong the useful life of an *Acremonium*-based plant defense system. That insect populations have the ability to evolve and overcome specific plant resistance factors is well known.[44] Clay[25] and Siegel et al.[93] first broached the subject of potential insect resistance to endophyte toxins. Recently, Pless et al.[75] provided the first experimental evidence, albeit preliminary, that an insect (*D. melanogaster*) can develop resistance to toxic factors associated with the *Acremonium* endophyte.

IV. ENDOPHYTES AS MICROBIAL GERMPLASM

This chapter is testimony to the many advances made in recent years in expanding our knowledge of the role played by anamorphic endophytes in conferring biotic stress tolerance to grasses. As a result, there is a growing awareness among plant scientists that anamorphic endophytes infecting grasses may constitute a valuable genetic resource for use in improving turf and crop grasses. Indeed, the discovery of natural strains of *A. lolii* that produce little or no detectable levels of animal toxins (i.e., lolitrem B), but that produce anti-insect alkaloids (i.e., peramine),[57] and the discovery of different *Acremonium* species and strains in grass germplasm, including cereal grass taxa (*Hordeum, Aegilops* L.) (see Section II), foretells the probable existence of a vast endophyte resource. This resource may provide the pool of *Acremonium* and alkaloid diversity needed for future exploitation.

Plant exploration is a primary mechanism for finding and adding new germplasm to seedbanks. As indicated earlier, this approach is being used to find new sources of endophytic fungi. Once endophyte-infected material has been placed in a germplasm repository, and surveys have identified infected accessions within a stored collection, it is important that the endophytic fungi (microbial germplasm) be preserved. Having a large diversity of endophytes as pest-resistance sources to draw from may be as beneficial as having a diversity of plant genetic resistance.

Germplasm maintenance procedures might have to be modified to maintain endophyte viability and diversity during seed regeneration and storage at repositories.[110] It might be necessary, for example, to regenerate seeds from endophyte-infected accessions on a frequent basis to refresh endophyte inoculum in seeds and to improve seed viability. Although guidelines on optimal seed storage conditions for endophyte-infected seeds of diverse taxa have not been developed, the available information indicates that seeds should not be stored at high temperatures (20 to 30°C) and a seed moisture content greater than 15%.[102] Seeds at the WRPIS are stored at 4°C, 30% RH, and a dew point of −10°C. These conditions are suitable for preserving endophytes in seeds of some grasses, as Hare et al.[50] found that viability of *A. lolii* in seeds of perennial ryegrass was maintained for 5 years at 0°C/30% RH and −15°C/90% RH. In addition, Do Valle Ribeiro[35] reported no loss of *A. lolii* viability in perennial ryegrass seeds stored for 6 to 7 years at 3 to 5°C (10% seed moisture) and −20°C (6 to 8% seed moisture). It is anticipated that future research will identify optimal storage conditions for the preservation of viable endophytes in seeds of diverse grass taxa.

V. CONCLUSIONS

Cultivars of *Acremonium*-infected perennial ryegrass, tall fescue, and hard and Chewings fescues are being marketed for enhanced resistance to webworms, billbugs, chinch bugs, and other pests of turf grasses.[41,77,92] The use of endophyte-infected turf grasses for insect resistance will likely expand, particularly if cultivars of Kentucky bluegrass (*Poa pratensis* L.), the most widely used turf grass species in the U.S., are developed with endophytes that impart resistance to insects. To date, however, researchers have had no success in locating anamorphic endophytes in *P. pratensis*.[99] Recent efforts to introduce *Acremonium* endophyte into Kentucky bluegrass were unsuccessful due to host-plant inhibition to the endophyte.[99] Perhaps, as suggested by Sun and Breen,[99] (1) the genetic resources of *P. pratensis* may contain germplasm without inhibition to nonchoke-inducing endophytes, and (2) *Acremonium* strains may be found that can be used to create *P. pratensis–Acremonium* associations for pest resistance. These points underscore the importance of grass and *Acremonium* genetic resources.

Researchers have discussed the possibility of introducing *Acremonium* fungi or genetically engineered forms of these fungi into crop grasses for insect control.[24,32,63] Recent discoveries of nonchoke-inducing endophytes in *Aegilops* and wild *Hordeum*[63,110] give some cause for optimism. Looking ahead, however, it is hard to predict if endophyte-infected cereals for insect control will become available. The development of such plants is fraught with potential problems and difficulties, including, but not limited to, the potential accumulation in seeds of metabolites toxic to humans, and possible incompatibility between modern cereal cultivars and natural or genetically modified endophyte strains. However, if research alleviates these problems so that grass–endophyte combinations can be tailored for insect resistance, pest managers could have a new method to strengthen the defensive capabilities of crop grasses.

There is also evidence that *Acremonium*-associated metabolites and their derivatives could be used as insecticides.[80,86] Also, metabolites from endophytes and grass–endophyte associations may have utility in pharmaceutical applications.[80]

In a span of 15 years, researchers have developed a body of information on the influence of *Acremonium* infection on grass susceptibility to a variety of insect herbivores. As more is learned about the diversity of endophytes in grass germplasm, and the ecological and biochemical factors governing grass–endophyte–insect relationships, the likelihood is increased that biotechnological approaches will succeed in producing new grass–endophyte associations for insect resistance and other specific purposes.

REFERENCES

1. **Ahmad, S., J. M. Govindarajan, C. R. Funk, and J. M. Johnson-Cicalese,** Fatality of house crickets on perennial ryegrasses infected with a fungal endophyte, *Entomol. Exp. Appl.,* 39:183–190, 1985.

2. **Ahmad, S., J. M. Johnson-Cicalese, W. K. Dickson, and C. R. Funk,** Endophyte-enhanced resistance in perennial ryegrass to the bluegrass billbug, *Sphenophorus parvulus, Entomol. Exp. Appl.,* 41:3–10, 1986.

3. **Ahmad, S., J. M. Govindarajan, J. M. Johnson-Cicalese, and C. R. Funk,** Association of a fungal endophyte in perennial ryegrass with antibiosis to larvae of the southern armyworm, *Spodoptera eridania, Entomol. Exp. Appl.,* 43:287–294, 1987.

4. **Bacon, C. W. and J. De Battista,** Endophytic fungi of grasses, In: D. K. Arora, B. Rai, K. G. Mikerji, and G. R. Knudson, Eds., *Handbook of Applied Mycology,* Vol. 1, *Soil and Plants,* Marcel Dekker, New York, 1991, 231.

5. **Bacon, C. W. and M. R. Siegel,** The endophyte of tall fescue, *J. Prod. Agric.,* 1:45–55, 1988.

6. **Ball, O. J.-P. and R. A. Prestidge,** The effect of the endophyte fungus *Acremonium lolii* on adult black beetle (*Heteronychus arator*) feeding, *Proc. N.Z. Plant Prot. Conf.,* 45:201–204, 1992.

7. **Barker, G. M.,** Grass host preferences of *Listronotus bonariensis* (Coleoptera:Curculionidae), *J. Econ. Entomol.,* 82:1807–1816, 1989.

8. **Barker, G. M., R. A. Prestidge, and R. P. Pottinger,** Role of *Acremonium lolii* in the population ecology of *Listronotus bonariensis* in New Zealand ryegrass pastures, In: S. S. Quisenberry, and R. E. Joost, Eds., *Proc. Int. Symp.* Acremonium/*Grass Interactions,* Louisiana Agriculture Experiment Station, Baton Rouge, 1990, 7.

9. **Barker, G. M., R. P. Pottinger, and P. G. Addison,** Effect of tall fescue and ryegrass endophytes on Argentine stem weevil, *Proc. N.Z. Weed Pest Control Conf.,* 36:216–219, 1983.

10. **Barker, G. M., R. P. Pottinger, and P. G. Addison,** Flight behavior of *Listronotus bonariensis* (Coleoptera:Curculionidae) in the Waikato, New Zealand, *Environ. Entomol.,* 18:996–1005, 1989.

11. **Barker, G. M., R. P. Pottinger, P. J. Addison, and R. A. Prestidge,** Effect of *Lolium* endophyte fungus infections on behavior of adult Argentine stem weevil, *N.Z. J. Agric. Res.,* 27:271–277, 1984.

12. **Barker, G. M., R. P. Pottinger, and P. J. Addison,** Effect of *Lolium* endophyte fungus infections on survival of larval Argentine stem weevil, *N.Z. J. Agric. Res.,* 27:279–281, 1984.

13. **Bettencourt, E. and J. Konopka,** Directory of germplasm collections, III. Cereals, International Board for Plant Genetic Resources, Rome, Italy, 1990.

14. **Breen, J. P.,** Temperature and seasonal effects on expression of *Acremonium* endophyte-enhanced resistance to *Schizaphis graminum* (Homoptera:Aphididae), *Environ. Entomol.,* 21:68–74, 1992.

15. **Breen, J. P.,** Enhanced resistance to fall armyworm (Lepidoptera:Noctuidae) in *Acremonium* endophyte-infected turfgrasses, *J. Econ. Entomol.,* 86:621–629, 1993.

16. **Bruehl, G. W., W. J. Kaiser, D. G. Lester, C. M. Davitt, and S. L. Clement,** Diversity of fungal endophytes in grass germplasm, In: D. E. Hume, G. C. M. Latch, and H. S. Easton, Eds., *Proc. 2nd Int. Symp.* Acremonium/*Grass Interactions,* AgResearch, Grasslands Research Centre, Palmerston North, New Zealand, 1993, 36.

17. **Buckley, R. J., P. M. Halisky, and J. P. Breen,** Variation in feeding deterrence of the corn leaf aphid related to *Acremonium* endophytes in grasses, *Phytopathology,* 81:120, 1991.

18. **Cheplick, G. P. and K. Clay,** Acquired chemical defences in grasses: the role of fungal endophytes, *Oikos,* 52:309–318, 1988.

19. **Christensen, M. J., G. C. M. Latch, and B. A. Tapper,** Variation within isolates of *Acremonium* endophytes from perennial rye-grasses, *Mycol. Res.,* 95:918–923, 1991.

20. **Christensen, M. J. and G. C. M. Latch,** Variation among isolates of *Acremonium* endophytes (*A. coenophialum* and possibly *A. typhinum*) from tall fescue (*Festuca arundinacea*), *Mycol. Res.,* 95:1123–1126, 1991.

21. **Clay, K.,** Fungal endophytes of grasses: a defensive mutualism between plants and fungi, *Ecology,* 69:10–16, 1988a.

22. **Clay, K.,** Clavicipitaceous fungal endophytes of grasses: coevolution and the change from parasitism to mutualism, In: K. A. Pirozynski and D. L. Hawksworth, Eds., *Coevolution of Fungi with Plants and Animals,* Academic Press, London, 1988b, 79.

23. **Clay, K.,** Clavicipitaceous endophytes of grasses: their potential as biocontrol agents, *Mycol. Res.,* 92:1–12, 1989.

24. **Clay, K.,** Fungal endophytes of grasses, *Annu. Rev. Ecol. Syst.,* 21:275– 297, 1990.

25. **Clay, K.,** Fungal endophytes, grasses, and herbivores, In: P. Barbosa, V. A. Krischik, and C. G. Jones, Eds., *Microbial Mediation of Plant-Herbivore Interactions,* John Wiley & Sons, New York, 1991, 199.

26. **Clay, K. and G. P. Cheplick,** Effect of ergot alkaloids from fungal endophyte-infected grasses on fall armyworm (*Spodoptera frugiperda*), *J. Chem. Ecol.,* 15:169–182, 1989.

27. **Clay, K., T. N. Hardy, and A. M. Hammond, Jr.,** Fungal endophytes of grasses and their effects on an insect herbivore, *Oecologia,* 66:1–5, 1985a.

28. **Clay, K., T. N. Hardy, and A. M. Hammond, Jr.,** Fungal endophytes of *Cyperus* and their effect on an insect herbivore, *Am. J. Bot.,* 72:1284–1289, 1985b.

29. **Clement, S. L., W. J. Kaiser, G. W. Bruehl, and R. C. Johnson,** Endophytic fungi in grass germplasm, In: *Agronomy Abstracts,* American Society of Agronomy, Madison, WI, 1992, 201.

30. **Clement, S. L., D. G. Lester, A. D. Wilson, R. C. Johnson, and J. H. Bouton,** Effect of fungal endophyte in tall fescue and wild barley on aphid performance, In: D. E. Hume, G. C. M. Latch, and H. S. Easton, Eds., *Proc. 2nd Int. Symp.* Acremonium/*Grass Interactions,* AgResearch, Grasslands Research Centre, Palmerston North, New Zealand, 1993, 155.

31. **Clement, S. L., K. S. Pike, W. J. Kaiser, and A. D. Wilson,** Resistance of endophyte-infected plants of tall fescue and perennial ryegrass to the Russian wheat aphid (Homoptera:Aphididae), *J. Kansas Entomol. Soc.,* 63:646–648, 1990.

32. **Clement, S. L., D. G. Lester, A. D. Wilson, and K. S. Pike,** Behavior and performance of *Diuraphis noxia* (Homoptera:Aphididae) on fungal endophyte-infected and uninfected perennial ryegrass, *J. Econ. Entomol.,* 85:583–588, 1992.

33. **Cole, A. M., C. D. Pless, and K. D. Gwinn,** Survival of *Drosophila melanogaster* (Diptera: Drosophilidae) on diets containing roots or leaves of *Acremonium*-infected or non-infected tall fescue, In: S. S. Quisenberry and R. E. Joost, Eds., *Proc. Int. Symp.* Acremonium/*Grass Interactions,* Louisiana Agriculture Experiment Station, Baton Rouge, 1990, 128.

34. **Dahlman, D. L., H. Eichenseer, and M. R. Siegel,** Chemical perspectives on endophyte-grass interactions and their implications to insect herbivory, In: P. Barbosa, V. A. Krischik, and C. G. Jones, Eds., *Microbial Mediation of Plant-Herbivore Interactions,* John Wiley & Sons, New York, 1991, 227.

35. **Do Valle Ribeiro, M. A. M.,** Transmission and survival of *Acremonium* and the implications for grass breeding, *Agric. Ecosyst. Environ.,* 44:195–213, 1993.

36. **Dubis, E. N., L. B. Brattsten, and L. B. Dungan,** Effects of the endophyte-associated alkaloid peramine on southern armyworm microsomal cytochrome P450, In: C. A. Mullin and J. G. Scott, Eds., *Molecular Mechanisms Insecticide Resistance,* ACS Symp. Ser. 505, American Chemical Society, Washington, D.C., 1992, 125.

37. **Dymock, J. J., D. D. Rowan, and I. R. McGee,** Effects of endophyte-produced mycotoxins on Argentine stem weevil and the cutworm, *Graphiana mutans*, In: P. P. Stahle, Ed., *Proc. 5th Aust. Conf. Grasslands Invert. Ecol.*, D & D Printing, Victoria, Australia, 1988, 35.

38. **Eichenseer, H. and D. L. Dahlman,** Antibiotic and deterrent qualities of endophyte-infected tall fescue to two aphid species (Homoptera:Aphididae), *Environ. Entomol.*, 21:1046–1051, 1992.

39. **Eichenseer, H., D. L. Dahlman, and L. P. Bush,** Influence of endophyte infection, plant age and harvest interval on *Rhopalosiphum padi* survival and its relation to quantity of N-formyl and N-acetyl loline in tall fescue, *Entomol. Exp. Appl.*, 60:29–38, 1991.

40. **Frost, W. E., P. E. Quigley, and P. J. Cunningham,** Research into the effect of *Acremonium lolii* on the feeding behavior and nutrition of principal Australian pests of perennial ryegrass, In: S. S. Quisenberry and R. E. Joost, Eds., *Proc. Int. Symp.* Acremonium/*Grass Interactions*, Louisiana Agriculture Experiment Station, Baton Rouge, 1990, 144.

41. **Funk, C. R., P. M. Halisky, and R. H. Hurley,** Implications of endophytic fungi in breeding for insect resistance, In: *Proc. Forage Turf-Grass Endophyte Workshop*, Oregon State University, Corvallis, 1983, 67.

42. **Funk, C. R., P. M. Halisky, M. C. Johnson, M. R. Siegel, A. V. Stewart, S. Ahmad, R. H. Hurley, and I. C. Harvey,** An endophytic fungus and resistance to sod webworms: association in *Lolium perenne* L., *Biotechnology*, 1:189–191, 1983.

43. **Funk, C. R., R. H. White, and J. P. Breen,** Importance of *Acremonium* endophytes in turfgrass breeding and management, *Agric. Ecosyst. Environ.*, 44:215–232, 1993.

44. **Gould, F.,** The evolutionary potential of crop pests, *Am. Sci.*, 79:496–507, 1991.

45. **Guy, P. L.,** Incidence of *Acremonium lolii* and lack of correlation with barley yellow dwarf viruses in Tasmanian perennial ryegrass pastures, *Plant Pathol.*, 41:29–34, 1992.

46. **Gwinn, K. D., C. A. Blank, A. M. Cole, and C. D. Pless,** Resistance of endophyte-infected tall fescue seedlings to pathogens and pests, In: B. P. Riechert, Ed., *Theme Issue: Tall Fescue and the Fungal Endophyte,* Tenn. Farm and Home Science, No. 160, Tennessee Agriculture Experiment Station, Knoxville, 1991, 72.

47. **Halisky, P. M., D. C. Saha, B. O. Bayaa, and C. R. Funk,** Fungal endophytes in *Festuca* and *Lolium*, *Can. J. Plant Pathol.*, 8:350, 1986.

48. **Hardy, T. N., K. Clay, and A. M. Hammond, Jr.,** Fall armyworm (Lepidoptera:Noctuidae): a laboratory assay and larval preference study for the fungal endophyte of perennial ryegrass, *J. Econ. Entomol.*, 78:571–575, 1985.

49. **Hardy, T. N., K. Clay, and A. M. Hammond, Jr.,** Leaf age and related factors affecting endophyte-mediated resistance to fall armyworm (Lepidoptera:Noctuidae) in tall fescue, *Environ. Entomol.*, 15:1083–1089, 1986.

50. **Hare, M. D., M. P. Rolston, and K. K. Moore,** Viability of *Lolium* endophyte fungus in seed and germination of *Lolium perenne* seed during five years of storage, In: S. S. Quisenberry and R. E. Joost, Eds., *Proc. Int. Symp.* Acremonium/*Grass Interactions*, Louisiana Agriculture Experiment Station, Baton Rouge, 1990, 147.

51. **Hoveland, C. S.,** Importance and economic significance of the *Acremonium* endophytes to performance of animals and grass plants, *Agric. Ecosyst. Environ.*, 44:3–12, 1993.

52. **Johnson, M. C., D. L. Dahlman, M. R. Siegel, L. P. Bush, G. C. M. Latch, D. A. Potter, and D. R. Varney,** Insect feeding deterrents in endophyte-infected tall fescue, *Appl. Environ. Microbiol.*, 49:568–571, 1985.

53. **Johnson-Cicalese, J. M. and R. H. White,** Effect of *Acremonium* endophytes on four species of billbug found on New Jersey turfgrasses, *J. Am. Soc. Hort. Sci.*, 115:602–604, 1990.

54. **Kindler, S. D., J. P. Breen, and T. L. Springer,** Reproduction and damage by Russian wheat aphid (Homoptera:Aphididae) as influenced by fungal endophytes and cool-season turfgrasses, *J. Econ. Entomol.*, 84:685–692, 1991.

55. **Kirfman, G. W., R. L. Brandenburg, and G. B. Garner,** Relationship between insect abundance and endophyte infestation level in tall fescue in Missouri, *J. Kansas Entomol. Soc.*, 59:552–554, 1986.

56. **Latch, G. C. M.,** Facilitator's comments, In: D. E. Hume, G. C. M. Latch, and H. S. Easton, Eds., *Proc. 2nd Int. Symp.* Acremonium/*Grass Interactions: Plenary Papers*, AgResearch, Grasslands Research Centre, Palmerston North, New Zealand, 1993, 41.

57. **Latch, G. C. M.,** Physiological interactions of endophytic fungi and their hosts. Biotic stress tolerance imparted to grasses by endophytes, *Agric. Ecosyst. Environ.*, 44:143–156, 1993.

58. **Latch, G. C. M. and M. J. Christensen,** Artificial infection of grasses with endophytes, *Ann. Appl. Biol.,* 107:17–24, 1985.

59. **Latch, G. C. M., M. J. Christensen, and D. L. Gaynor,** Aphid detection of endophyte infection in tall fescue, *N.Z. J. Agric. Res.,* 28:129–132, 1985.

60. **Latch, G. C. M., M. J. Christensen, and G. J. Samuels,** Five endophytes of *Lolium* and *Festuca* in New Zealand, *Mycotaxon,* 20:535–550, 1984.

61. **Latch, G. C. M., L. R. Potter, and B. F. Tyler,** Incidence of endophytes in seeds from collections of *Lolium* and *Festuca* species, *Ann. Appl. Biol.,* 111:59–64, 1987.

62. **Lewis, G. C. and R. O. Clements,** A survey of ryegrass endophyte (*Acremonium loliae*) in the U. K. and its apparent ineffectuality on a seedling pest, *J. Agric. Sci. (Camb.),* 107:633–638, 1986.

63. **Marshall, D., L. R. Nelson, and B. Tunali,** The occurrence of *Acremonium* and other endophytic fungi in the indigenous wild cereals of Turkey, In: D. E. Hume, G. C. M. Latch, and H. S. Easton, Eds., *Proc. 2nd Int. Symp.* Acremonium/*Grass Interactions,* AgResearch, Grasslands Research Centre, Palmerston North, New Zealand, 1993, 8.

64. **Mathias, J. K., R. H. Ratcliffe, and J. L. Hellman,** Association of an endophytic fungus in perennial ryegrass and resistance to the hairy chinch bug (Hemiptera:Lygaeidae), *J. Econ. Entomol.,* 83:1640–1646, 1990.

65. **Mathias, J. K., J. L. Hellman, and R. H. Ratcliffe,** Endophyte-enhanced resistance to the bluegrass webworm (Lepidoptera:Pyralidae) in perennial ryegrass, *J. Econ. Entomol.,* in press, 1993.

66. **Morgan-Jones, G. and W. Gams,** Notes on Hyphomycetes. XLI. An endophyte of *Festuca arundinacea* and the anamorph of *Epichloe typhina,* new taxa in one of two new sections of *Acremonium, Mycotaxon,* 15:311–318, 1982.

67. **Morgan-Jones, G., R. A. Phelps, and J. F. White, Jr.,** Systematic and biological studies in the Balansieae and related anamorphs. I. Prologue, *Mycotaxon,* 63:401–415, 1992.

68. **Muegge, M. A., S. S. Quisenberry, G. E. Bates, and R. E. Joost,** Influence of *Acremonium* infection and pesticide use on seasonal abundance of leafhoppers and froghoppers (Homoptera: Cicadellidae; Cercopidae) in tall fescue, *Environ. Entomol.,* 20:1531–1536, 1991.

69. **Murphy, J. A., S. Sun, and L. L. Betts,** Endophyte-enhanced resistance to billbug (Coleoptera:Curculionidae), sod webworm (Lepidoptera:Pyralidae) and white grub (Coleoptera: Scarabaeidae) in tall fescue, *Environ. Entomol.,* 22:699–703, 1993.

70. **Murray, F. R., G. C. M. Latch, and D. B. Scott,** Surrogate transformation of perennial ryegrass, *Lolium perenne,* using genetically modified *Acremonium* endophyte, *Mol. Gen. Genet.,* 233:1–9, 1992.

71. **Nelson, L. R., D. Marshall, and B. Tunali,** Exploration for fungal endophyte in *Lolium* and other grass species of central Turkey, In: D. E. Hume, G. C. M. Latch, and H. S. Easton, Eds., *Proc. 2nd Int. Symp.* Acremonium/*Grass Interactions,* AgResearch, Grasslands Research Centre, Palmerston North, New Zealand, 1993, 11.

72. **Oliver, J. B., C. D. Pless, and K. D. Gwinn,** Effect of endophyte, *Acremonium coenophialum,* in 'Kentucky 31' tall fescue, *Festuca arundinacea,* on survival of *Popillia japonica,* In: S. S. Quisenberry and R. E. Joost, Eds., *Proc. Int. Symp.* Acremonium/*Grass Interactions,* Louisiana Agriculture Experiment Station, Baton Rouge, 1990, 173.

73. **Patterson, C. G., D. A. Potter, and F. F. Fannin,** Feeding deterrency of alkaloids from endophyte-infected grasses to Japanese beetle grubs, *Entomol. Exp. Appl.,* 61:285–289, 1991.

74. **Pearson, W. D.,** The pasture mealy bug, *Balanoccus poae* (Maskell) in Canterbury: a preliminary report, In: P. P. Stahle, Ed., *Proc. 5th Aust. Conf. Grasslands Invert. Ecol.,* D & D Printing, Victoria, Australia, 1988, 297.

75. **Pless, C. D., K. D. Gwinn, A. M. Cole, D. B. Chalkley, and V. C. Gibson,** Development of resistance by *Drosophilia melanogaster* (Diptera:Drosophilidae) to toxic factors in powdered *Acremonium*-infected tall fescue, In: D. E. Hume, G. C. M. Latch, and H. S. Easton, Eds., *Proc. 2nd Int. Symp.* Acremonium/*Grass Interactions,* AgResearch, Grasslands Research Centre, Palmerston North, New Zealand, 1993, 170.

76. **Popay, A. J., R. A. Mainland, and C. J. Saunders,** The effect of endophytes in fescue grass on growth and survival of third instar grass grub larvae, In: D. E. Hume, G. C. M. Latch, and H. S. Easton, Eds., *Proc. 2nd Int. Symp.* Acremonium/*Grass Interactions,* AgResearch, Grasslands Research Centre, Palmerston North, New Zealand, 1993, 174.

77. **Potter, D. A. and S. K. Braman,** Ecology and management of turfgrass insects, *Annu. Rev. Entomol.,* 36:383–406, 1991.

198

78. **Potter, D. A., C. G. Patterson, and C. T. Redmond,** Influence of turfgrass species and tall fescue endophyte on feeding ecology of Japanese beetle and southern masked chafer grubs (Coleoptera:Scarabaeidae), *J. Econ. Entomol.,* 85:900–909, 1992.
79. **Pottinger, R. P., G. M. Barker, and R. A. Prestidge,** A review of the relationships between endophytic fungi of grasses (*Acremonium* spp.) and Argentine stem weevil [*Listronotus bonariensis* (Kuschel)], In: R. B. Chapman, Ed., *Proc. 4th Aust. Conf. Grasslands Invert. Ecol.,* Caxton Press, Christchurch, New Zealand, 1985, 322.
80. **Powell, R. G. and R. J. Petroski,** Alkaloid toxins in endophyte-infected grasses, *Nat. Toxins,* 1:163–170, 1992.
81. **Prestidge, R. A. and R. T. Gallagher,** Endophyte fungus confers resistance to ryegrass: Argentine stem weevil larval studies, *Ecol. Entomol.,* 13:429–435, 1988.
82. **Prestidge, R. A., D. R. Lauren, S. G. van der Zijpp, and M. E. di Menna,** Isolation of feeding deterrents to Argentine stem weevil in cultures of endophytes of perennial ryegrass and tall fescue, *N.Z. J. Agric. Res.,* 28:87–92, 1985.
83. **Prestidge, R. A., R. P. Pottinger, and G. M. Barker,** An association of *Lolium* endophyte with ryegrass resistance to Argentine stem weevil, *Proc. N.Z. Weed Pest Control Conf.,* 35:119–122, 1982.
84. **Prestidge, R. A. and O. J.-P. Ball,** The role of endophytes in alleviating plant biotic stress in New Zealand, In: D. E. Hume, G. C. M. Latch, and H. S. Easton, Eds., *Proc. 2nd Int. Symp.* Acremonium/*Grass Interactions: Plenary Papers,* AgResearch, Grasslands Research Centre, Palmerston North, New Zealand, 1993, 141.
85. **Quigley, P., X. Li, G. McDonald, and A. Noske,** Effects of *Acremonium lolii* on mixed pastures and associated insect pests in south-eastern Australia, In: D. E. Hume, G. C. M. Latch, and H. S. Easton, Eds., *Proc. 2nd Int. Symp.* Acremonium/*Grass Interactions,* AgResearch, Grasslands Research Centre, Palmerston North, New Zealand, 1993, 177
86. **Riedell, W. E., R. E. Kieckhefer, R. J. Petroski, and R. G. Powell,** Naturally-occurring and synthetic loline alkaloid derivatives: insect feeding behavior modification and toxicity, *J. Entomol. Sci.,* 26:122–129, 1991.
87. **Rowan, D. D., J. J. Dymock, and M. A. Brimble,** Effect of fungal metabolite peramine and analogs on feeding and development of Argentine stem weevil (*Listronotus bonariensis*), *J. Chem. Ecol.,* 16:1683–1695, 1990.
88. **Saha, D. C., J. M. Johnson-Cicalese, P. M. Halisky, M. I. van Heemstra, and C. R. Funk,** Occurrence and significance of endophytic fungi in the fine fescues, *Plant Dis.,* 71:1021–1024, 1987.
89. **Schmidt, D.,** La quenouille rend-elle le fourrage toxique?, *Revue Suisse Agric.,* 18:329–332, 1986.
90. **Schmidt, D.,** Effects of *Acremonium uncinatum* and a *Phialophora*-like endophyte on vigour, insect and disease resistance of meadow fescue, In: D. E. Hume, G. C. M. Latch, and H. S. Easton, Eds., *Proc. 2nd Int. Symp.* Acremonium/*Grass Interactions,* AgResearch, Grasslands Research Centre, Palmerston North, New Zealand, 1993, 185.
91. **Siegel, M. R., G. C. M. Latch, and M. C. Johnson,** Fungal endophytes of grasses, *Annu. Rev. Phytopathol.,* 25:293–315, 1987.
92. **Siegel, M. R., D. L. Dahlman and L. P. Bush,** The role of endophytic fungi in grasses: new approaches to biological control of pests, In: A. R. Leslie and R. L. Metcalf, Eds., *Intergrated Pest Management for Turfgrasses and Ornamentals,* U.S. Environmental Protection Agency, Washington, D.C., 1989, 169.
93. **Siegel, M. R., G. C. M. Latch, L. P. Bush, F. F. Fannin, D. D. Rowan, B. A. Tapper, C. W. Bacon, and M. C. Johnson,** Fungal endophyte-infected grasses: alkaloid accumulation and aphid response, *J. Chem. Ecol.,* 16:3301–3315, 1990.
94. **Siegel, M. R. and C. L. Schardl,** Fungal endophytes of grasses: detrimental and beneficial associations, In: J. H. Andrews and S. S. Hirano, Eds., *Microbial Ecology of Leaves,* Springer-Verlag, New York, 1991, 198.
95. **Smith, C. M.,** *Plant Resistance to Insects: A Fundamental Approach,* John Wiley & Sons, New York, 1989.
96. **Springer, T. L. and S. D. Kindler,** Endophyte-enhanced resistance to the Russian wheat aphid and the incidence of endophytes in fescue species, In: S. S. Quisenberry and R. E. Joost, Eds., *Proc. Int. Symp.* Acremonium/*Grass Interactions,* Louisiana Agriculture Experiment Station, Baton Rouge, 1990, 194.

97. **Stewart, A. V.,** Perennial ryegrass seedling resistance to Argentine stem weevil, *N.Z. J. Agric. Res.,* 28:403–407, 1985.

98. **Stovall, M. E. and K. Clay,** Adverse effects on fall armyworm feeding on fungus-free leaves of fungus-infected plants, *Ecol. Entomol.,* 16:519–523, 1991.

99. **Sun, S. and J. P. Breen,** Inhibition of *Acremonium* endophyte by Kentucky bluegrass, In: D. E. Hume, G. C. M. Latch and H. S. Easton, Eds., *Proc. 2nd Int. Symp.* Acremonium/*Grass Interactions,* AgResearch, Grasslands Research Centre, Palmerston North, New Zealand, 1993, 19.

100. **van Heeswijck, R. and G. McDonald,** *Acremonium* endophytes in perennial ryegrass and other pasture grasses in Australia and New Zealand, *Aust. J. Agric. Res.,* 43:1683–1709, 1992.

101. **von Bothmer, R., N. Jacobsen, C. Baden, R. B. Jorgensen, and I. Linde-Laursen,** An Ecogeographical Study of the Genus *Hordeum,* International Board for Plant Genetic Resources, Rome, Italy, 1991.

102. **Welty, R. E., M. D. Azevedo, and T. M. Cooper,** Influence of moisture content, temperature, and length of storage on seed germination and survival of endophytic fungi in seeds of tall fescue and perennial ryegrass, *Phytopathology,* 77:893–900, 1987.

103. **West, C. P., M. McConnell, S. Saidi, F. Ben Jeddi, and G. Charmet,** Collection of endophyte-infected *Festuca* germplasm in western Mediterranean countries, In: *Agronomy Abstracts,* American Society of Agronomy, Madison, WI, 1992, 207.

104. **West, C. P. and K. D. Gwinn,** Role of *Acremonium* in drought, pest, and disease tolerances of grasses, In: D. E. Hume, G. C. M. Latch, and H. S. Easton, Eds., *Proc. 2nd Int. Symp.* Acremonium/*Grass Interactions: Plenary Papers,* AgResearch, Grasslands Research Centre, Palmerston North, New Zealand, 1993, 131.

105. **West, C. P., T. L. Holder, K. E. Turner, M. E. McConnell, and E. L. Piper,** Endophyte infection status of *Festuca* germplasm, In: D. E. Hume, G. C. M. Latch, and H. S. Easton, Eds., *Proc. 2nd Int. Symp.* Acremonium/*Grass Interactions,* AgResearch, Grasslands Research Centre, Palmerston North, New Zealand, 1993, 222.

106. **White, J. F.,** *Epichloë* and *Acremonium* endophytes of grasses: systematic and experimental studies, In: D. E. Hume, G. C. M. Latch, and H. S. Easton, Eds., *Proc. 2nd Int. Symp.* Acremonium/*Grass Interactions: Plenary Papers,* AgResearch, Grasslands Research Centre, Palmerston North, New Zealand, 1993, 47.

107. **White, J. F., A. C. Morrow, and G. Morgan-Jones,** Endophyte-host associations in forage grasses. XII. A fungal endophyte of *Trichachne insularis* belonging to *Pseudocercosporella, Mycologia,* 82:218–226, 1990.

108. **White, J. F., A. C. Morrow, G. Morgan-Jones, and D. A. Chambless,** Endophyte-host associations in forage grasses. XIV. Primary stromata formation and seed transmission in *Epichloë typhina*: developmental and regulatory aspects*, Mycologia,* 82:72–81, 1991.

109. **White, J. F., G. Morgan-Jones, and A. C. Morrow,** Taxonomy, life cycle reproduction and detection of *Acremonium* endophytes, *Agric. Ecosyst. Environ.,* 44:13–37, 1993.

110. **Wilson, A. D., S. L. Clement, and W. J. Kaiser,** Endophytic fungi in a *Hordeum* germplasm collection, *FAO/IBPGR Plant Genet. Res. Newsl.,* 87:1–4, 1991.

111. **Wilson, A. D., S. L. Clement, and W. J. Kaiser,** Survey and detection of endophytic fungi in *Lolium* germ plasm by direct staining and aphid assays, *Plant Dis.,* 75:169–173, 1991.

112. **Yates, S. G., J. G. Fenster, and R. J. Bartelt,** Assay of tall fescue seed extracts, fractions, and alkaloids using the large milkweed bug, *J. Agric. Food Chem.,* 37:354–357, 1989.

Chapter 14

Role of Endophytes in Grasses Used for Turf and Soil Conservation

C. Reed Funk, Faith C. Belanger, and James A. Murphy

CONTENTS

I. INTRODUCTION

Grass is the forgiveness of nature — her constant benediction. Fields trampled with battle, saturated with blood, torn with the ruts of cannon, grow green again with grass, and carnage is forgotten. Streets abandoned by traffic become grass-grown like rural lanes, and are obliterated. Forests decay, harvests perish, flowers vanish, but grass is immortal. Beleaguered by the sullen hosts of winter, it withdraws into the impregnable fortress of its subterranean vitality, and emerges upon the first solicitation of spring. Sown by the winds, by wandering birds, propagated by the subtle horticulture of the elements which are its ministers and servants, it softens the rude outline of the world. Its tenacious fibers hold the earth in its place, and prevent its soluble components from washing into the wasting sea. It invades the solitude of deserts, climbs the inaccessible slopes and forbidding pinnacles of mountains, modifies climates, and determines the history, character, and destiny of nations. Unobtrusive and patient, it has immortal vigor and aggression. Banished from the thoroughfare and the field, it bides its time to return, and when vigilance is relaxed, or the dynasty has perished, it silently resumes the throne from which it never abdicates. It bears no blazonry or bloom to charm the senses with fragrance or splendor, but its homely hue is more enchanting than the lily or the rose. It yields no fruit in earth or air, and yet should its harvest fail for a single year, famine would depopulate the world.[13]

Agronomists have long recognized the universal importance of grass in soil development, stabilization, and improvement; erosion control; food production; and providing recreation and environmental enhancement. Many of the world's most fertile and productive soils were developed under grasslands. Soils eroded and depleted by unsound agricultural and developmental activities have been restored to productivity by grasses and associated forbes and legumes. Coastal beaches are protected from wind and waves. Turfgrasses are used to beautify the earth, enrich our lives, and provide recreation and enjoyment for mankind.

Scientists are becoming increasingly aware of the role and importance of fungal endophytes, as well as other microbial associations, in the persistence and performance of grasses used for turf and soil preservation. These beneficial associations are especially evident in native and naturalized grasslands and are abundant in many old turfs that receive limited, if any, supplemental fertilizer, pesticides, or irrigation. Plant breeders, agronomists, microbiologists, ecologists, and other scientists have much to learn from studies of old low-maintenance turfs and other grasslands.

Our major cool-season turfgrass species originated and evolved in the cool, moist summer climates of northern Europe and the British Isles or the hot, dry summer climates of the Mediterranean region. In

0-8493-6276-8/94/$0.00+$.50

many respects, they are poorly adapted to the warm, humid summers, and the disease and insect pests, common to the mid-Atlantic region and transition zone of the U.S. Adequate tolerance of the temperature extremes found in the more continental climates of North America is also lacking. Early colonists brought their grasses to America as hay and bedding, as well as seeds to establish their hay fields and pastures. Eventually millions of kilograms of seeds were used to establish permanent pastures, conservation plantings, sports fields, and lawns. Birds and other animals carried grass seeds to many sites. Through decades or centuries of natural selection, the best populations became better adapted to our soils, pests, and climates. These naturalized turfs normally consisted of mixtures of many species, with significant genetic diversity within each species, and frequently contained endophytes and other beneficial microbial associations.

All currently used turfgrass cultivars trace their original ancestry to plants selected from old turfs. Many of our best-adapted cultivars were developed from a very limited number of plants selected from old low-maintenance turfs of the U.S. A few other cultivars were introduced from Europe. Some contain germplasm from both American and European collections. Population improvement programs have been very effective in improving many, but not all, characteristics useful for turf. However, there is concern about the limited germplasm presently being used in our major turfgrass species. Moreover, we should be concerned about the limited diversity of the endophytes used in current turfgrass improvement programs.

II. PERENNIAL RYEGRASS

Scientists in New Zealand[19] were the first to recognize an association of a fungal endophyte (*Acremonium lolii* Latch, Christensen, and Samuels) with field resistance of perennial ryegrass (*Lolium perenne* L.) to an important insect pest. Stewart[27] subsequently showed that endophyte-enhanced resistance to the Argentine stem weevil, *Listronotus bonariensis* Kuschel, was essential to the persistence and performance of perennial ryegrass used for turf in many parts of New Zealand. It is now recognized that many of the improved turf-type perennial ryegrasses developed in the U.S. prior to the discovery of endophyte-enhanced insect resistance had high levels of the *Acremonium* Link endophyte. Most of the attractive persistent ryegrasses selected from old turfs contained an *Acremonium* endophyte. "Manhattan" perennial ryegrass, released in 1967, was developed from endophyte-infected plants surviving in old lawns located near the sheep meadow of Central Park in New York City. "Pennfine," released in 1970, was developed from endophyte-infected plants selected from old turfs in Pennsylvania. "Pennant," released in 1980, and "All*Star," released in 1983, contained germplasm and endophytes from plants selected from old turfs located in College Park and Baltimore, MD, respectively. Presently, a very high percentage of all turf-type perennial ryegrass germplasm and endophytes used in North America trace their origin to the above cultivars and closely related plants selected from the same old turfs.

The developers of Manhattan and Pennfine were originally unaware of the presence and benefits of *Acremonium* endophytes, as well as the potential for loss of endophyte viability during prolonged seed storage. As a result, many, if not all, current seed lots of these cultivars are free of *Acremonium* endophytes. It is now apparent that much of the persistence and success of these early cultivars, under insect predation and summer stress, was associated with their high levels of endophytes.

Turfgrass breeders, seed companies, and knowledgeable consumers rapidly became aware of the benefits of endophytes in perennial ryegrass turfs, during the mid-1980s. Striking examples of endophyte-enhanced performance were observed in many turf trials. These included enhanced resistance to many lepidopterous species of sod webworm,[10] chinch bugs, *Blissus leucopterus* hirtus Montandon,[11,17] the Argentine stem weevil,[19] and billbugs, *Sphenophorus* spp.[1] Plant breeders found that many of the plants in leading cultivars, germplasm collections, and advanced breeding populations contained endophytes. This made it feasible to rapidly develop endophyte-enhanced cultivars. Seed companies generally recognized and implemented practices needed to produce and market seeds with high levels of viable endophytes. Turf professionals and other well-informed consumers started to demand such seeds. As a result, most of the best performing new turf-type perennial ryegrasses released during the past few years contain high levels of endophytes. Currently available ryegrasses with many seed lots containing high percentages of endophyte include "All*Star," "Pennant," "Regal," "Repell," "Repell II," "Manhattan II (E)," "Yorktown III," "Palmer II," "Prelude II," "Sherwood," "Gettysburg," "SR-4000," "SR-4100," "SR-4200," "SR-4300," "Brightstar," "Advent," "APM," "Assure," "Legacy," "Dandy," "Elf," "Saturn,"

"Quickstart," "Prizm," "Pinnacle," "Affinity," "Seville," "Duet," "Target," and "Delaware Dwarf." Most of the best new turf-type ryegrasses now being developed in the U.S. will contain high levels of endophytes.

III. TALL FESCUES

Tall fescue (*Festuca arundinacea* Schreb.) has been extensively used for pastures, soil conservation, and turf, throughout the transition zone of the U.S. since the release of "Kentucky 31" in 1943. Kentucky 31 tall fescue originated as a naturalized ecotype discovered on a hillside pasture in Menifee County, KY. Through many decades of natural selection, the better-adapted plants of an unknown introduced seed source survived and reproduced. Unpublished data and observations at Rutgers University show that Kentucky 31 has better cold hardiness, improved summer performance, and increased resistance to the large brown patch disease incited by *Rhizoctonia solani* Kühn, compared with most other tall fescues obtained from foreign sources. It is now apparent that a high percentage of endophyte-infected plants were present in most plantings of Kentucky 31 and contributed to its vigor, pest resistance, and persistence, in many environments. Unfortunately, endophyte infection was associated with many instances of poor animal performance.[2,3] The use of endophyte-free tall fescues may increase known or unknown pests currently being held in check by biologically active *Acremonium* endophytes. Increased efforts should be made to (1) develop tall fescues for forage, having genetically improved persistence, pest resistance, and stress tolerance; (2) identify and use endophytes associated with increased stress tolerance and pest resistance, but without serious deleterious effects on livestock; and (3) use genetically improved turf-type tall fescues containing the most useful endophytes for turf and conservation plantings.

Largely due to the success of Kentucky 31, turfgrass breeders are devoting considerable effort to the genetic improvement of tall fescue. Essentially all of the germplasm used in the development of improved turf-type tall fescues traces its origin to attractive, persistent plants selected from old turfs in the U.S. "Rebel" tall fescue, released in 1980, traces much of its parental germplasm to a single plant collected in 1961 from an old lawn in Princeton, NJ. Three cycles of both phenotypic and genotypic recurrent selection over a 19-year period were used to develop Rebel.[9] The Rebel breeding program demonstrated that it was feasible to develop turf-type tall fescues with a much darker green color, a lower growth profile, finer leaves, good seed yields, greater turf density and improved resistance to net blotch caused by *Drechslera dictyoides* (Drechs.) Shoemaker. As a result, turfgrass breeders spent thousands of hours examining old turfs throughout the U.S. for attractive turf-type tall fescues. Following clonal evaluation, the best of these selections were intercrossed with other selections from old turf, or with plants related to Rebel, to develop new cultivars or initiate population improvement programs.

Most attractive tall fescue plants selected from old turfs contained *Acremonium* endophytes. "Titan" tall fescue, released in 1987,[20] was developed as a high-endophyte cultivar. Endophyte-infected plants selected from old turfs in Georgia, Mississippi, Maryland, Kansas, Ohio, Texas, Virginia, Pennsylvania, North Carolina, Kentucky, and New Jersey were used in the breeding of Titan.

In spite of the widespread availability of *Acremonium* endophytes in elite turfgrass germplasm, only a few of the newer turf-type tall fescues have high percentages of endophyte-infected plants. The limited use of *Acremonium* endophytes in turf-type tall fescues is due to a number of factors, including (1) the potential misuse of seeds of these cultivars, for the establishment of pastures; (2) concerns about grazing seed fields and the use of forage produced as a byproduct in seed production; and (3) limited knowledge of which endophytes are the most desirable for enhancing stress tolerance, persistence, and pest resistance. Programs to select, identify, and utilize desirable endophytes are needed, in addition to more effective programs in seed testing and labeling.

Research and observations show that many strains of *Acremonium* endophytes can enhance the persistence and performance of tall fescue. This appears to be especially significant under severe stress.[5] Recent studies in New Jersey[18] showed that billbugs (*Sphenophorus* spp.) and sod webworms (Lepidoptera:Pyralidae) can cause extensive damage to endophyte-free tall fescues. Endophyte-free plots averaged 54 billbug and 52 sod webworm larvae per square meter, whereas comparable endophyte-infected plots averaged 4 billbug and 23 sod webworm larvae. Endophyte-infected tall fescues showed more rapid and complete recovery from a severe summer drought and produced a denser, lusher, brighter, but lighter-colored, turf compared with the duller, darker, more open turf of comparable endophyte-free tall fescues. Subsequent observations showed a significant reduction in number of, and damage by, white grubs (Coleoptera:Scarabeidae) in endophyte-infected tall fescues. It is possible that adults preferred the darker, duller, thinner turfs of the endophyte-free tall fescues, for oviposition of eggs during mid summer.

Pythium blight can be a serious disease of lush, dense, overfertilized tall fescue turfs under hot, humid, wet conditions. Observations at Rutgers University have shown that at least some strains of *Acremonium* endophytes have been associated with increased susceptibility to *Pythium* blight in some turf trials. This may be an indirect effect of some endophytes promoting a denser, lusher turf that would enhance disease development. Reduced nitrogen fertility during hot, wet seasons might be expected to offset endophyte-enhanced disease susceptibility. Research is underway to identify endophyte strains and host–endophyte combinations that are not associated with increased susceptibility to *Pythium* blight.

IV. FINE FESCUES

The fine fescues are receiving increased attention and use by the turfgrass industry because of the development of genetically improved cultivars possessing tolerance of acid soils, moderate shade, and low soil fertility. Fine fescues include a number of fine-leaved species and subspecies of *Festuca* L. having fine, tough, bristle-like leaves and a low-growing, leafy, turf-type growth habit. They are especially useful for medium-to-low maintenance turfs in moderate shade, on poor soils, and in regions with cool summer climates. Useful fine fescues include Chewings fescue (*F. rubra* L. subsp. *commutata* Gaud.), strong creeping red fescue (*F. rubra* L. subsp. *rubra*), hard fescue (*F. longifolia* Thuill.), blue fescue (*F. glauca* Lam.), and slender creeping red fescue (*F. rubra* L. subsp. *litoralis* Meyer Auquier).

Endophyte infection of fine fescues is common in old turfs of many species of fine fescue. Nearly 80% of the fine fescue seed lots collected in Iceland showed high levels of endophytes. Fine fescues growing in undisturbed sites, including hedgebanks, stone walls, and cliff walls, in southwest England were normally infected with endophytes, whereas endophyte-infected plants were infrequently encountered in nearby meadow-type habitats.[33] A total of 92 endophyte-infected plants were found in a collection of 2873 fine fescues, made throughout the U.S.[21,30] It appears that endophytes are very common in old native grasslands. However, loss of endophyte viability during storage of commercial seed lots, during overseas shipments, and in germplasm repositories has reduced endophyte frequency in most commercial seed sources and new plantings.

Acremonium-endophyte-infected fine fescues have frequently shown dramatic improvements in turf performance, persistence, and resistance to some diseases and insect pests, in turf trials at Rutgers University. They often produce a brighter, lusher, denser turf, under certain conditions of summer stress. Field resistance to the hairy chinch bug (*Blissus leucopterus hirtus* Montandon) has been observed in many tests. Laboratory studies[7] showed that no fall armyworms (*Spodoptera frugiperda* J. E. Smith) fed endophyte-infected hard and Chewings fescues survived to pupation, whereas 40 and 100% survived on endophyte-free hard and Chewings fescues, respectively.

Many instances of *Acremonium*-endophyte-enhanced resistance to the dollar spot disease (caused by *Sclerotina homoeocarpa* F. T. Bennett) have been observed in turf trials of strong creeping red fescue, Chewings fescue, and hard fescues, at both Adelphia and North Brunswick, NJ. Endophyte-enhanced resistance was manifest by both dramatic reductions of fungal mycelia and disease damage. Endophyte-infected perennial ryegrasses in similar tests did not exhibit enhanced resistance to this disease. The selection and/or development of endophyte strains that enhance field resistance to important diseases infecting turfgrasses merits increased attention. This should be an especially exciting and productive area of research.

Turfgrass breeders are currently devoting considerable resources to the incorporation of beneficial endophytes into genetically improved fine fescues. Several cultivars of endophyte-infected hard fescues ("SR-3000," "SR-3100," "Reliant," "Aurora," and "Discovery") and Chewings fescues ("Jamestown II," "Victory," "Tiffany," "SR-5000," "SR-5100," and "Shadow") are in commercial production and use. Other promising endophyte-containing cultivars are being developed and increased. However, much additional research is needed before we can fully utilize the potential benefits of endophytes in the various species of fine fescue. Four areas of needed research include (1) controlling choke expression in many *Festuca–Acremonium* associations, (2) developing methods of prolonging endophyte viability during seed storage, (3) evaluating the effects of endophyte-infected fine fescues on animal herbivores, and (4) determining the genetic resources and variability of endophytes of fine fescue.

Epichloë typhina (Per.:Fr.) Tul., the teleomorph or sexual stage of *A. typhinum*, often develops a stroma on the panicles of Chewings, hard, and strong creeping red fescues. The stroma may prevent, or partially prevent, the emergence of flower panicles, resulting in a disease called "choke."[23,25] Choke can reduce seed yields and quality,[22] thereby limiting the potential commercial production and use of a cultivar. Sound management practices, including optimum rates and timing of nitrogen fertilization, can reduce or eliminate choke formation in endophyte-infected fine fescues.[28] It also appears possible to select fine-fescue genotypes with increased genetic resistance to choke development and to identify strains of endophytes with reduced potential for choke expression. Current endophyte-infected cultivars and leading experimental selections of hard fescue and blue fescue have a very low frequency of choke expression. Choke is more common in current cultivars of endophyte-infected Chewings fescue. However, acceptable yields of high-quality seeds are being reported. Nevertheless, there is concern that unpredictable and uncontrollable environmental effects might increase choke expression to the extent that seed production would be reduced. Current breeding populations of strong creeping red fescues occasionally show an unacceptable level of choke expression. However, new strains of endophytes and selection for host-plant resistance to choke expression make it likely that useful endophyte-infected cultivars will soon be developed and produced. Choke stroma have been observed in mowed turfs, especially when turfs have been thinned and are growing under low fertility. However, the stroma are very small and would not generally be noticed by a casual observer. They are present for only a few days during seed-head formation in late spring.

Fine-fescue seeds appear to maintain endophyte viability for a shorter time than do seeds of tall fescue and perennial ryegrass. Endophyte viability is also more likely to be lost under adverse storage conditions. This partial or complete loss of endophyte viability during prolonged seed storage can reduce or eliminate the benefits of endophyte-enhanced turf performance. Storing seeds under cool, dry conditions can prolong both seed germination and endophyte viability. Under normal commercial conditions, high levels of viable endophytes can be found in endophyte-infected seeds in the fall and spring following a June or July harvest. Endophyte viability can be lost rather quickly during the following summer if storage conditions are warm and humid. Research is needed to develop improved seed conditioning, packaging, and storage practices, directed towards maintaining adequate levels of viable endophytes.

Considerable research has been conducted on the effects of endophyte-infected perennial ryegrass and tall fescue forage, on animal health and performance. Very little is known about the possible effects of endophyte-infected fine fescue. Turf-type species of fescues are seldom recommended or used as forage grasses in the U.S., but are often grazed in some other countries, usually as components of naturalized grasslands growing on poorer soils. As a minor component of a mixed diet, possible adverse effects of endophyte-infected plants might be of little importance or go unnoticed. Most serious problems associated with endophyte-infected forage grasses generally occur in monocultures and where endophyte-infected grasses make up the sole or major portion of the animal diet. However, until more information is available, it would not be wise to use endophyte-infected fine fescues for animal consumption.

V. REDTOP AND BENTGRASSES

The genus *Agrostis* L. includes a number of species suitable for many turfs, especially in temperate climates with cool, moist summers. Bentgrasses are generally tolerant of acid soils and can thrive at lower levels of soil fertility relative to that needed for good performance of Kentucky bluegrass or perennial ryegrass. Future breeding work could significantly expand the areas of adaptation and usefulness of many of these grasses.

The creeping bentgrasses, *A. palustris* Huds. and *A. stolonifera* L., are soft, fine-textured, low-growing grasses with a vigorous stoloniferous growth habit. They are capable of producing a very attractive, dense, compact turf tolerant of very close mowing. They are widely used for golf course putting greens, tees, and closely mowed fairways, in the U.S. and Canada. Such turfs normally receive very high maintenance inputs, including frequent irrigation and mowing, aerification and topdressing for thatch reduction, and pesticides for disease and insect control. Colonial bentgrasses, *A. tenuis* Sibth., can also produce exceptionally attractive turfs at moderately close mowing heights. They are widely used for golf turfs and closely mowed lawns, often in mixtures with fine fescue, in regions with cool summers. Currently available cultivars are very susceptible to diseases prevalent in regions with warm humid summers.

Velvet bentgrass, *A. caniva* L., is an exceptionally fine-textured, low-growing grass, but of limited commercial use at present. Dryland bentgrass, *A. castellana* Boiss. and Reuter, is a hardy rhizomatous species frequently found in old low-maintenance turfs. Redtop, *A. gigantea* Roth, is a strongly rhizomatous, moderately robust species adapted to acid soils of low fertility. It was once widely used in both forage and turfgrass mixtures. Significant genetic diversity exists in each of the above species, with significant opportunities for substantial genetic improvement.

Non-choke-inducing *Acremonium* endophytes have been found in a number of species of *Agrostis*, including Redtop, but do not appear to be readily available in creeping, colonial, dryland, or velvet bentgrasses.[32] The discovery or development of beneficial non-choke-inducing strains of endophytes for use in each of these species could prove to be very valuable to the turfgrass industry.

Endophyte-infected plants of creeping and colonial bentgrasses are often found in old, low-maintenance turfs of the British Isles. Bradshaw[6] found that they were more vigorous, and produced more tillers, than nearby endophyte-free plants. Abundant choke makes them of little or no value for commercial use. However, opportunities might exist to hybridize these endophytes with non-choke-inducing endophytes from other species, or otherwise modify them to produce endophyte strains of value to turfgrass breeders. It is also possible that useful non-choke-inducing endophytes exist in nature. An expanded effort should be made to search for such endophytes.

VI. KENTUCKY BLUEGRASSES

Kentucky bluegrass (*Poa pratensis* L.) is a hardy, persistent, attractive grass with extensive rhizomes. It is adapted to a wide range of climates and soils, partly as a result of the tremendous range of genetic diversity within the species. Introduced from the old world by early colonists to North America, Kentucky bluegrass became naturalized throughout much of the U.S. and Canada. It provides nutritious and highly palatable forage on millions of hectares of pastures throughout the northern half of the U.S. As a premier turfgrass, it is found alone, or in mixtures, on more than 40 million lawns throughout temperate North America. Many of these old pastures and turfs were never seeded by man, but were established from seeds carried from birds and other animals. In many instances, individual seedlings have persisted, spread, and colonized to produce attractive turfs many meters in diameter.

Kentucky bluegrass is a facultative apomict with complex cytological and embryological characteristics and a highly variable chromosome number ($x = 7$; $2n = 28$ to 154). It has the ability to hybridize with, and absorb, entire genomes from other species of *Poa*. The ability of apomictic reproduction to facilitate fertility of superior hybrid genotypes largely accounts for the great diversity and widespread distribution of the species. The highly complex and variable genome constitutions in Kentucky bluegrass may also help explain the apparent lack of compatible *Acremonium* endophyte strains capable of existing in a mutualistic relationship.

Many species of *Poa* L. native to both North America and Eurasia are hosts of *Acremonium* spp.[31] Sprague[26] and Kohlmeyer and Kohlmeyer[15] reported Kentucky bluegrass as a host of *E. typhina* in plants collected from both the U.S. and Europe. Living plants of Kentucky bluegrass with viable endophytes were not maintained. Collection efforts by turfgrass breeders in the U.S. have been unsuccessful, to date, in finding a Kentucky bluegrass with a useful non-choke-inducing *Acremonium* endophyte. It would be very useful to expand these collection efforts into areas of Europe and Asia, where different sources of genetic diversity exist.

Apomictic reproduction has many advantages for endophyte research and utilization. Nearly all seeds of an endophyte-infected, highly apomictic plant would contain the endophyte and would be genetically identical to each other. After storage or heat treatment, performance of the resulting endophyte-free seedlings could be compared with genetically identical endophyte-infected seedlings. Various endophytes inoculated into different seedlings of the same apomictic cultivar could be compared in the same host genotype(s). This would be equivalent to inoculating clonal propagules in other grasses.

Most Kentucky bluegrass cultivars originated as the progeny of a single, highly apomictic plant. Nearly all plants of a particular cultivar are genetically identical to each other and to their maternal parent. The successful inoculation of a desirable, compatible endophyte into only one plant would be sufficient to incorporate a desirable endophyte into a Kentucky bluegrass cultivar. This contrasts with the need to successfully inoculate many plants of a highly heterogeneous and highly heterozygous cultivar of perennial ryegrass, tall fescue, fine fescue, or bentgrass. However, because of the highly heterozygous genotypes characteristic of all Kentucky bluegrasses, the back-cross technique cannot be used to transfer an endophyte into an existing cultivar. Hybridization and a modified back-cross technique could be used to generate highly heterogeneous and highly heterozygous populations of Kentucky bluegrass containing useful endophytes.[4] Elite, highly apomictic, endophyte-enhanced cultivars could be selected from such populations.

The authors are unaware of a successful compatible transfer of an *Acremonium* endophyte from other grass genera, or another species of *Poa*, into Kentucky bluegrass, although a number of efforts have been made to transfer an endophyte into Kentucky bluegrass through crossing with other *Acremonium*-endophyte-infected species of *Poa*. "Service," an endophyte-infected, highly apomictic cultivar of big bluegrass (*P. ampla* Merr.), was used as a female parent and pollinated with a group of Kentucky bluegrass cultivars and selections.[29] Over 200 F[1] hybrids showing rhizomes and other characteristics of their Kentucky bluegrass parent were selected from large populations of seedlings. These seedlings contained the *Acremonium* endophyte a few weeks after germination, but were free of the endophyte after spending the fall, winter, and spring in a spaced-plant nursery. Additional endophyte-infected *Poa* species are being collected for additional studies with hybridization and inoculation.

To date, the many attempts to incorporate *Acremonium* endophytes into *P. pratensis*, through breeding or direct inoculation, have not been successful. In general, it has been found that there is limited compatibility between various *Acremonium* endophyte isolates and grasses other than their natural hosts.[24] Most attempts to produce novel endophyte–host combinations have failed. Better understanding of the physiology of the interaction between the endophytes and their hosts may ultimately lead to new approaches to the incorporation of desired endophytes into *P. pratensis*. The existence of other endophyte-infected *Poa* spp. provides valuable material for such studies.

Recently a novel fungal protease was described that is expressed in the endophytic association of *A. typhinum* with *P. ampla* Merr.[16] This protease has some unusual characteristics, such as activity in high concentrations of detergent, and its expression appears to be specifically associated with the endophytic symbiosis. The protease is a membrane-associated serine endoprotease with an essential thiol group. Its amino acid specificity appears to be quite broad, since it degrades many different proteins to small peptides. Currently efforts are underway to further characterize the protease and to determine its *in vivo* function.

This protease is expressed by the fungus in the endophytic infection, but is not expressed on long-term culture of the fungus (Lindstrom and Belanger, unpublished). Since expression of this protease is not required for growth of the fungus in culture, its function may relate specifically to the grass–fungus interaction. Similar protease activity has been detected in all other endophyte-infected *Poa* spp. that have been examined. These are *P. ampla*, *P. autumnális* Muhl. ex Ell., *P. sylvestris* A. Gray, *P. palustris* L., and two other *Poa* spp. that have been tentatively identified as *P. interior* Rydb. and *P. cúrta* Rydb. Expression of such a fungal protease, however, does not appear to be a general feature of *Acremonium* infection of grasses, since similar activity was not detected in endophyte-infected perennial ryegrass or tall fescue. This raises the possibility that expression of this fungal protease may be a unique feature of endophyte infection of *Poa* spp.

In general, it is well established that there are specific genetic requirements for effective host–pathogen interactions and for the effective establishment of symbiotic and mutualistic relationships between plants and their microbial partners.[8,12,14] It may be that expression of this novel fungal protease is an important factor in the host–fungus compatibility in *Poa* spp. Information on such unique features of specific endophyte–grass associations may be important in future efforts to broaden the host range of desirable endophytes.

VII. SUMMARY: ENDOPHYTE RESEARCH AND EFFICIENCY IN PLANT BREEDING

It is essential to know the endophyte status of a plant or cultivar before we can evaluate its genetic potential.[11] Thus, an understanding of endophyte effects is of considerable value in increasing the efficiency, reliability, and effectiveness of plant breeding and cultivar evaluation programs. Grass breeders may want to remove endophytes from breeding populations, to more effectively select for nonendophytic sources of pest resistance and stress tolerance. The appropriate endophyte(s) could be added later to enhance performance. Use of the most effective endophyte(s) will require an increased knowledge of the genetic variation within endophytes, and their interactions with various hosts and environments. Turfgrass breeders will want to continue to work closely and synergistically with other scientists in endophyte research. Synergistic collaborators will have many opportunities for making contributions of value to grass improvement programs. Increased knowledge of the ecology of grasses and their associated symbiots will aid in restoring and enhancing the beauty, dependability, and usefulness of our pristine grasslands and other areas where turf is used for soil preservation and improvement, sports and recreation, pollution control, and the enhancement of our environment and quality of life.

ACKNOWLEDGMENTS

This is New Jersey Agricultural Experiment Station Publication no. F-12264-1-94 supported by State and U.S. Hatch Act Funds. Additional support was received from the U.S. Golf Association-Golf Course Superintendents Association of America Research Fund and from other grants and gifts.

REFERENCES

1. **Ahmad, S., J. M. Johnson-Cicalese, W. K. Dickson, and C. R. Funk,** Endophyte-enhanced resistance in perennial ryegrass to the bluegrass billbug, *Sphenophorus parvulus*, *Entomol. Exp. Appl.*, 41:3–10, 1986.

2. **Bacon, C. W., J. K. Porter, J. D. Robbins, and E. S. Luttrell,** *Epichloë typhina* from toxic tall fescue grasses, *Appl. Environ. Microbiol.*, 34:576, 1977.

3. **Ball, D. M., J. F. Pedersen, and G. D. Lacefield,** The tall-fescue endophyte, *Am. Sci.*, 81:370–379, 1993.

4. **Bashaw, E. C. and C. R. Funk,** Apomictic grasses, In: W. R. Fehr, Ed., *Principles of Cultivar Development*, Vol. 2, *Crop Species*, Macmillan, New York, 1987, 40.

5. **Bouton, J. H., R. N. Gates, D. P. Belesky, and M. Owsley,** Yield and persistence of tall fescue in the southeastern coastal plain after removal of its endophyte, *Agron. J.*, 85:52–55, 1993.

6. **Bradshaw, A. D.,** Population differentiation in *Agrostis tenuis* Sibth. II. The incidence and significance of infection by *Epichloë typhina*, *New Phytol.*, 58:310–315, 1959.

7. **Breen, J. P.,** Enhanced resistance to fall armyworm (Lepidoptera:Noctudiae) in Acremonium endophyte-infected turfgrasses, *J. Econ. Entomol.*, 86:621–629, 1993.

8. **de Wit, P. J. G. M.,** Molecular characterization of gene systems in plant-fungus interactions and the application of avirulence genes in control of plant pathogens, *Annu. Rev. Phytopathol.*, 30:391–418, 1992.

9. **Funk, C. R., R. E. Engel, W. K. Dickson, and R. H. Hurley,** Registration of Rebel tall fescue, *Crop Sci.*, 21:632, 1981.

10. **Funk, C. R., P. M. Halisky, M. C. Johnson, M. R. Siegel, A. V. Stewart, S. Ahmad, R. H. Hurley, and I. C. Harvey,** An endophytic fungus and resistance to sod webworms: association in *Lolium perene* L., *Biotechnology*, 1:189–191, 1983.

11. **Funk, C. R., P. M. Halisky, S. Ahmad, and R. H. Hurley,** How endophytes modify turfgrass performance and response to insect pests in turfgrass breeding and evaluation trials, In: F. Lemaire, Ed., *Proc. 5th Int. Turfgrass Res. Conf.*, INRA, Versailles, 1985, 137.

12. **Gabriel, D. W. and B. G. Rolfe,** Working models of specific recognition in plant-microbe interactions, *Annu. Rev. Phytopathol.*, 29:365–391, 1990.

13. **Ingalls, J. J.,** In praise of bluegrass, *Kansas Magazine*, 1872. (Reprinted in: *Grass, 1948 Yearbook of Agriculture*, U.S. Government Printing Office, Washington, D.C., 1948, 6.)

14. **Koide, R. D. and R. P. Schreiner,** Regulation of the vesicular arbuscular mycorrhizal symbiosis, *Annu. Rev. Plant Physiol. Plant Mol. Biol.*, 43:557–581, 1992.

15. **Kohlmeyer, J. and E. Kohlmeyer,** Distribution of *Epichloë typhina* (Ascomycetes) and its parasitic fly, *Mycology*, 66:77–86, 1974.

16. **Lindstrom, J. T., S. Sun, and F. C. Belanger,** A novel fungal protease expressed in endophytic infection of *Poa* species, *Plant Physiol.*, 102:645–650, 1993.

17. **Mathias, J. K.,** The relationship of endophytic fungi in perennial ryegrass and resistance to the hairy chinch bug and sod webworm, Ph.D. thesis, University of Maryland, College Park, 1988.

18. **Murphy, J. A., S. Sun, and L. L. Betts,** Endophyte-enhanced resistance to billbug (Coleoptera: Curculionidae), sod webworm (Lepidoptera:Pyralidae), and white grubs (Coleoptera:Scarabeidae) in tall fescue, *Environ. Entomol.*, 22:699–703, 1993.

19. **Prestidge, R. A., R. P. Pottinger, and G. M. Barker,** An association of *Lolium* endophyte with ryegrass resistance to Argentine stem weevil, *Weed Pest Control Conf.*, 35:119–122, 1982.

20. **Robinson, M. F., B. B. Clarke, R. W. Duell, and C. R. Funk,** Registration of Titan tall fescue, *Crop Sci.,* 30:956–957, 1990.
21. **Saha, D. C., J. M. Johnson-Cicalese, P. M. Halisky, M. I. van Heemstra, and C. R. Funk,** Occurrence and significance of endophytic fungi in the fine fescue, *Plant Dis.,* 71:1021–1024, 1987.
22. **Sampson, K.,** The systemic infection of grasses by *Epichloë* (Pers.) Tul., *Trans. Br. Mycol. Soc.,* 18:30–47, 1933.
23. **Siegel, M. R., G. C. M. Latch, and M. R. Johnson,** Fungal endophytes of grasses, *Ann. Rev. Phytopathol.,* 25:393–315, 1987.
24. **Siegel, M. R. and C. L. Schardl,** Fungal endophytes of grasses: detrimental and beneficial associations, In: J. H. Andrews and S. S. Hirano, Eds., *Microbial Ecology of Leaves,* Springer-Verlag, New York, 1991, 198.
25. **Smith, J. D., N. Jackson, and A. R. Woolhouse,** Fungal diseases of amenity turf grasses, 3rd ed., E. and F.N. Spon, New York, 1989.
26. **Sprague, R.,** *Diseases of Cereals and Grasses (Fungi, Except Smuts and Rusts),* Ronald Press, New York, 1950, 77.
27. **Stewart, A. V.,** The effect of endophyte on perennial ryegrass turf performance, *Rutgers Turfgrass Proc.,* 17:120–126, 1986.
28. **Sun, S., B. B. Clarke, and C. R. Funk,** Effect of fertilizer and fungicide applications on choke expression and endophyte transmission in Chewings fescue, In: S. S. Quisenberry and R. E. Joost, Eds., *Proc. Int. Symp.* Acremonium/*Grass Interactions,* Louisiana Agriculture Experiment Station, Baton Rouge, 1990, 62.
29. **Sun, S. and J. P. Breen,** Inhibition of *Acremonium* endophyte in Kentucky bluegrass, In: D. E. Hume, G. C. M. Latch, and H. S. Easton, Eds., *Proc. 2nd Int. Symp.* Acremonium/*Grass Interactions,* AgResearch, Grasslands Research Centre, Palmerston North, New Zealand, 1993a, 19.
30. **Sun, S., B. B. Clarke, D. R. Huff, L. L. Betts, D. A. Smith, and J. F. White, Jr.,** Enhanced performance and new sources of *Acremonium typhinum* in fine fescues, In: D. E. Hume, G. C. M. Latch, and H. S. Easton, Eds., *Proc. 2nd Int. Symp.* Acremonium/*Grass Interactions,* AgResearch, Grasslands Research Centre, Palmerston North, New Zealand, 1993b, 23.
31. **White, J. F., Jr.,** Widespread distribution of endophytes in the Poaceae, *Plant Dis.,* 71:340–342, 1987.
32. **White, J. F., Jr., P. M. Halisky, S. Sun, G. Morgan-Jones, and C. R. Funk,** Endophyte-host associations in grasses. XVI. Patterns of endophyte distribution in species of the tribe Agrostideae, *Am. J. Bot.,* 79:472–477, 1992.
33. **White, J. F., Jr., G. Lewis, S. Sun, and C. R. Funk,** A study of distribution and reproduction of red fescue endophytes in populations in southwest England, *Sydowia,* in press, 1994.

INDEX